Revolutions of Choice

For architecture, like most other disciplines, the COVID-19 pandemic is a mirror. The crisis reflects architecture's strengths and failures. It shows what is urgent and what is secondary. It makes one question what is important and what is superfluous. Can there ever be a 'back to normal'?

Barkow Leibinger's work assembled in this third edition of *Atlas* was made between 2013 and 2021. The selection corresponds closely to the exhibition *Revolutions of Choice*, realized during the pandemic, in 2020, at Haus am Waldsee in Berlin. It could have also been called *Experiment*. In fact, one phenomenon the crisis-as-mirror reflects is the idea of experiment. In the first decades of the 2000s, experimentation was marginalized within the field of architecture. Unlike in the 1970s, when the status of architects was often defined by what they did not build, today status is mostly defined built works. As a consequence, notions such as 'experiment' or 'utopia' have been overshadowed by more popular notions such as 'project' or 'strategy.'

Barkow Leibinger are admired and known for their interest in pragmatism. Yet *Atlas* shows their practice in a new light. It points to experimentation – acts of testing, failing, starting anew. Their experimentation has led to workshops and exchanges directly with manufacturers. And it addresses the unknown, moments of creation and imagination where the outcome is unclear and where compromises of realization are still far away.

The focus on the experiment is, of course, not caused by the pandemic. Most of the work presented here was realized before COVID-19. Yet the pandemic has shifted attention. The very emblem for this shift of attention is the labyrinthine bundle that Barkow Leibinger designed for the Serpentine Summer House Program in 2016, and which was exhibited on the back lawn of Haus am Waldsee. It had no clear purpose but was exactly what people needed. Its playful, beautiful structure accommodates different forms of use and is open for imagination. It is a plea for experimentation. And it demonstrates that there is no back to normal.

Philip Ursprung
Professor of the History of Art
and Architecture, ETH Zürich

Revolutions of Choice

1

This third compilation of research is an anthology of our experiments over the last eight years. We have shifted our priority to an experimentation that is increasingly less ornamental and more performative, and made through material economies, engineering, and sustainability.

2

In an existing and post-pandemic world, we have also prioritized working ever more locally with students, fabricators, manufacturers, consulting engineers, and artists. This has allowed an immediacy, a hands-on aspect to our work realized through trial and error.

3

Experimenting requires a work arena that can allow for success and failure, and that is unburdened from the everyday demands of budgets and strict deadlines. It must exist in parallel to our building and teaching practices.

4

This edition emerges from the 2020 exhibition *Revolutions of Choice* at Haus am Waldsee in Berlin. The work shown and described here corresponds with the six organizing principles of the exhibition: Bundles, Casting, Topographical, Liminal Facade, Ultrastructural, and Frames.

As with the two earlier editions of our *Atlas* series, the emphasis of this collection is on research – rather than on a specific or completed building project. Here, process is favored over product. Conceived as a manual, this book traces the trajectory of the continuing evolution of our work and research.

Bundles

7

Bundles are aggregate clusters of similar elements that together generate larger composite structures. The cellular character of an individual element allows, additively, larger and more complex structures to emerge. These systems are then combined into bundles to function as wholes.

8

But the whole is much more than the sum of its parts, and the elemental character of bundles is never subsumed by the whole. Cells vary with respect to geometry and form. Typically, these cells extrude into three-dimensional volumes, which are bound or fastened together mechanically.

9

These aggregates may operate at different scales and can be harnessed for various purposes – structural or tectonic (lighting, furniture, or surface). Plexiglass tubing, beer bottles, timber boxes, loops of cardboard – there is an inexhaustive list of bundle-structure materials.

10

How does one make history new? This question has guided many of our recent experiments with materiality. Three approaches, three ways of stacking structural elements, based on three distinct architectural histories, characterize our attempts at an answer.

For the 2017 Chicago Architecture Biennial, *Make New History*, we designed three affordable housing towers. The towers, presented as large-scale models, were similar in scope and program, but were differentiated by the historical models that informed them, suggesting new architectural solutions.

This is not a sentimental mining of history, but rather a pragmatic and performative one. As alternatives to conventional column grids (of the 7.3 × 7.3-m variety) or load-bearing walls, these bundle structures spatialize differently and offer alternative domestic experiences.

13

The advantage of these alternative spatializations lies in their references to the randomness of the natural world: they produce, in turn, spaces that can be experienced as clearings.

14

Confronted with the question of how to make history new, we interrogated obvious and unexpected historical models of spatializing; we studied the structural, visual, and typological formulas they offer; and we generated our own structures on the foundations of these bygone experiments.

15

Certainly, our creations are indebted to these models. But we avoid anachronistic creation; our spaces belong unmistakably in their place, in our present.

16

Stacks of peeled logs – informed by Heinrich Tessenow's design for the 1936 Festhalle in Rügen – characterize the first model. We stacked the colonnaded *halle* vertically on each respective floor, which resembled stacked forests rather than Tessenow's monumental interior.

17

Intended for affordable housing – which must accommodate a high density of people – these forests thin out as they rise in response to structural loading.

18

Tessellated wire (metal), which draws from Nancy Holt and Robert Smithson's 1971 film *Swamp*, characterizes the second. For this model, we appropriated the idea of a thicket as, once again, a series of stacked vertical volumes.

19

The ephemeral, dematerialized structure acts as a blurred visual field, complicating and intensifying the spatial experience of the space. We liked the idea that the thicket could offer the disorientation of a labyrinth.

20

The third prototype, made of infralight panels (concrete), responds to Eastern European socialist constructions (*Plattenbau*). It is another citation of a historical model (this time with relevant resonance as social housing).

21

We revisited a historical type in order to update it with modern, sustainable production methods and materials. We also exploited latent possibilities for sculptural formwork – formwork not contained in the historical model, but as a structural idea for self-supporting elements.

22

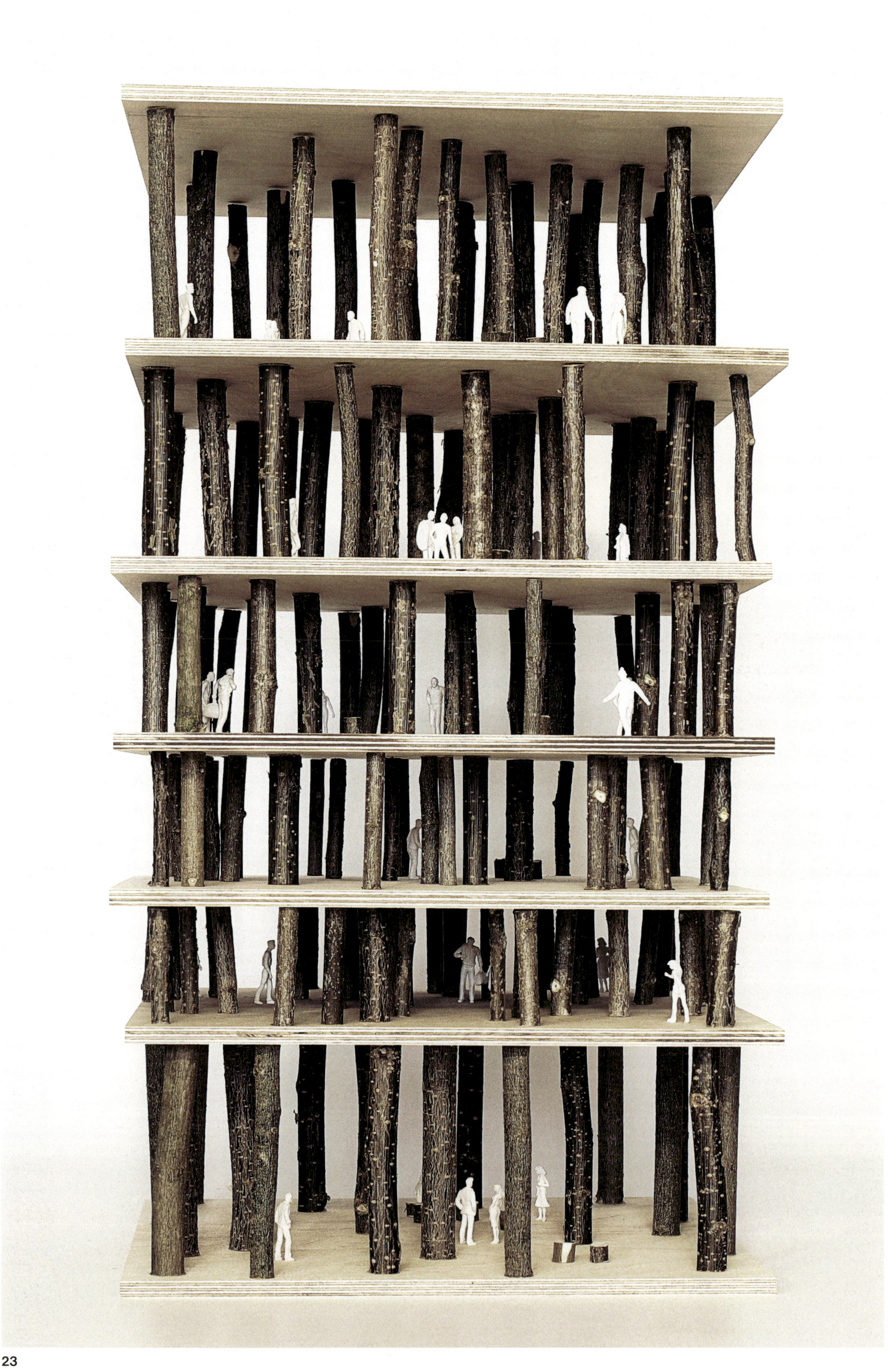

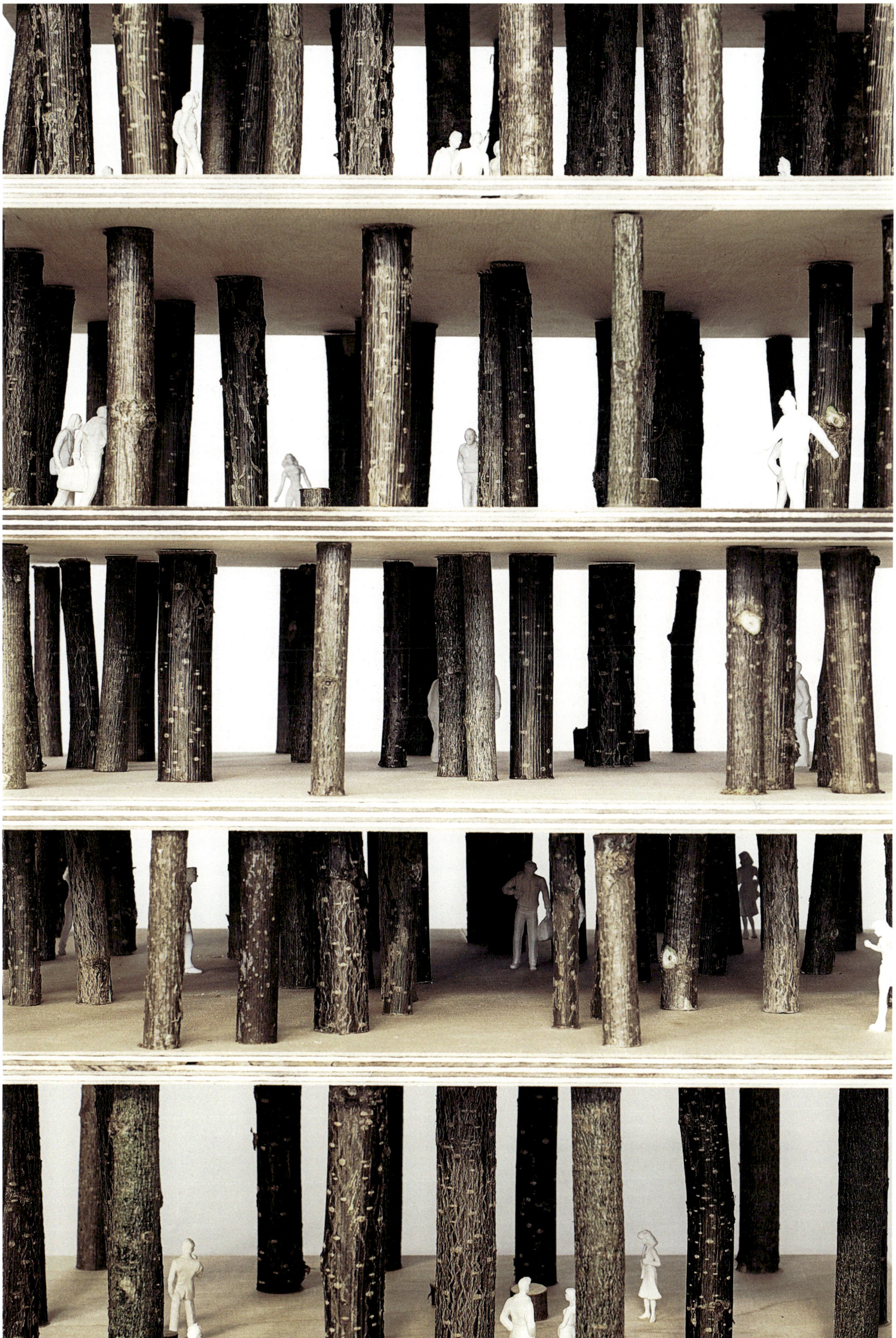

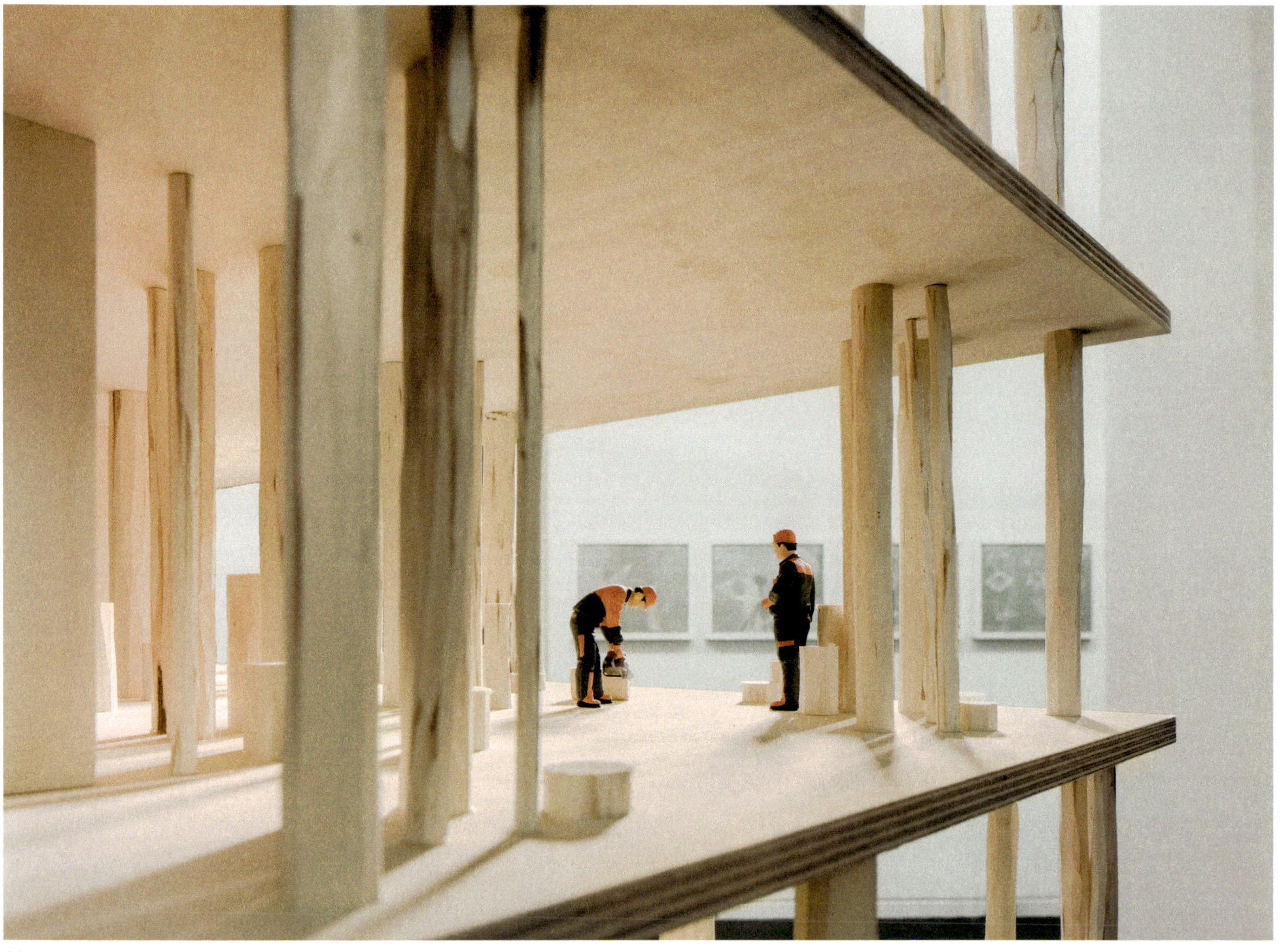

25

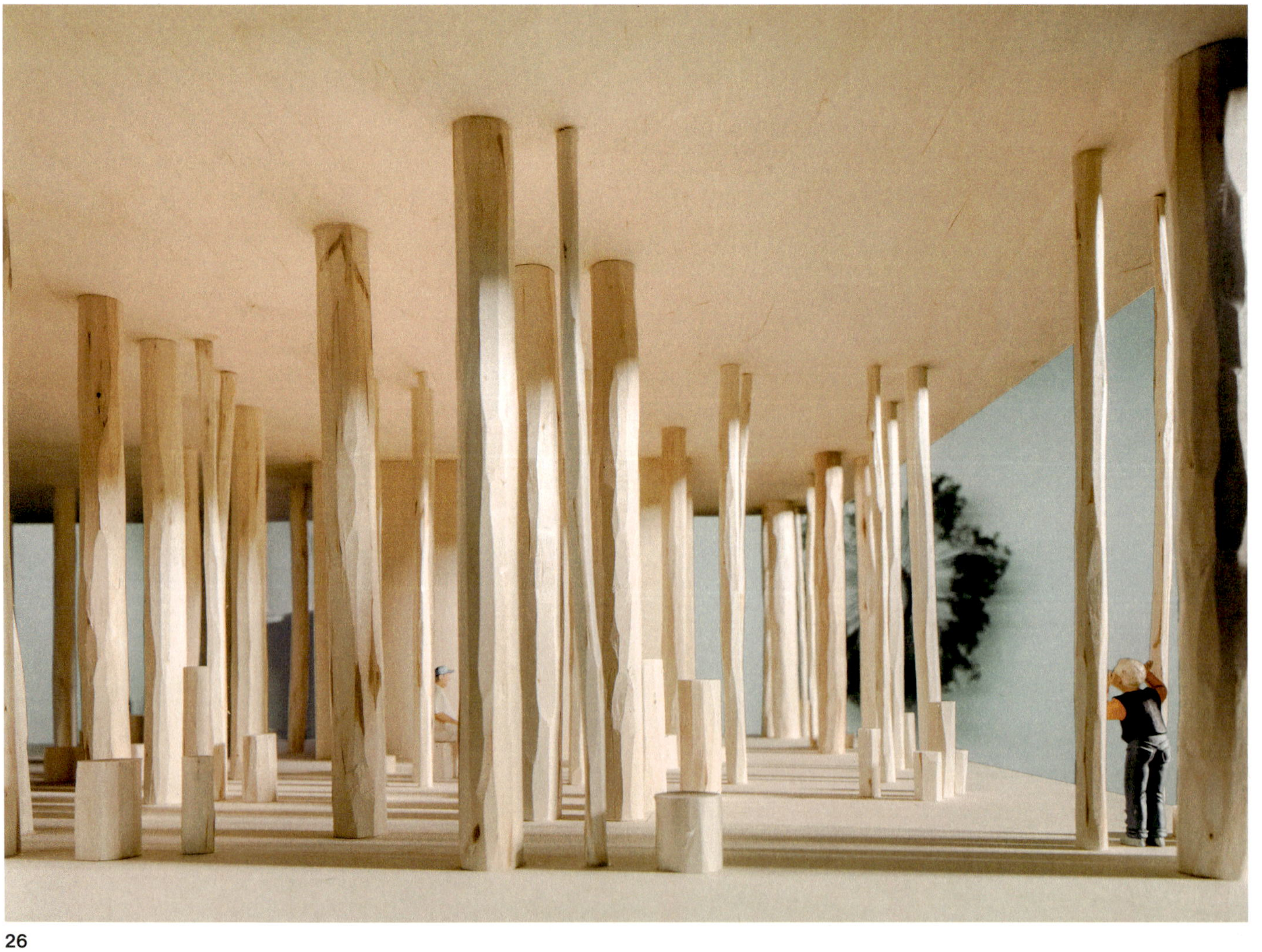

26

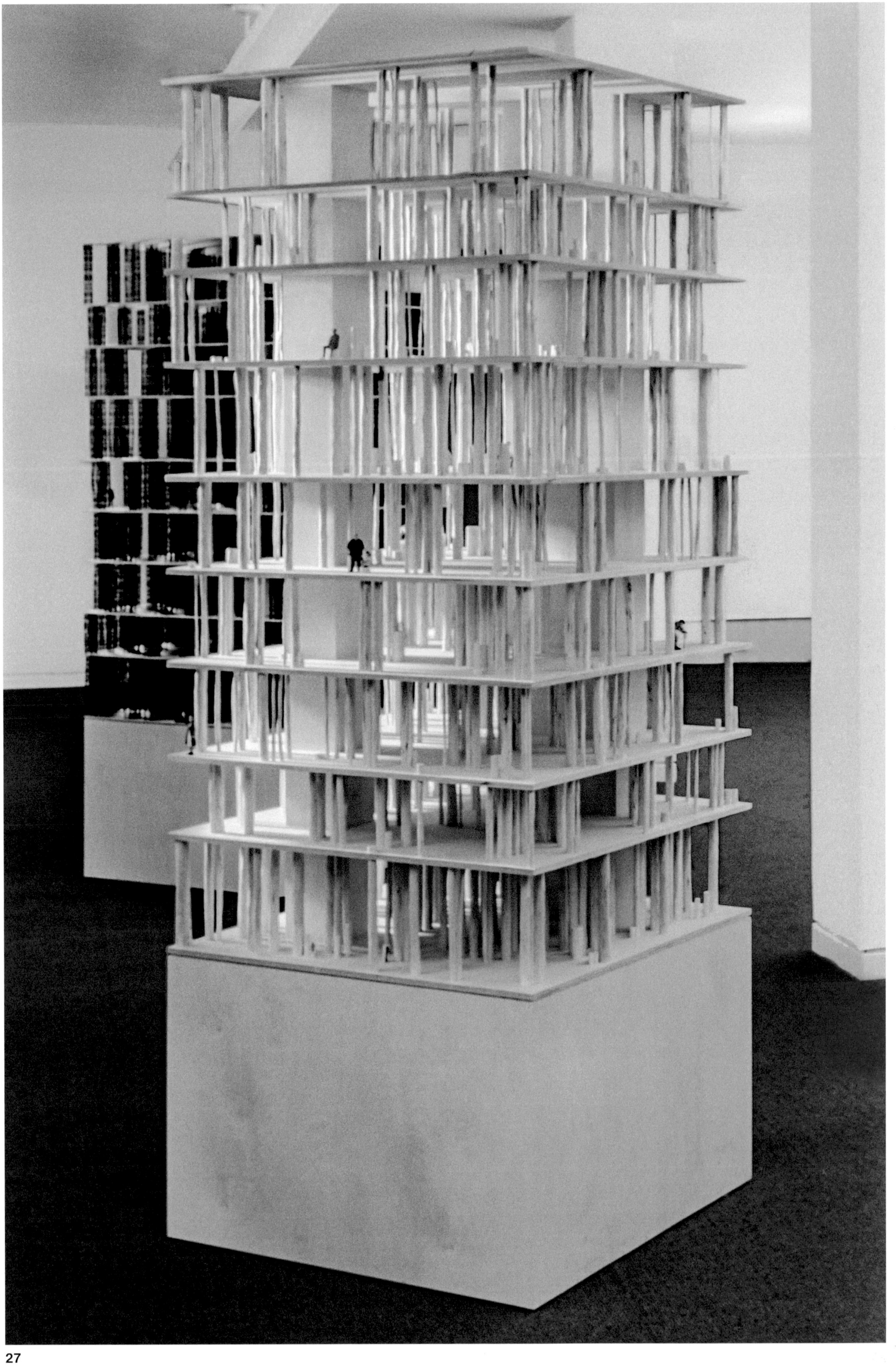

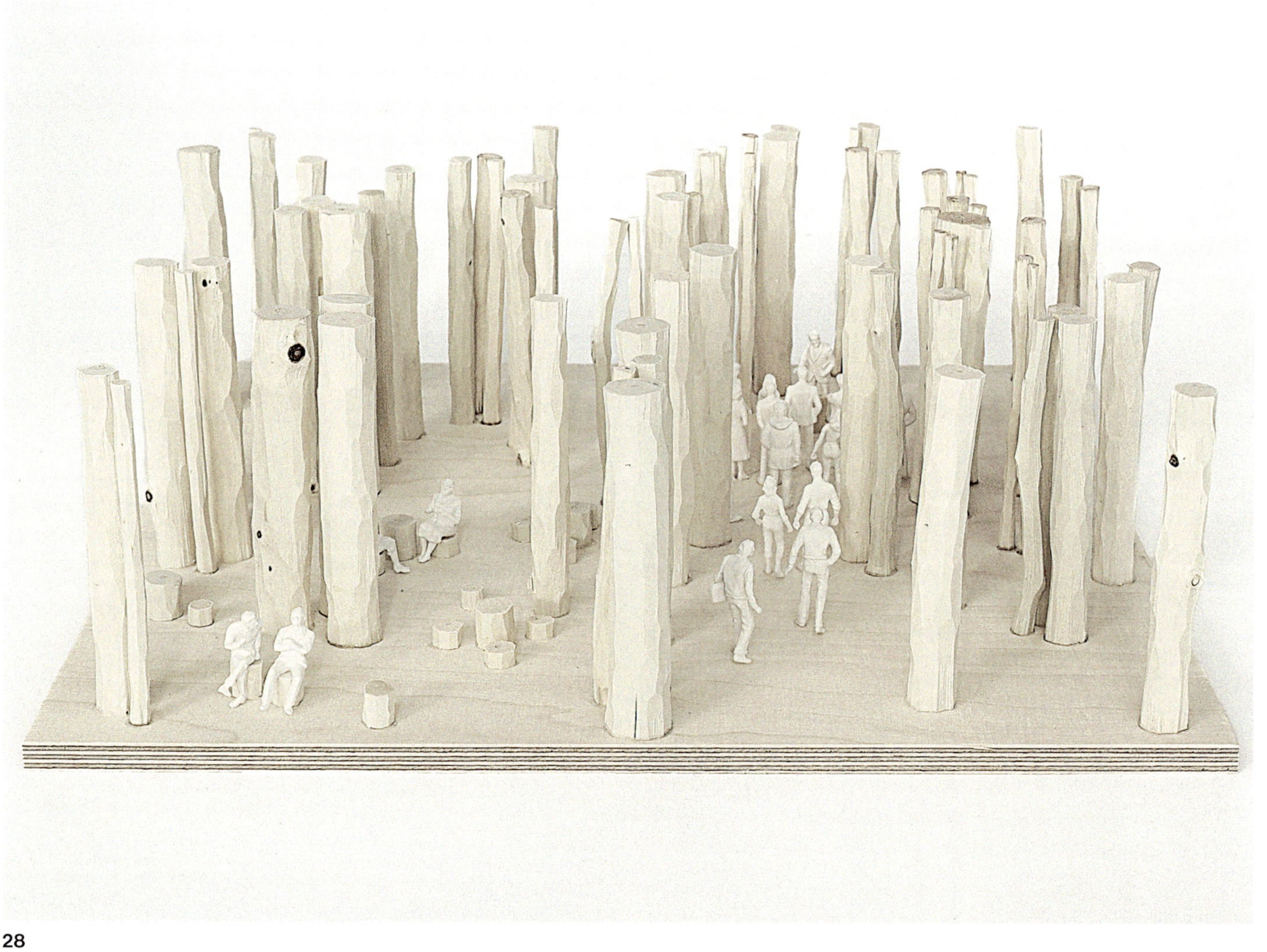

28

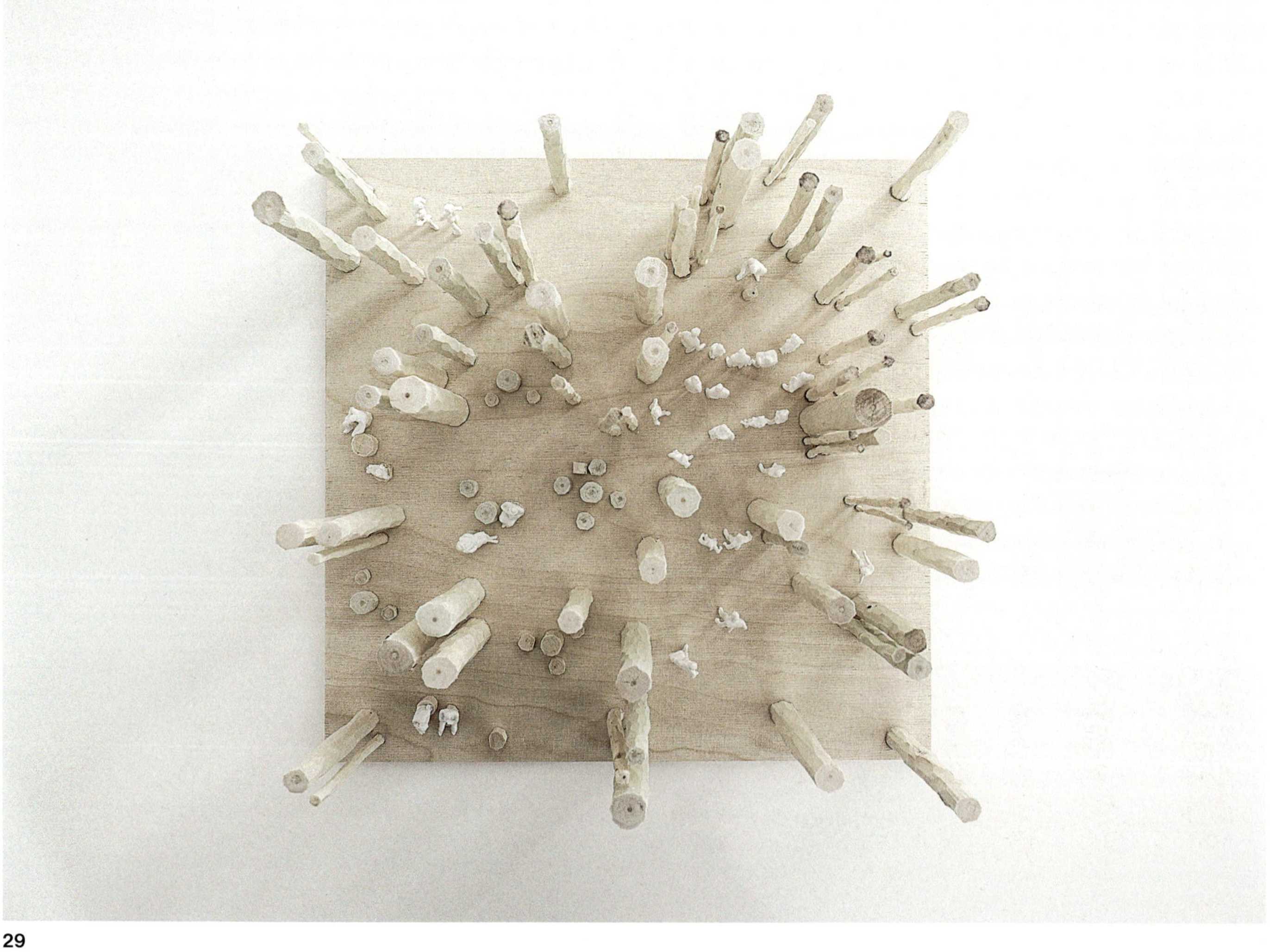

29

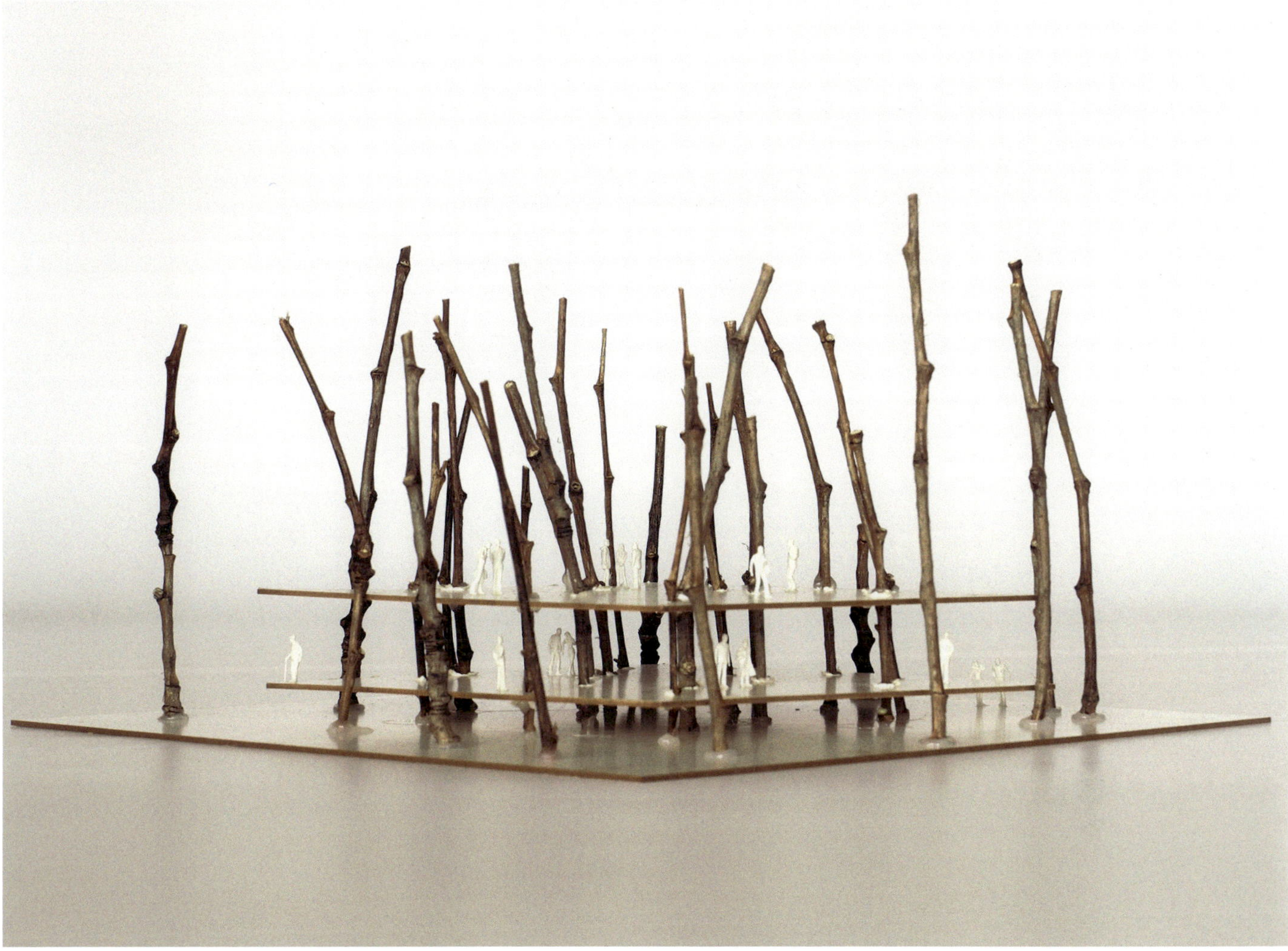

30

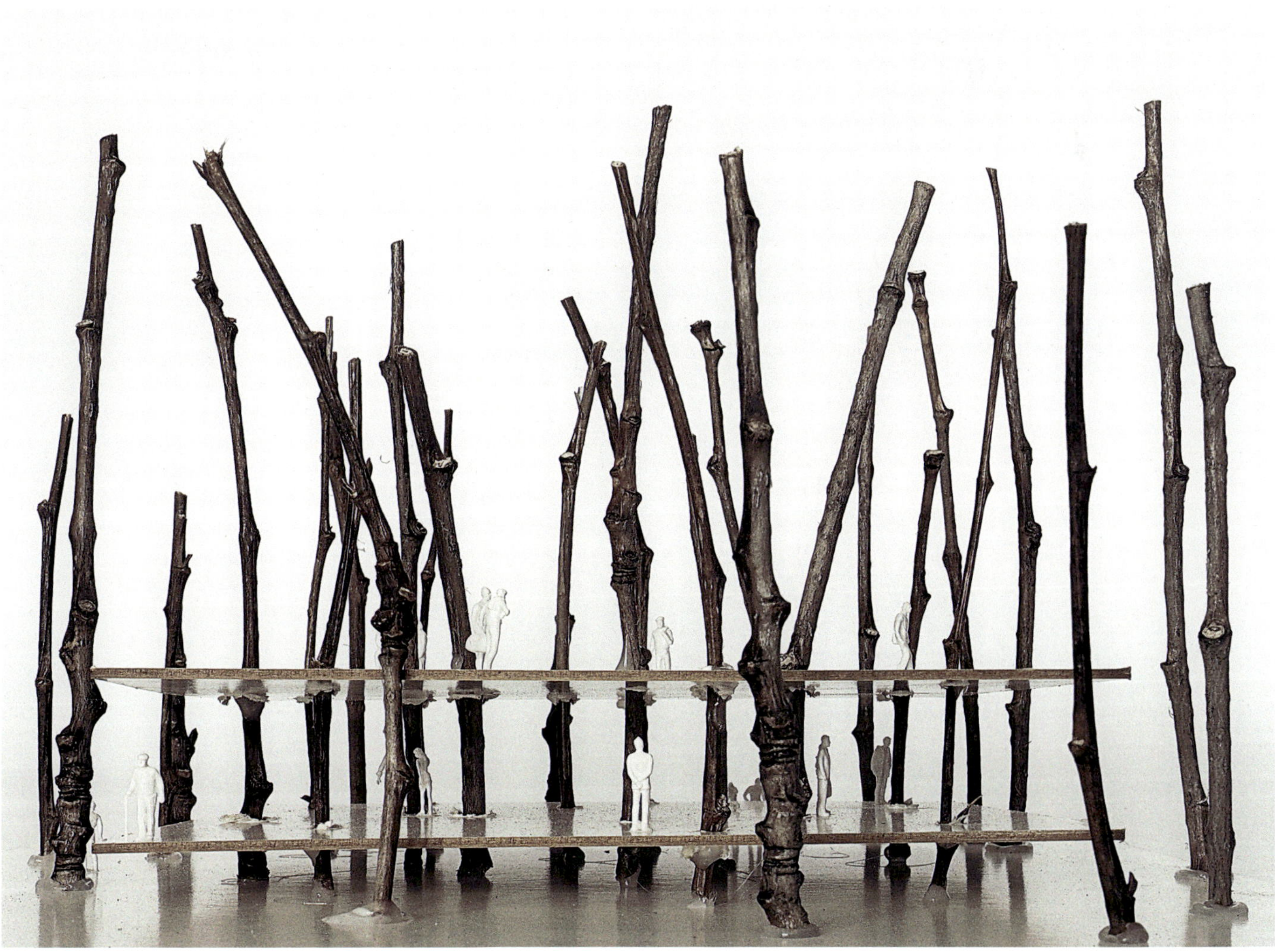

31

For Chicago's Smart Factory, we made a bundle at the scale of a large industrial chandelier. Reflective/polished stainless steel extruded arcs were seamlessly welded together to reflect and refract projected down light, at once sculptural and specific to the technologies of its making.

The sculptural quality of the chandelier demonstrates the operation of digital fabrication at varied scales and the harnessing of such techniques for the sake of capturing the play of light and material.

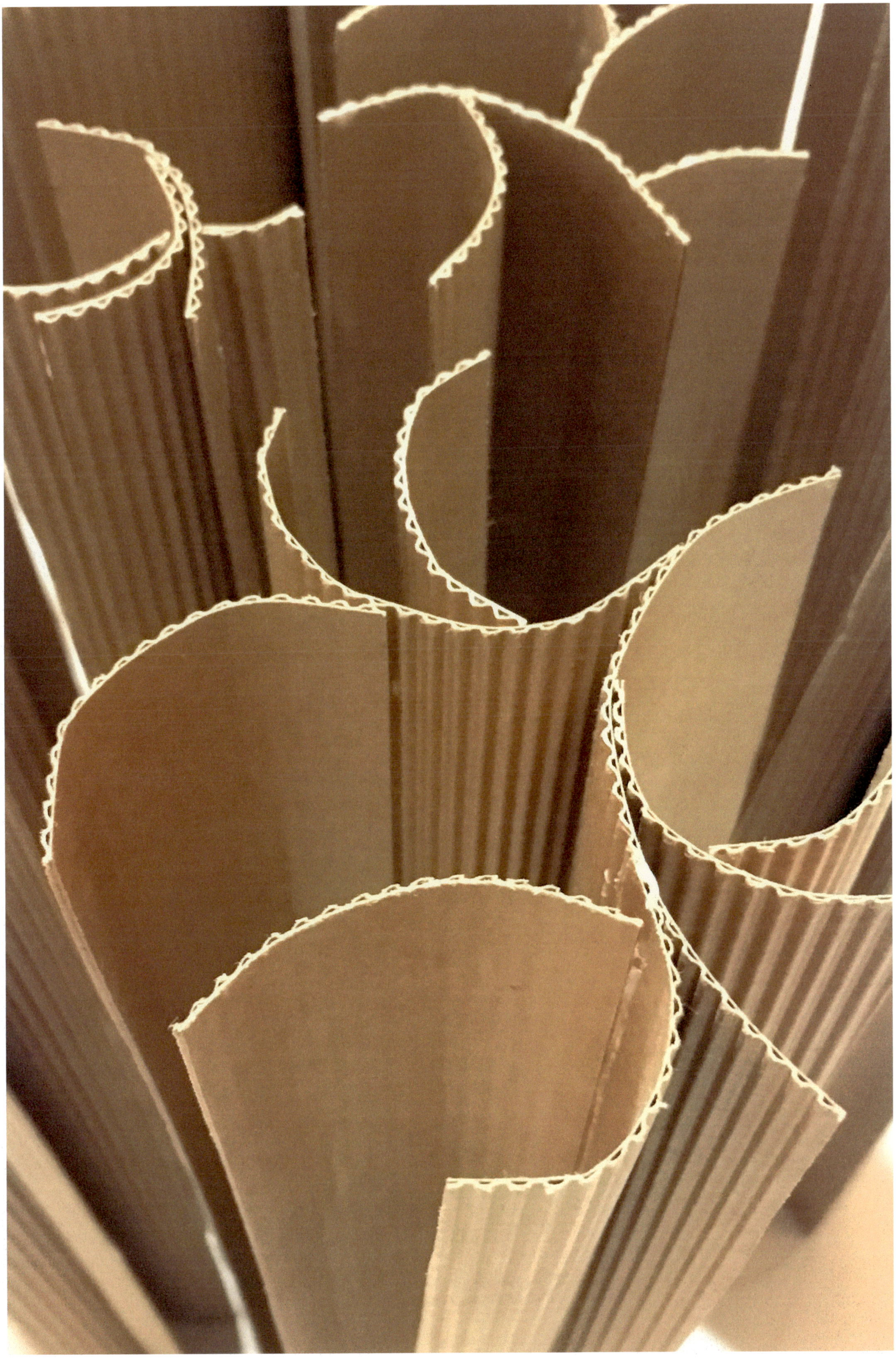

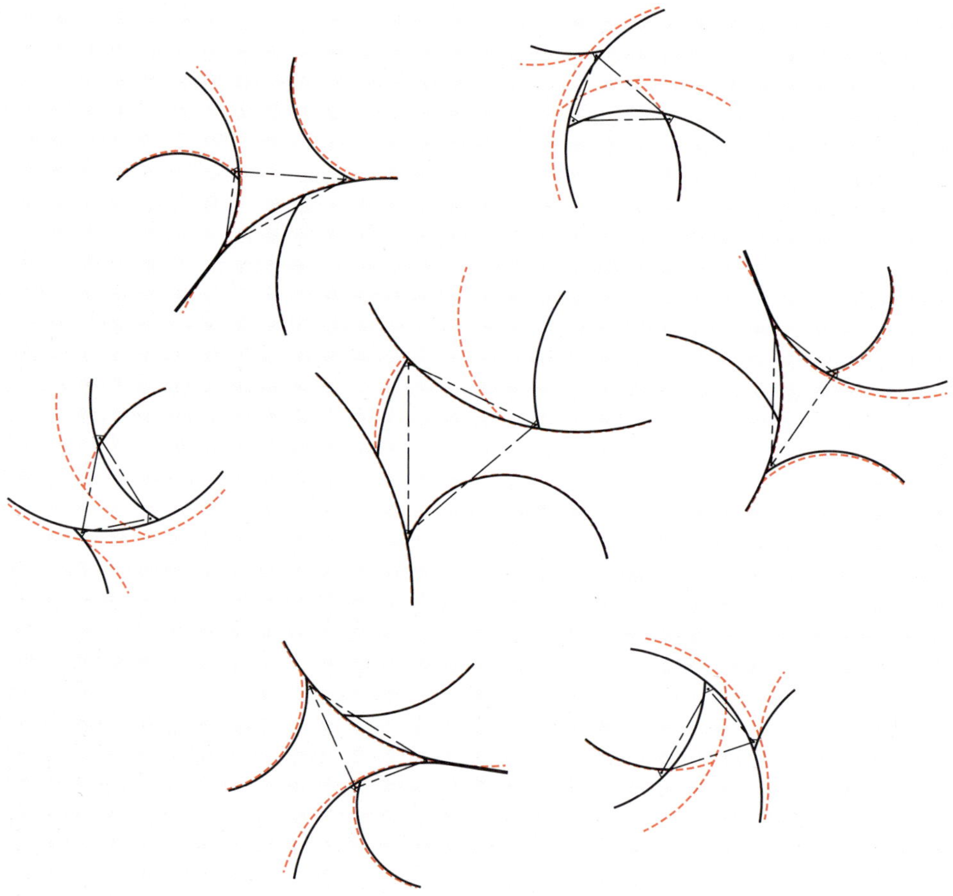

35

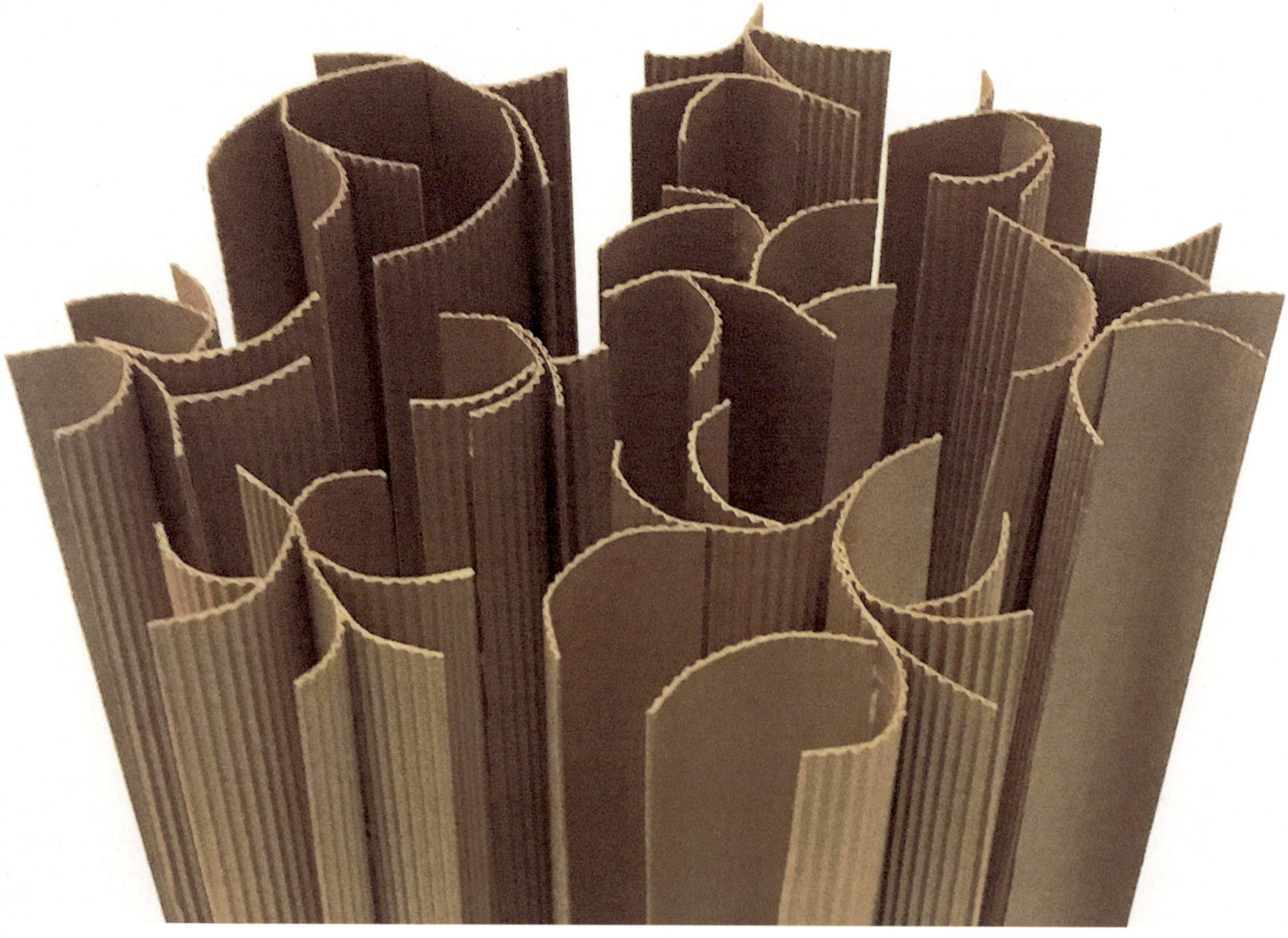

36

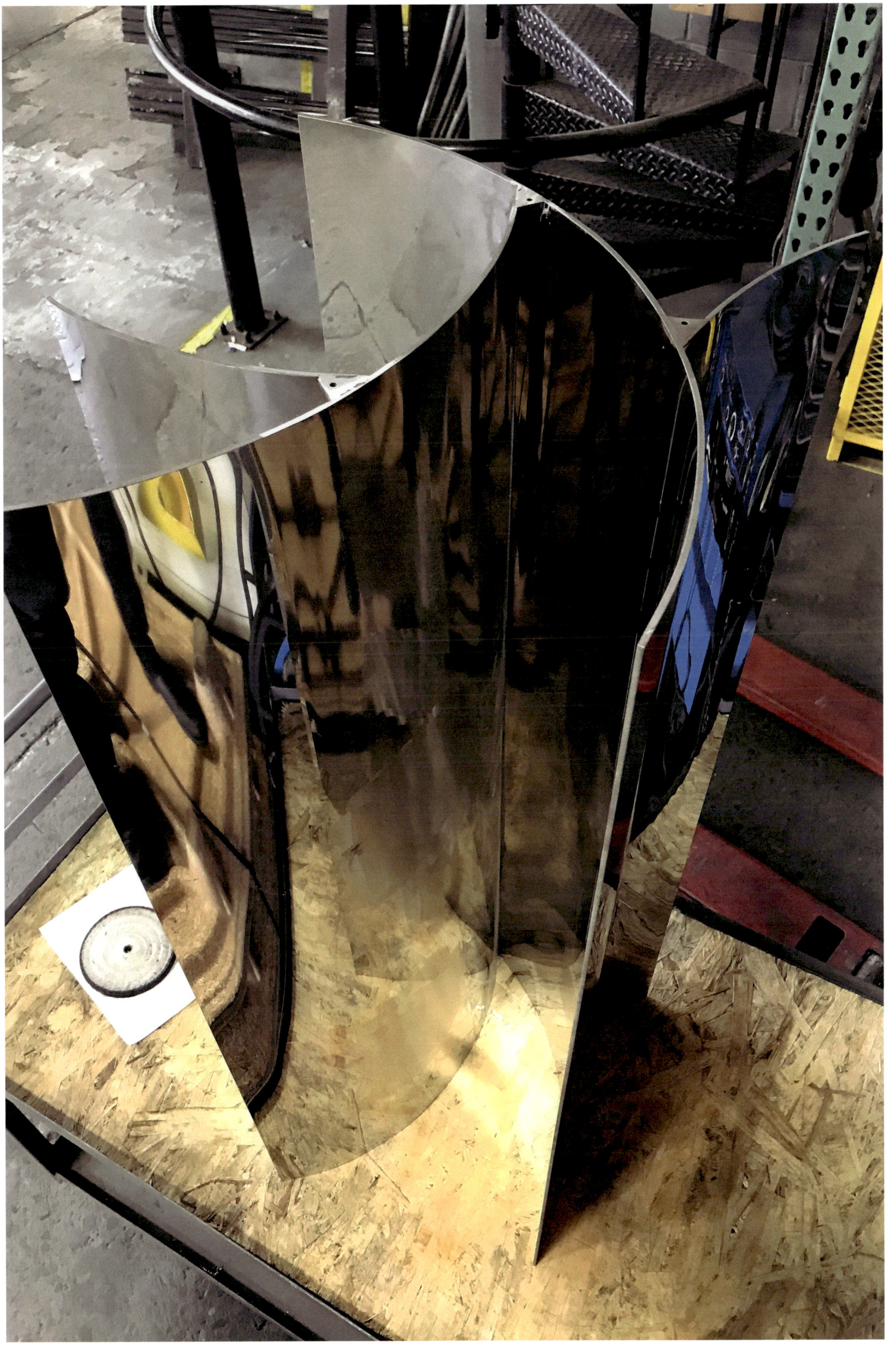

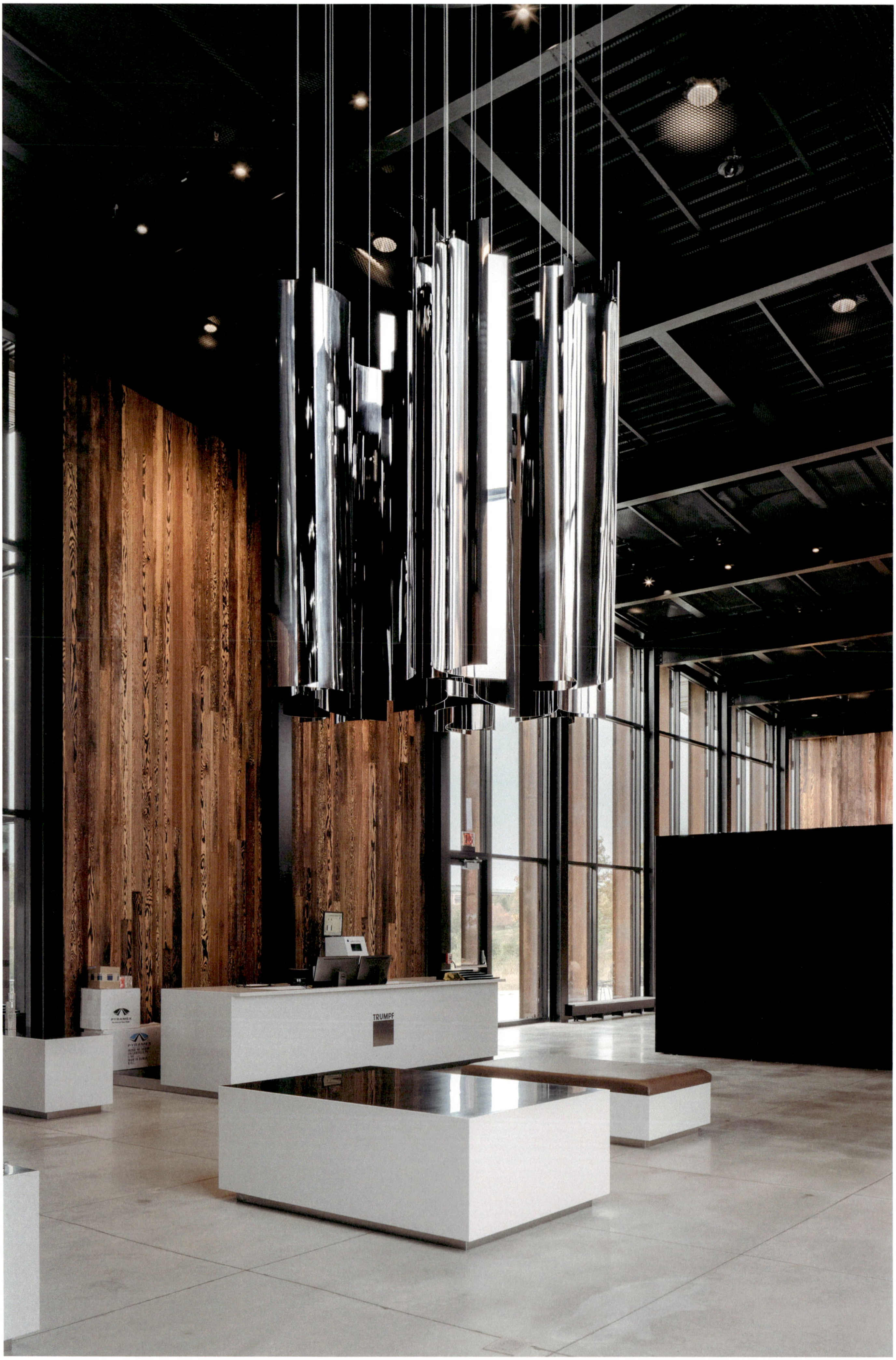
TRUMPF

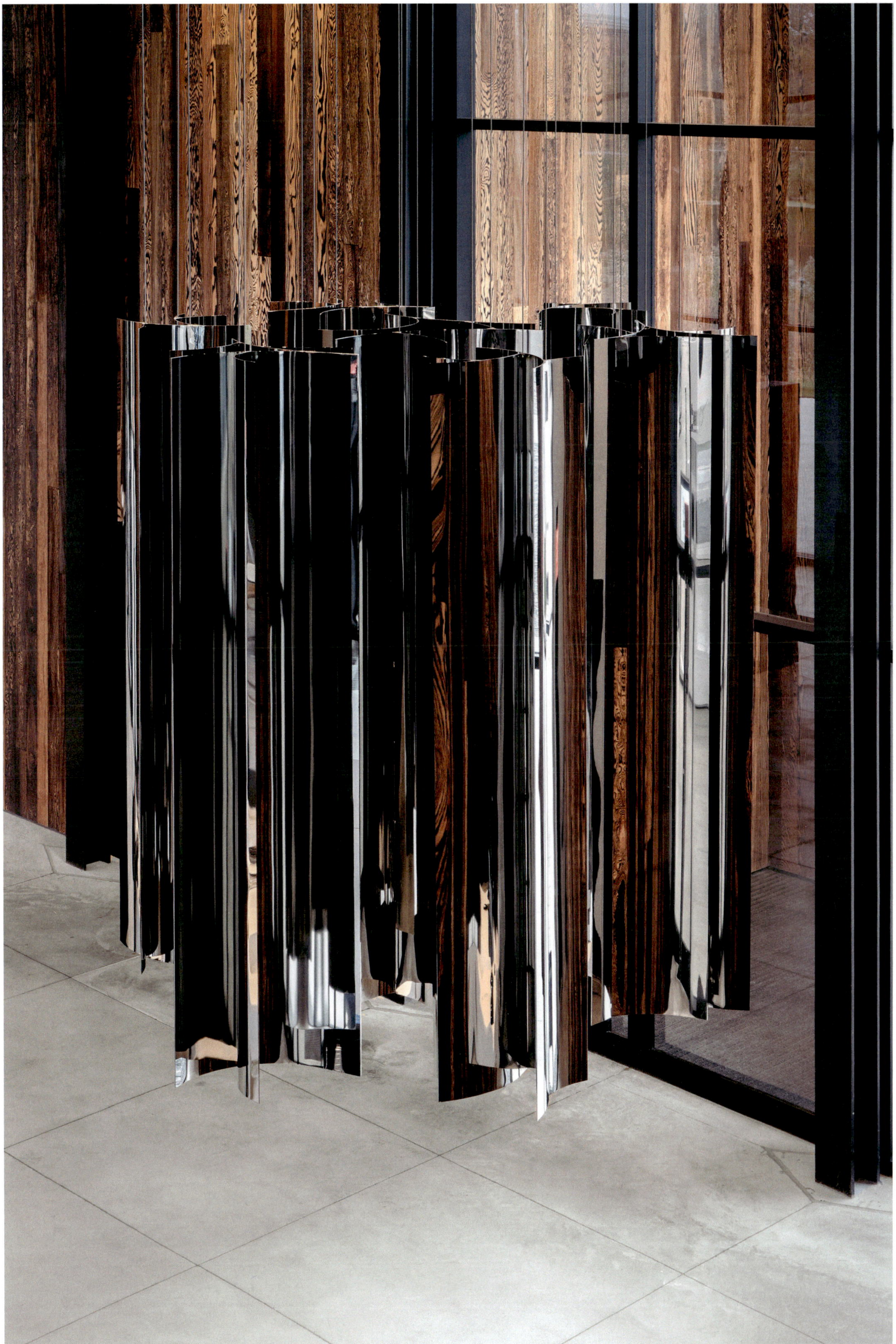

For the 2016 Serpentine Summer House Program, we conceived of a labyrinthine bundle. Constructed of steel tubes clad in 'bendy plywood' skin, the bands undulated and looped upon themselves, generating three bands of bench seating, walls, and a cantilevering roof.

Study models of cardboard were irregular: we strengthened them by looping the material over itself and fastening it. This looping technique remained instrumental in the realized bundle.

The temporary installation responded to a history of pavilion building within Kensington Gardens in London – specifically Queen Caroline's Temple (1735) by William Kent, as well as Kent's historical rotating pavilion, an early model of a picturesque viewing machine of sorts.

We conceived of the pavilion as a prototypical structure, built to be adapted and expanded. A counterpoint to Kent's structure, which asserts its solidity and longevity, our pavilion is organic – fluid, even. Despite updated means, our ends were nevertheless similar to Kent's: our structure, too, is meant to be a pavilion in the round.

This structure was relocated and reclad in a metal skin for the Haus am Waldsee exhibition. In this iteration, the new skin reflects sky, trees, water, and visitors, diffusing its form and shape into the gardens.

44

This site-specific iteration reflects, refracts, and dematerializes (thus tests) the architectural form within its new context. Where the Serpentine was haptic or opaque in its first iteration, self-referential in its materiality, it absorbs its new context in its reincarnation at Haus am Waldsee.

45

Barkow Leibinger's elegant, looping wooden project in Hyde Park is only the latest example in an expanding atlas of material research that they have undertaken for the past two decades.

What's harder to see, but evident in their own writing about this project, is the affiliative, experimental impulse that by now has its own historical genealogy within architecture and goes even deeper than that of conventional modern histories.

One of the most satisfying features of their new installation is the happenstance consequence of a couple of Berlin-based architects stacking together some stressed, looping wooden elements to invoke an idea of a classically arranged summer house by a foundational English classicist and master, now long forgotten.

Thanks to Barkow Leibinger's declaration of interest, this allows us to see much more than formal similarities, or shared context, but rather this: the enduring importance of affinity itself in any attempt to meaningly construct architecture as a interdisciplinary project today.

These days, experimental approaches like that of Barkow Leibinger should be seen as no less classical than they are modern, above all in their commitment to an architecture not of fleeting material beauty, but of enduring interdisciplinary continuity.

Brett Steele
Dean of the UCLA School
of the Arts and Architecture

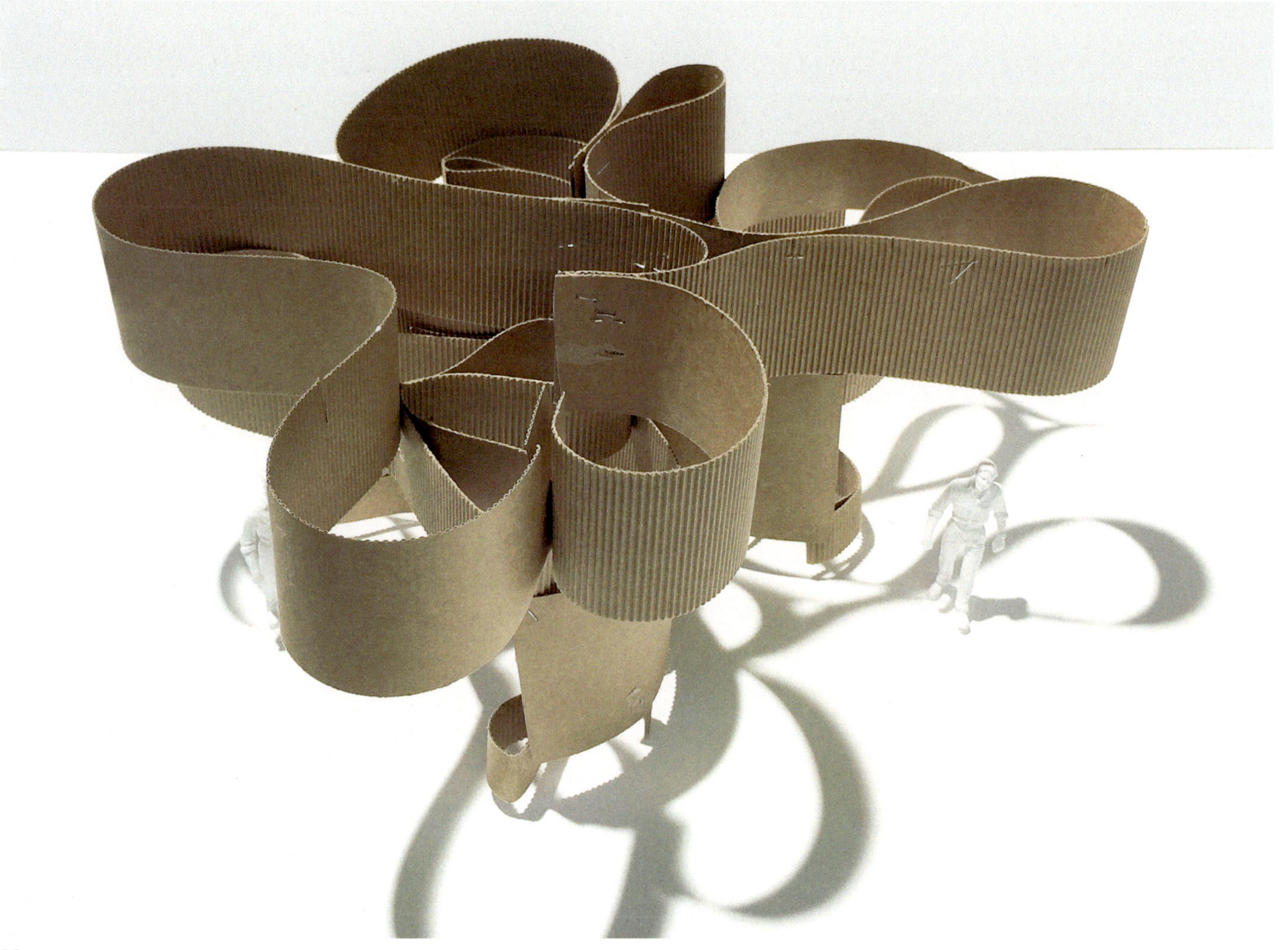

52

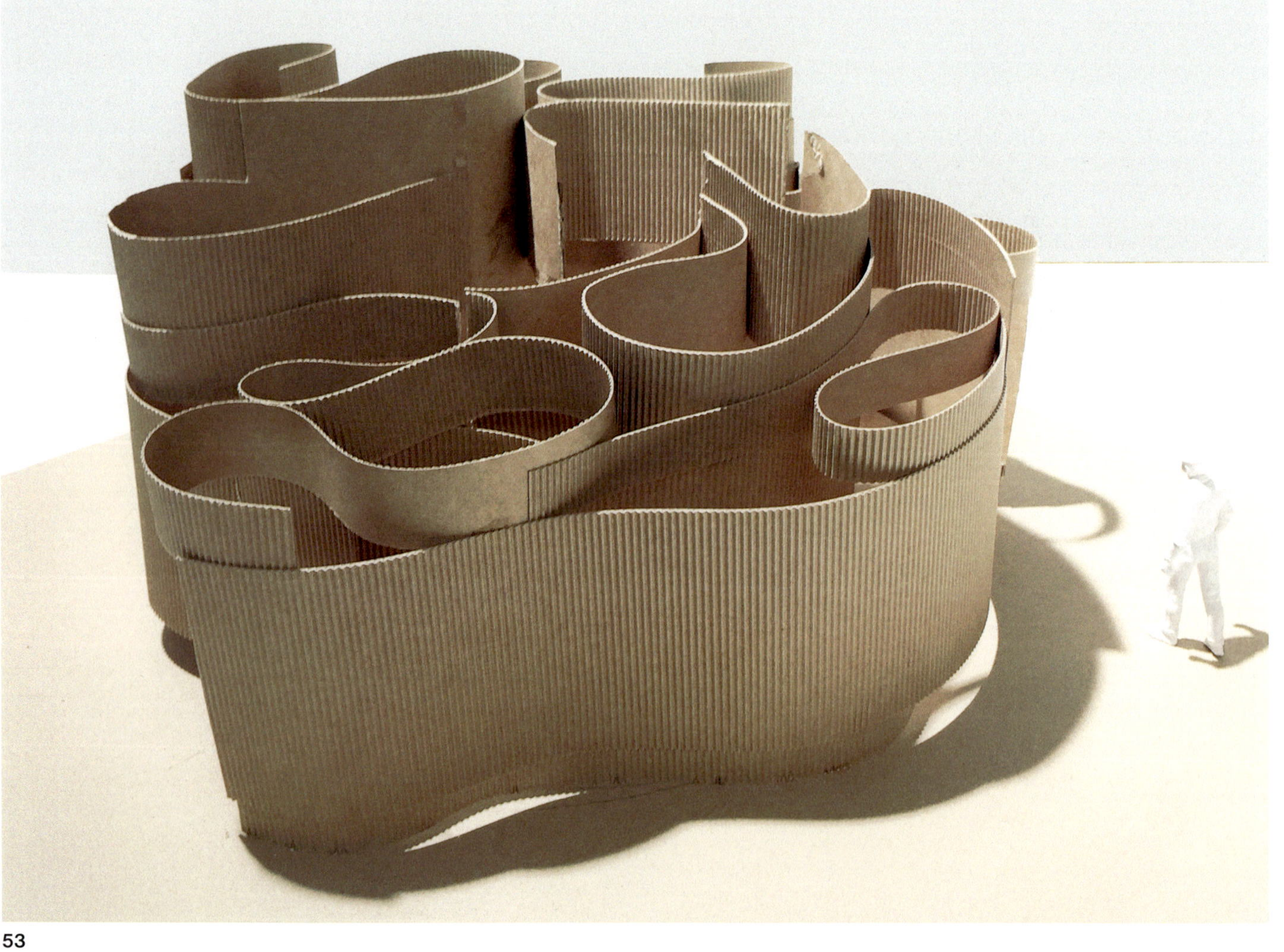

53

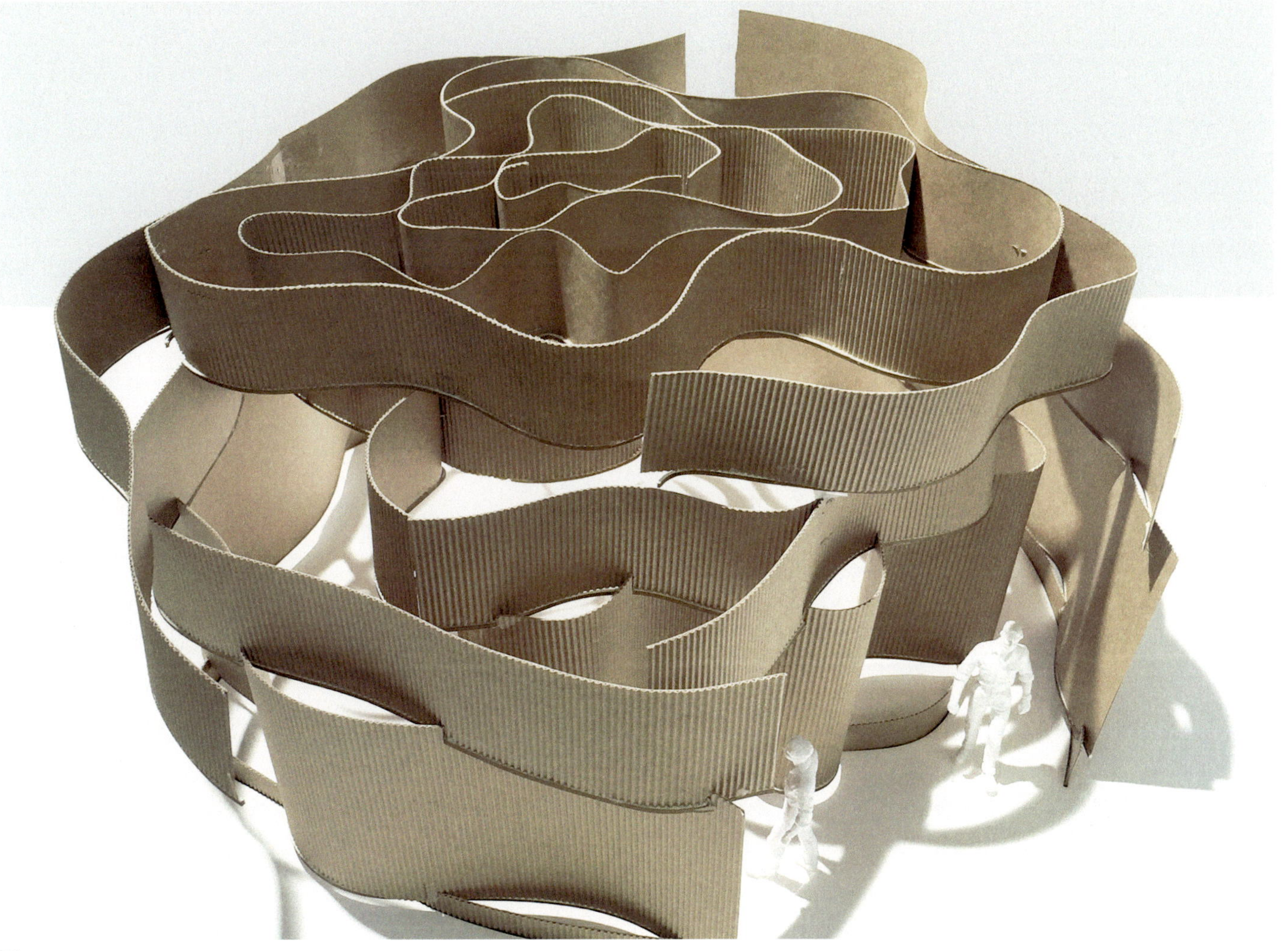

54

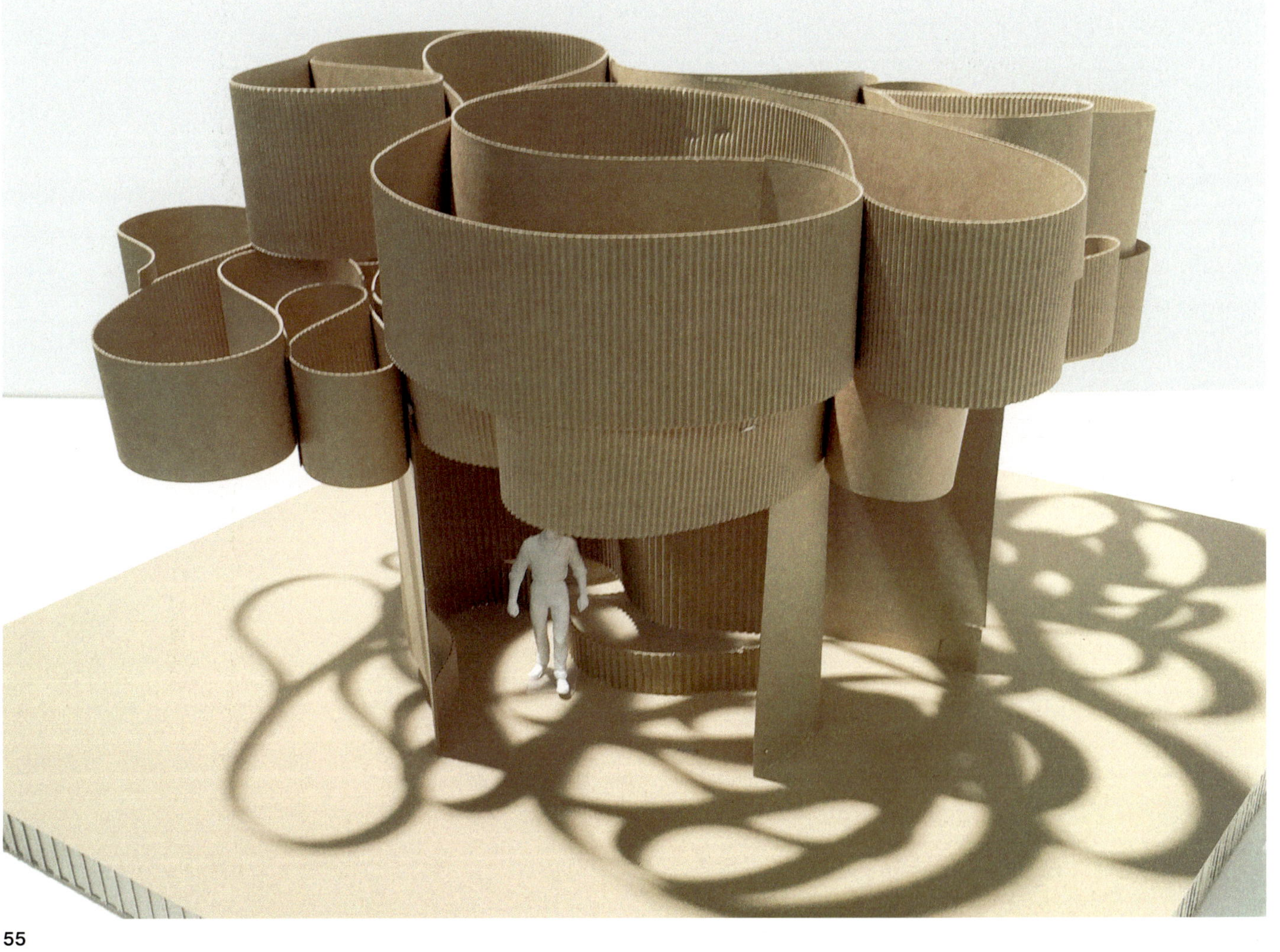

55

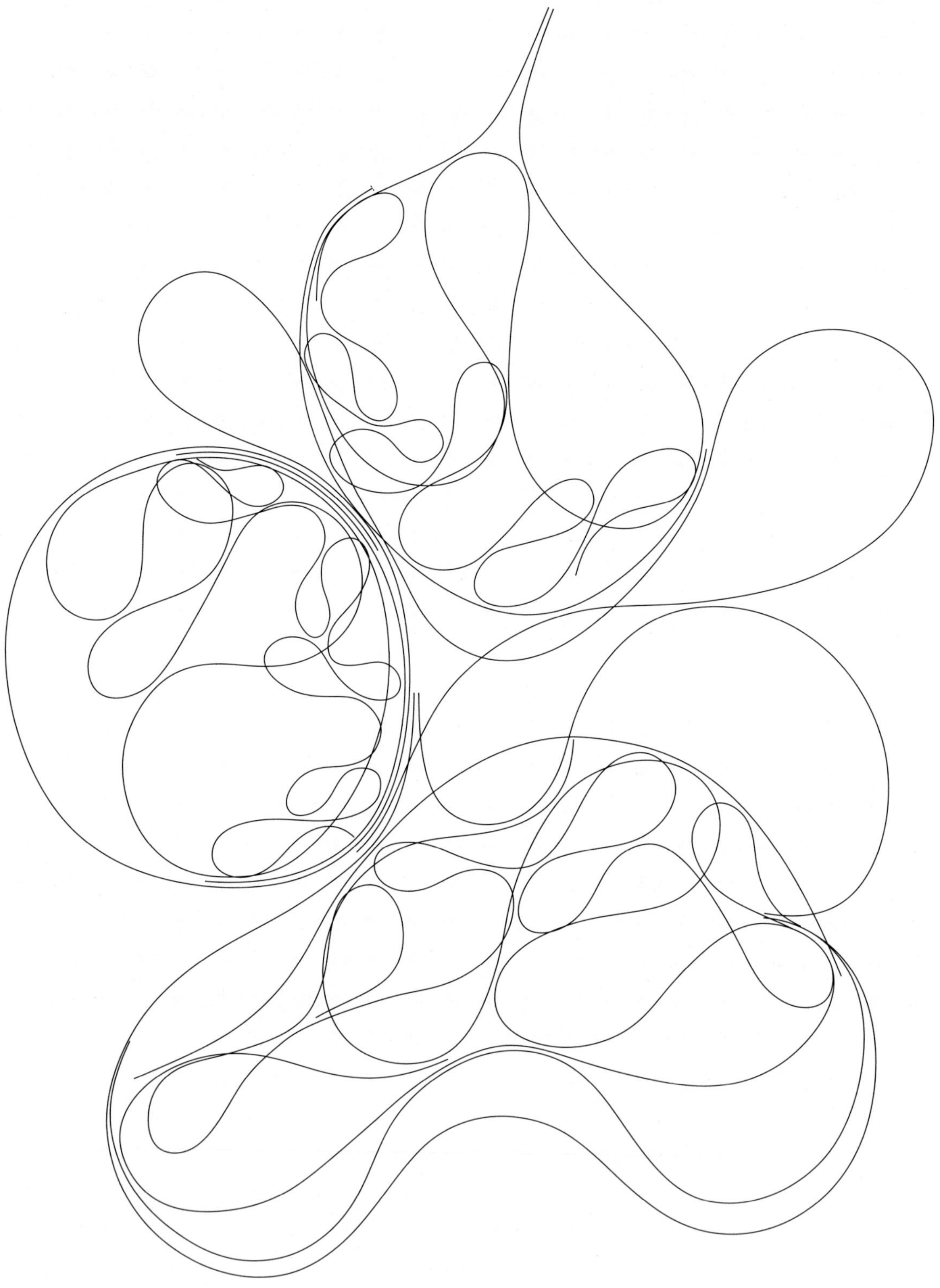

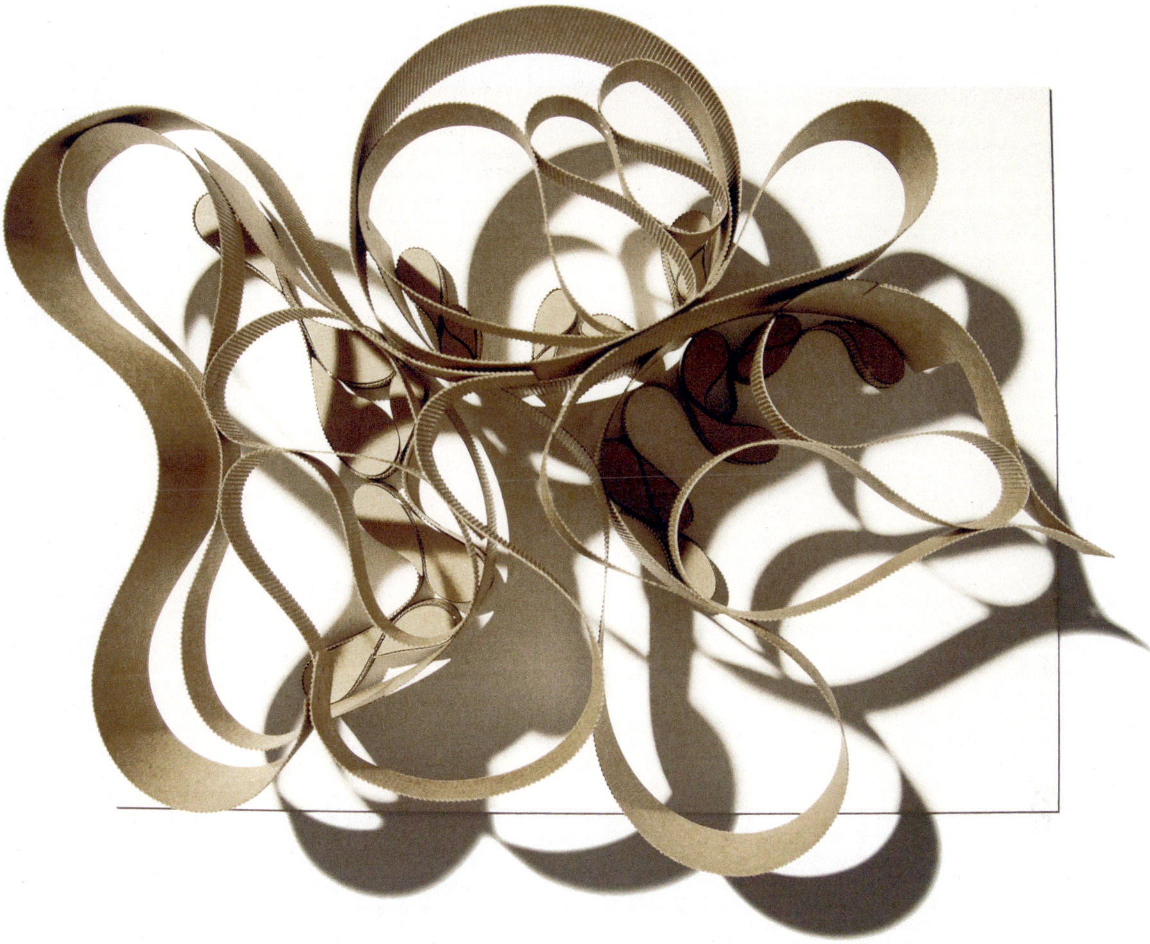

57

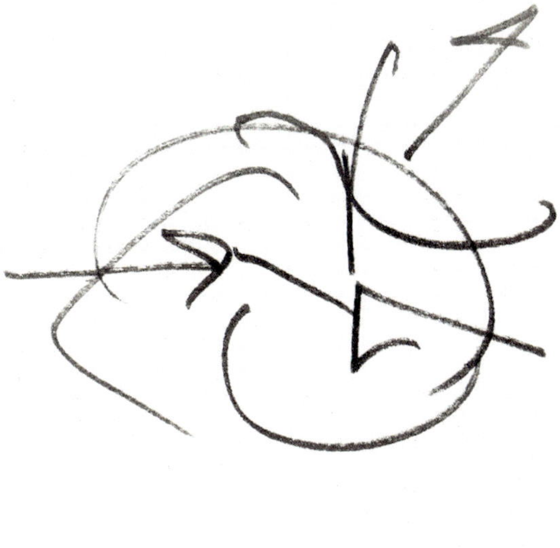

58

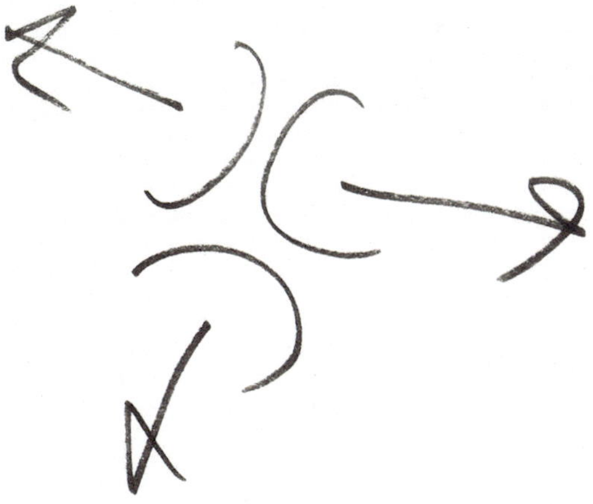

59

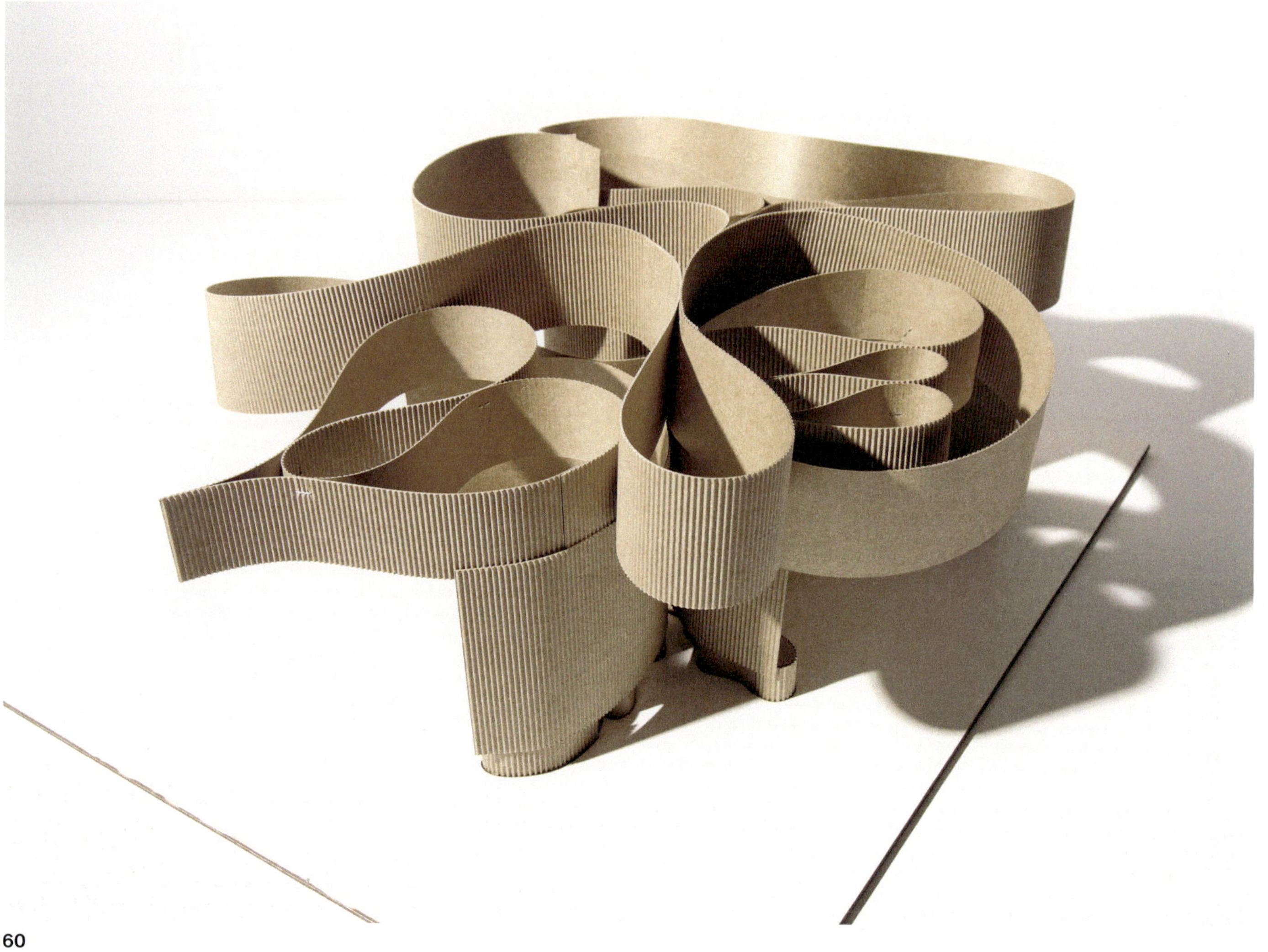

60

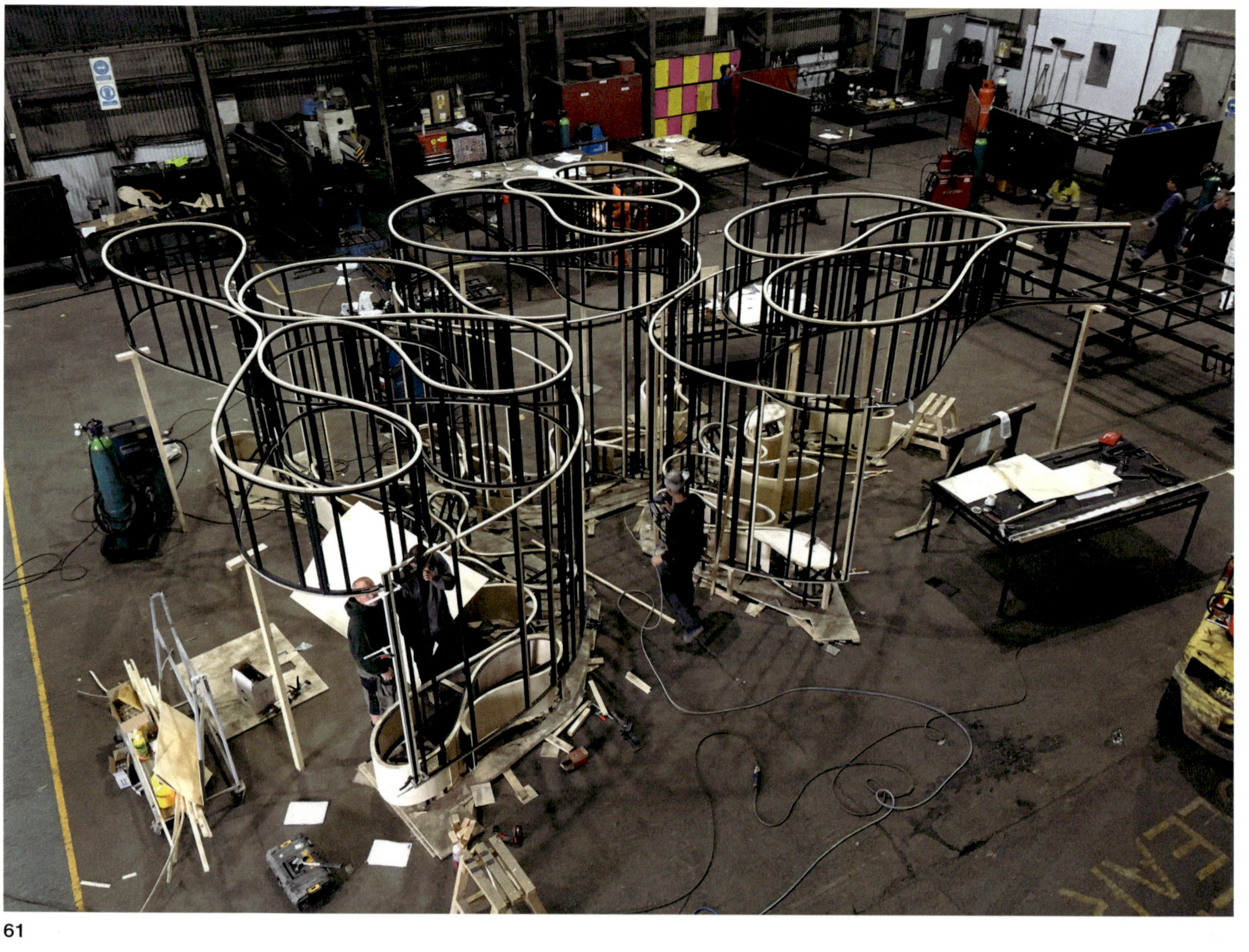

61

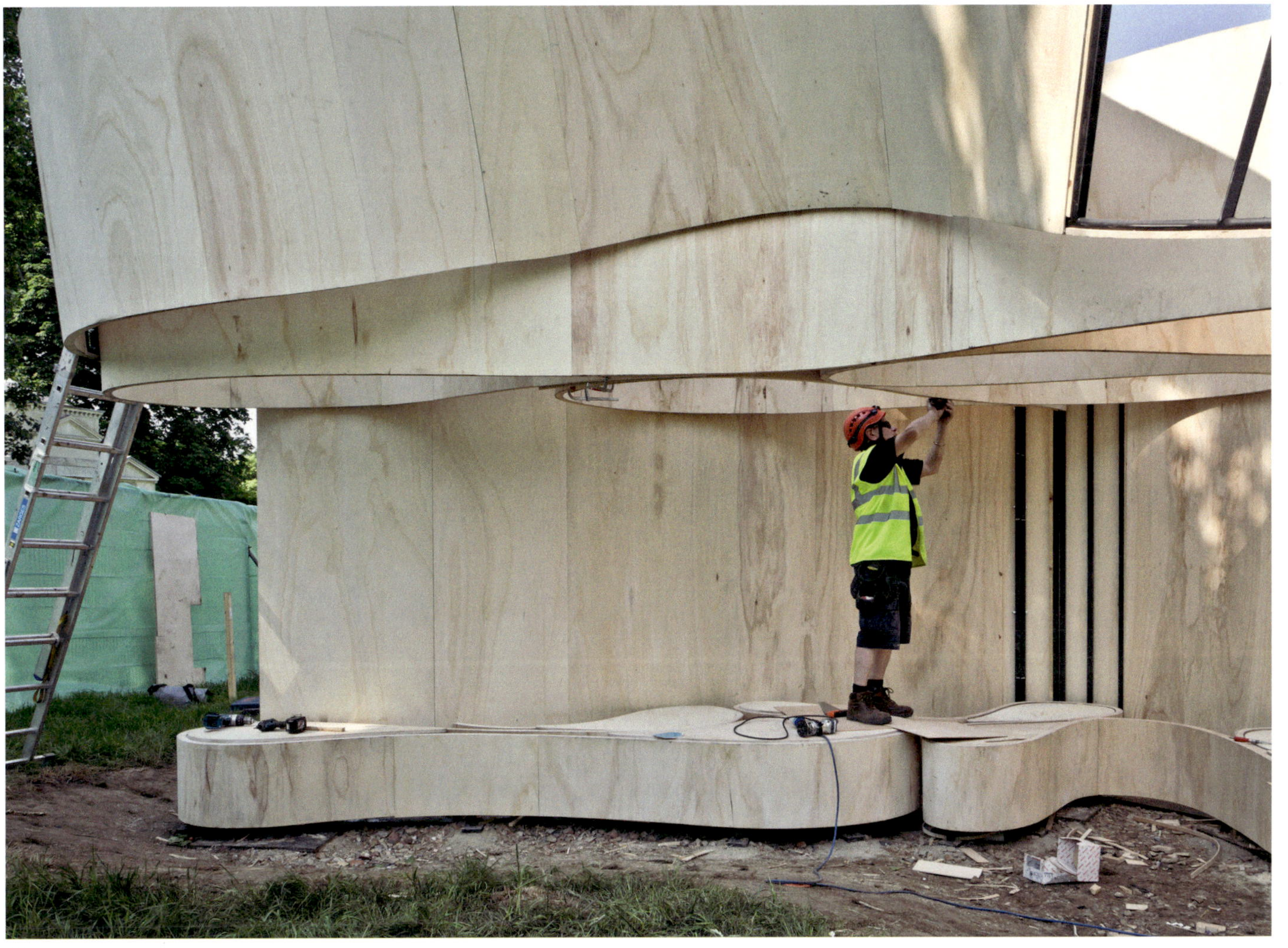

62

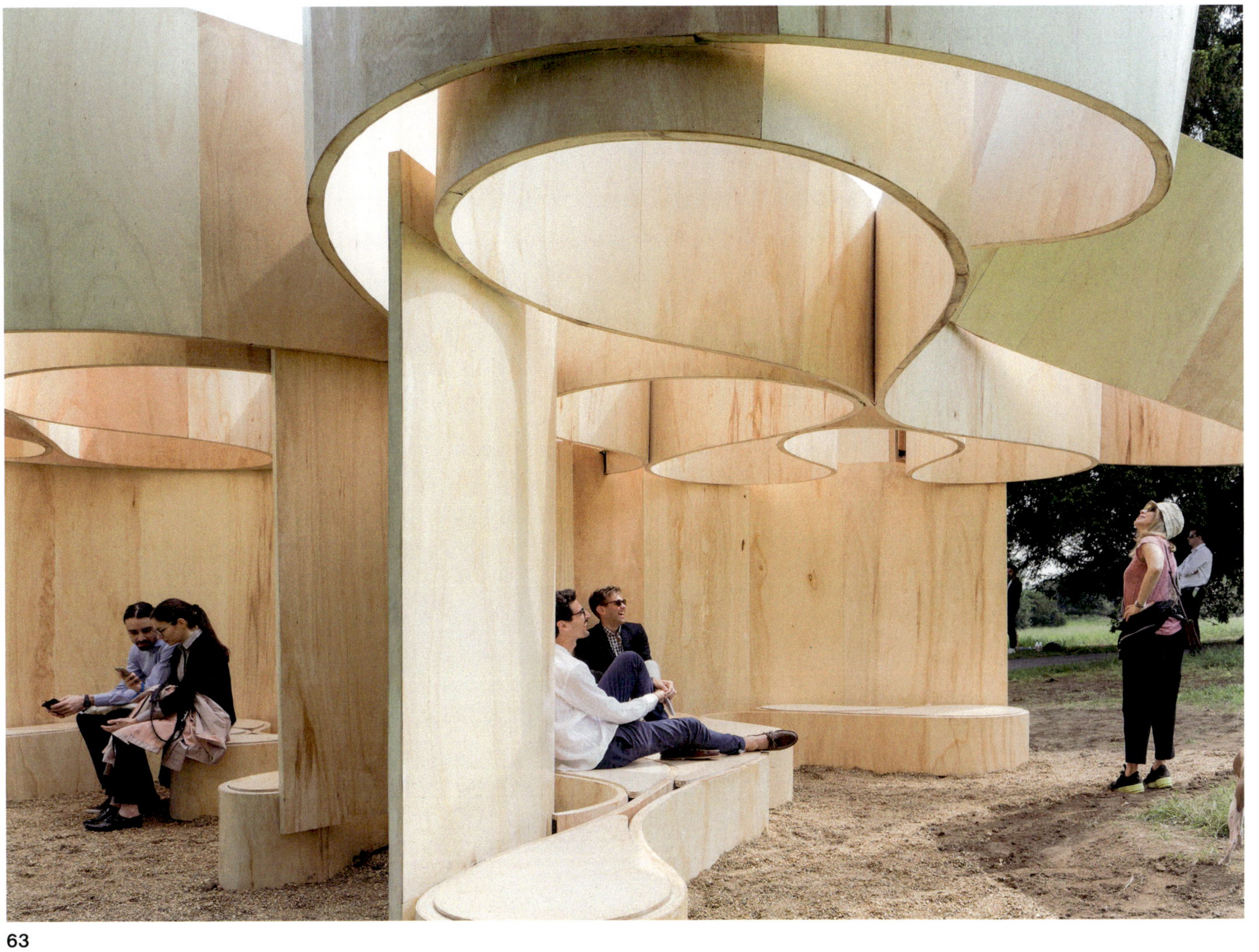

63

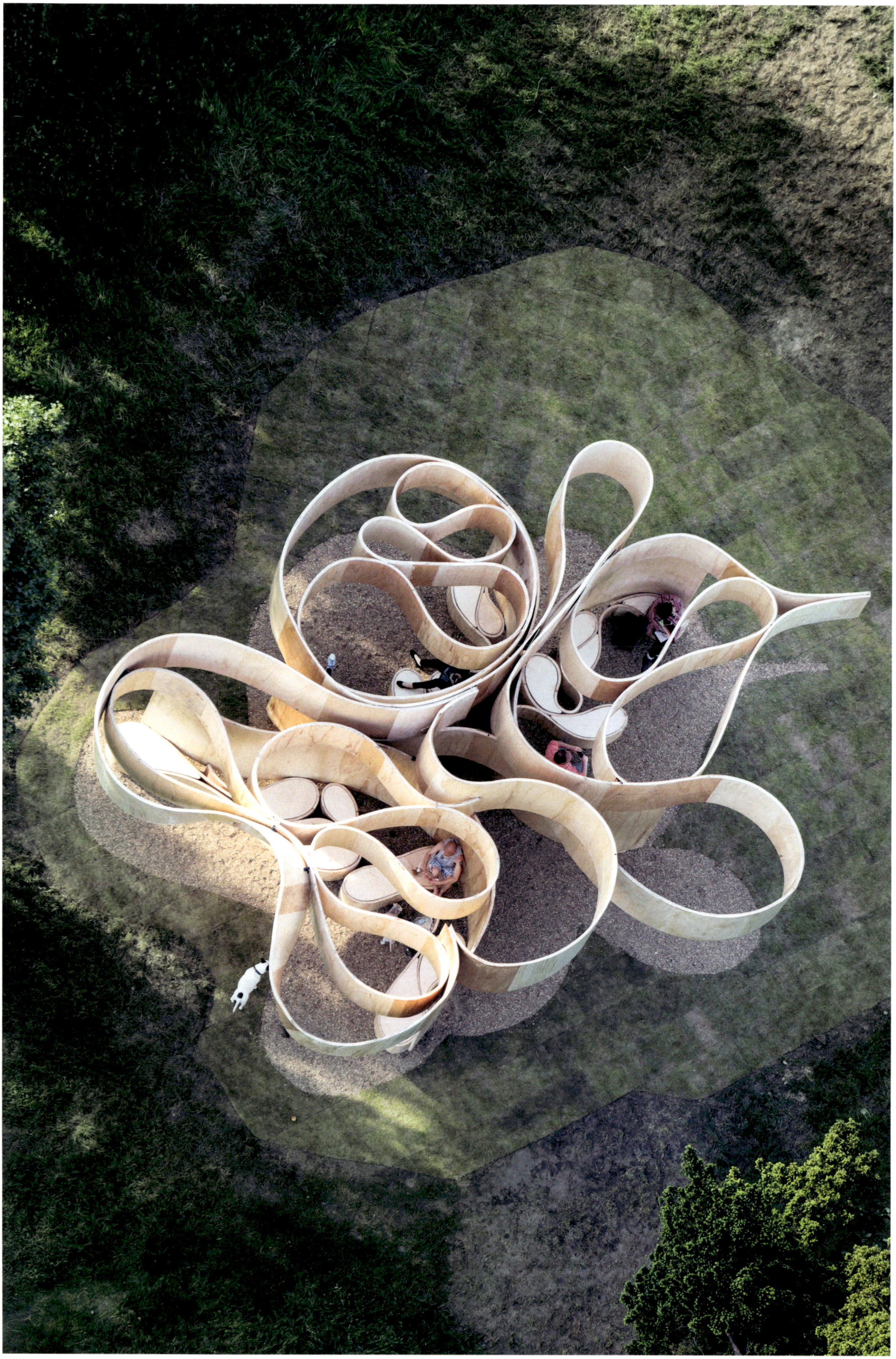

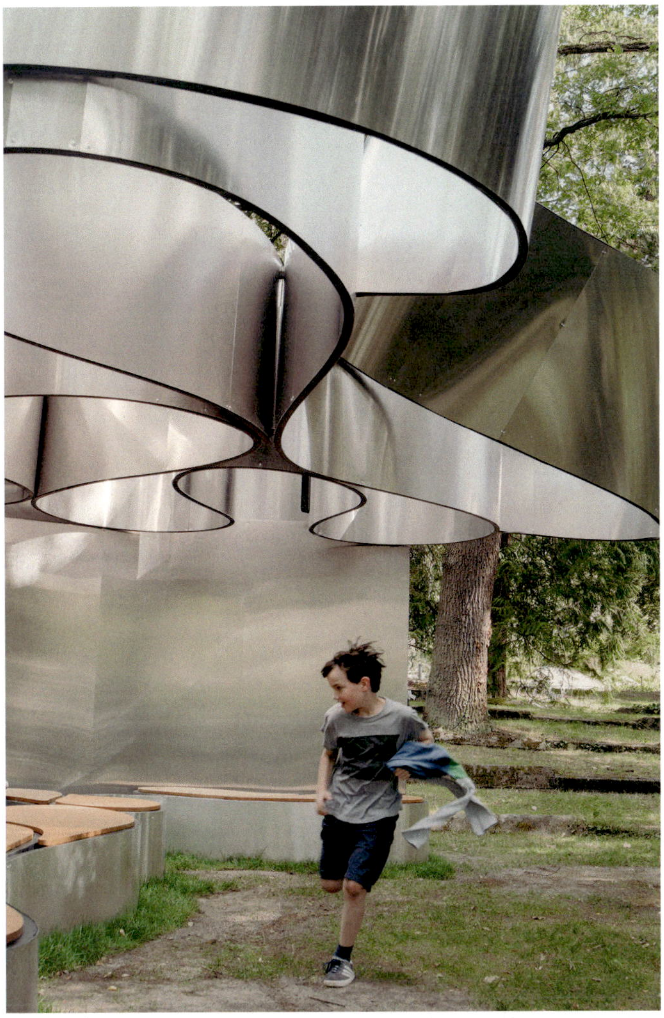

65

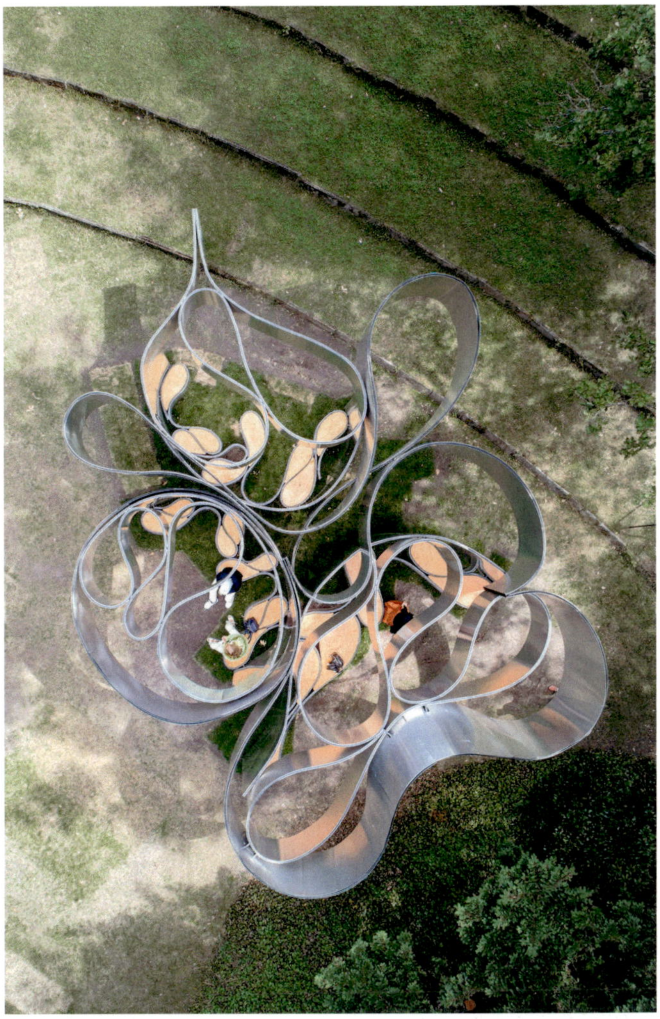

66

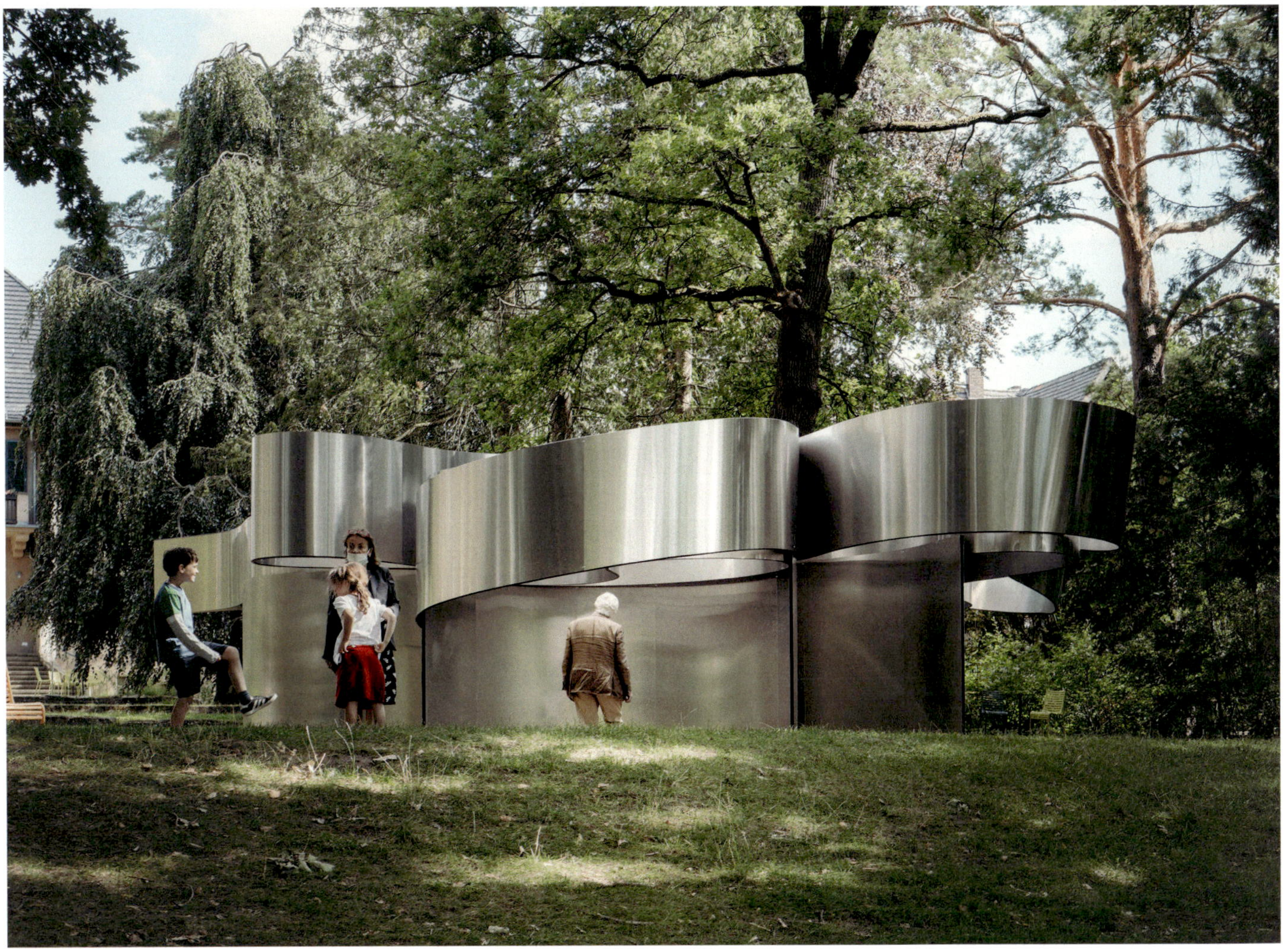

67

Thicket, a differently immersive bundle typology, was conceived by bundling together, with the use of zip ties, 2-mm stainless steel wires (digitally bent at a coat hanger factory) to form load-bearing columns.

Unlike conventional columns, these tessellated branch-like columns dematerialize the opacity of the vertical structure into layered and ephemeral lines – much like thickets in nature. The rods generate a series of spaces and passages through which visual and spatial densities coalesce, obscure, blur, and open.

Thicket blurs the boundaries between materiality and space making. This blurring of structure is a-perspectival and lent itself to the labyrinth-like space. The experience of disorientation, the effect of any labyrinth, is mimicked here through the evocation of tactile density – of a kind of fog.

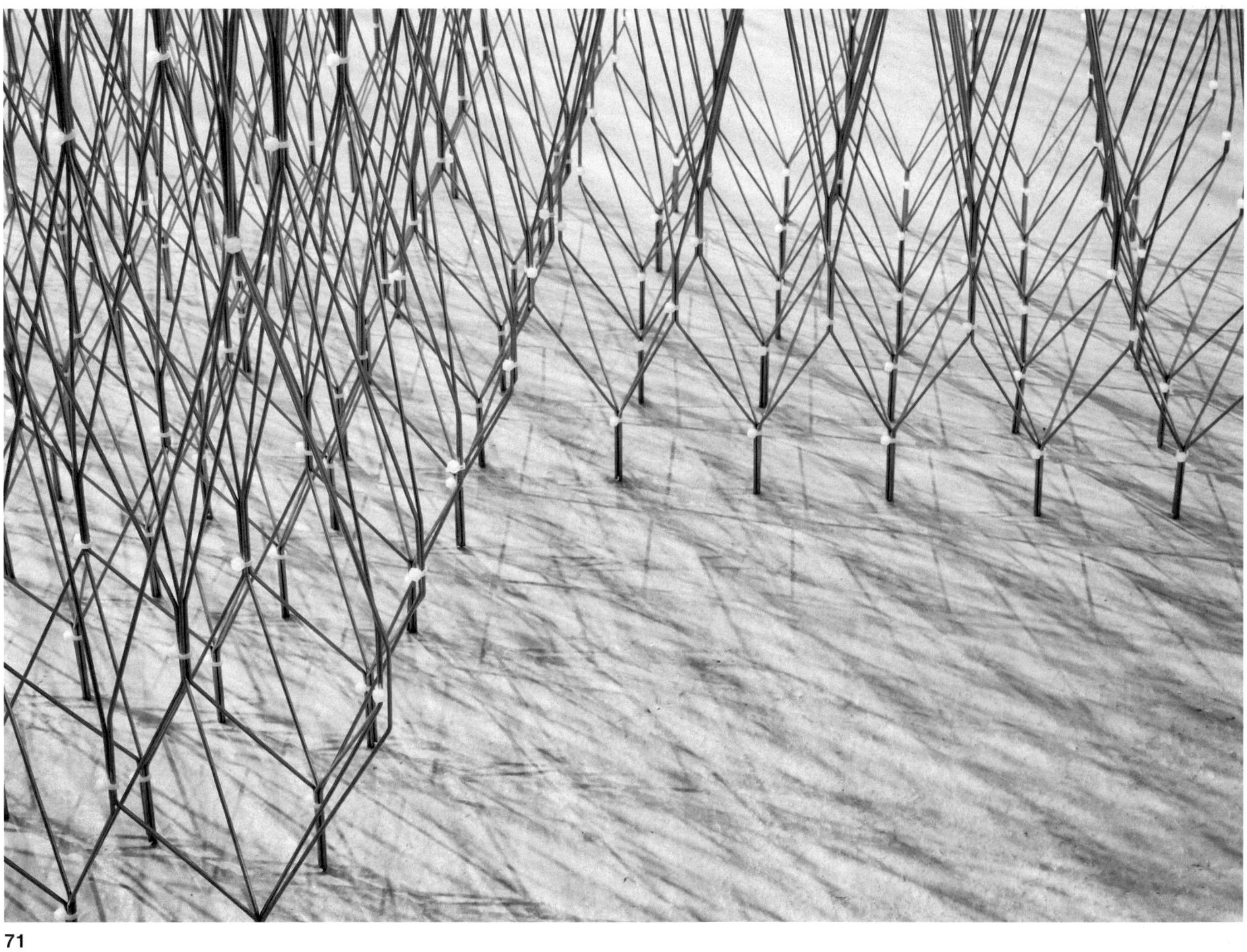

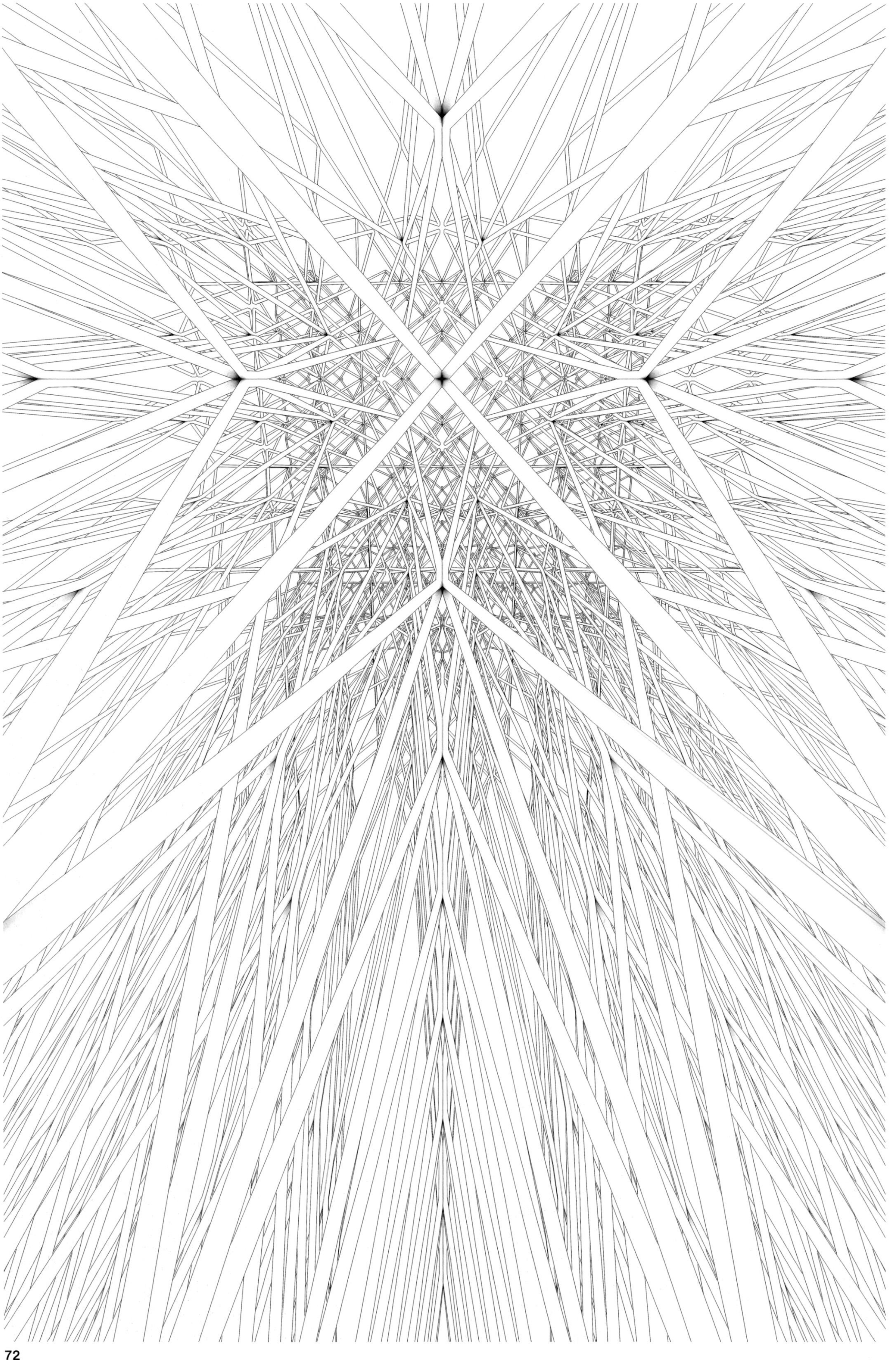

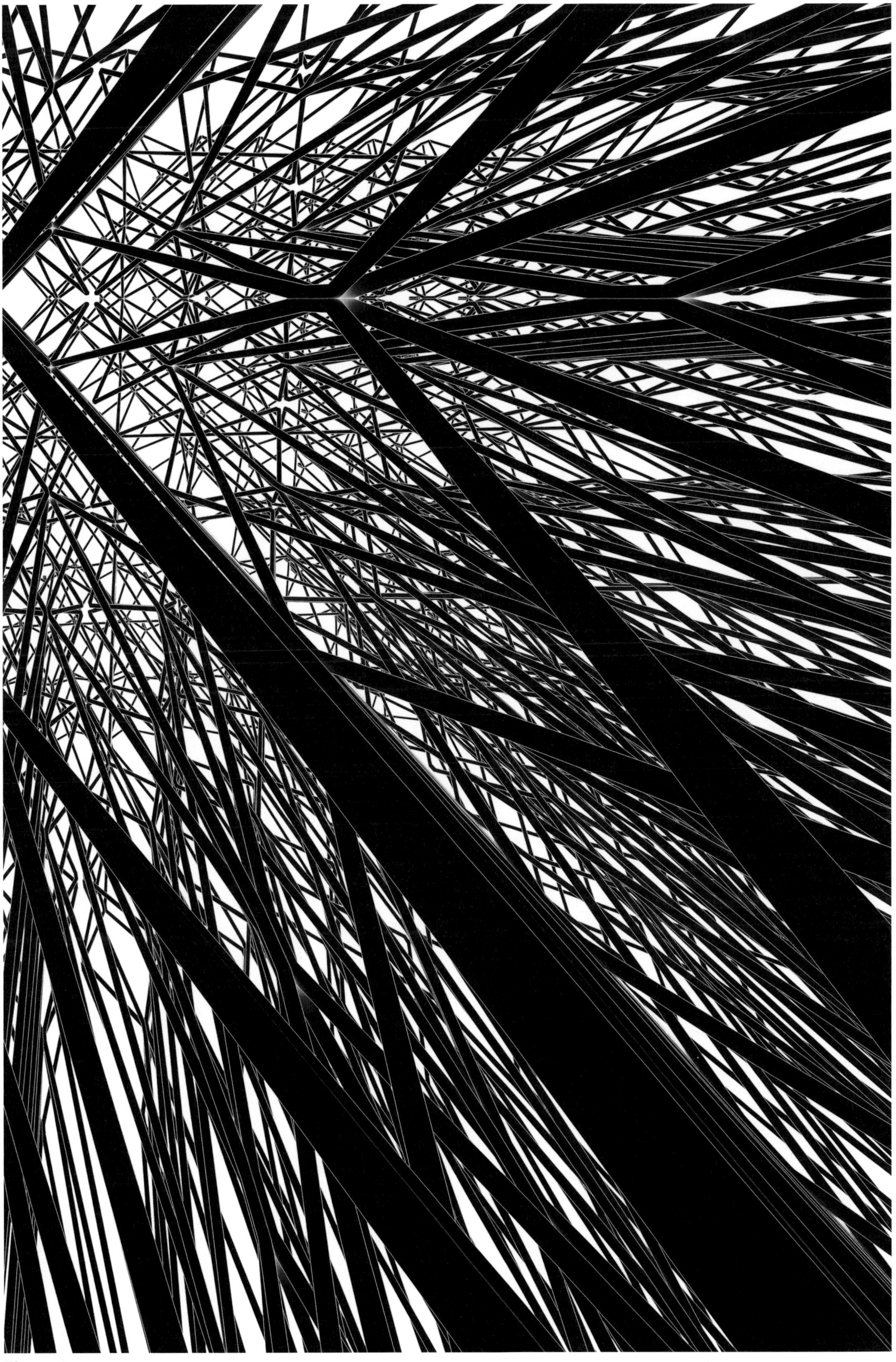

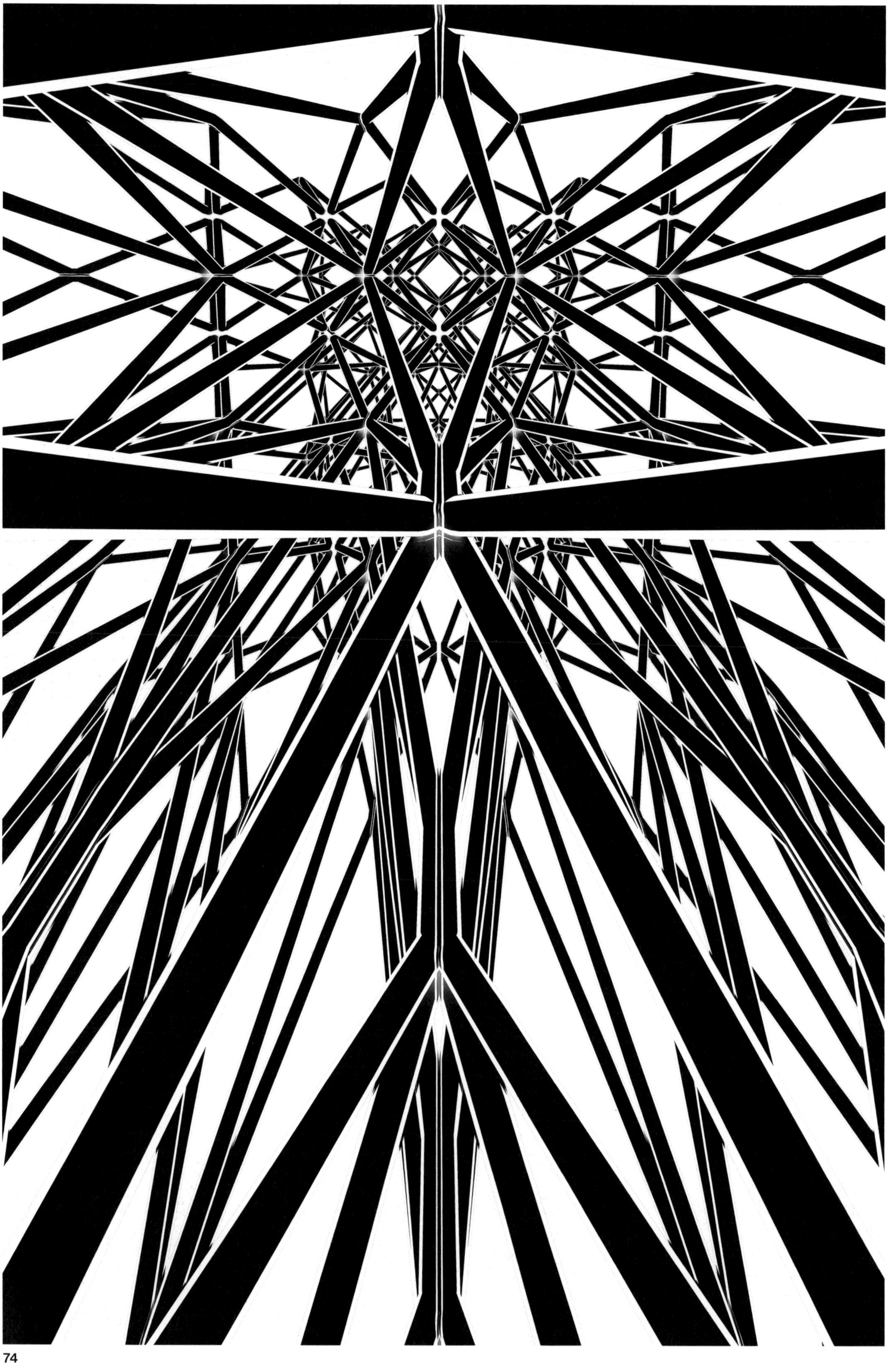

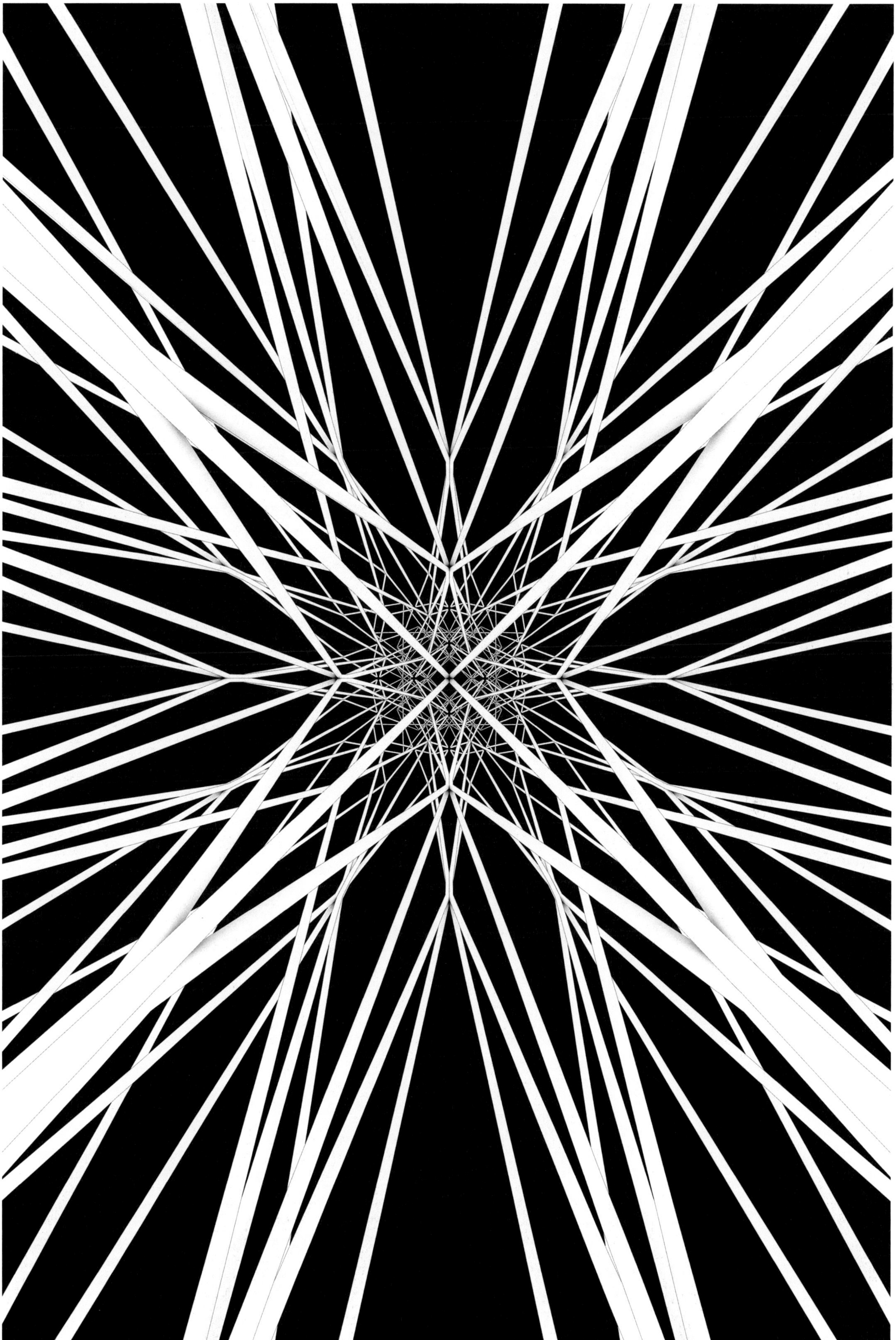

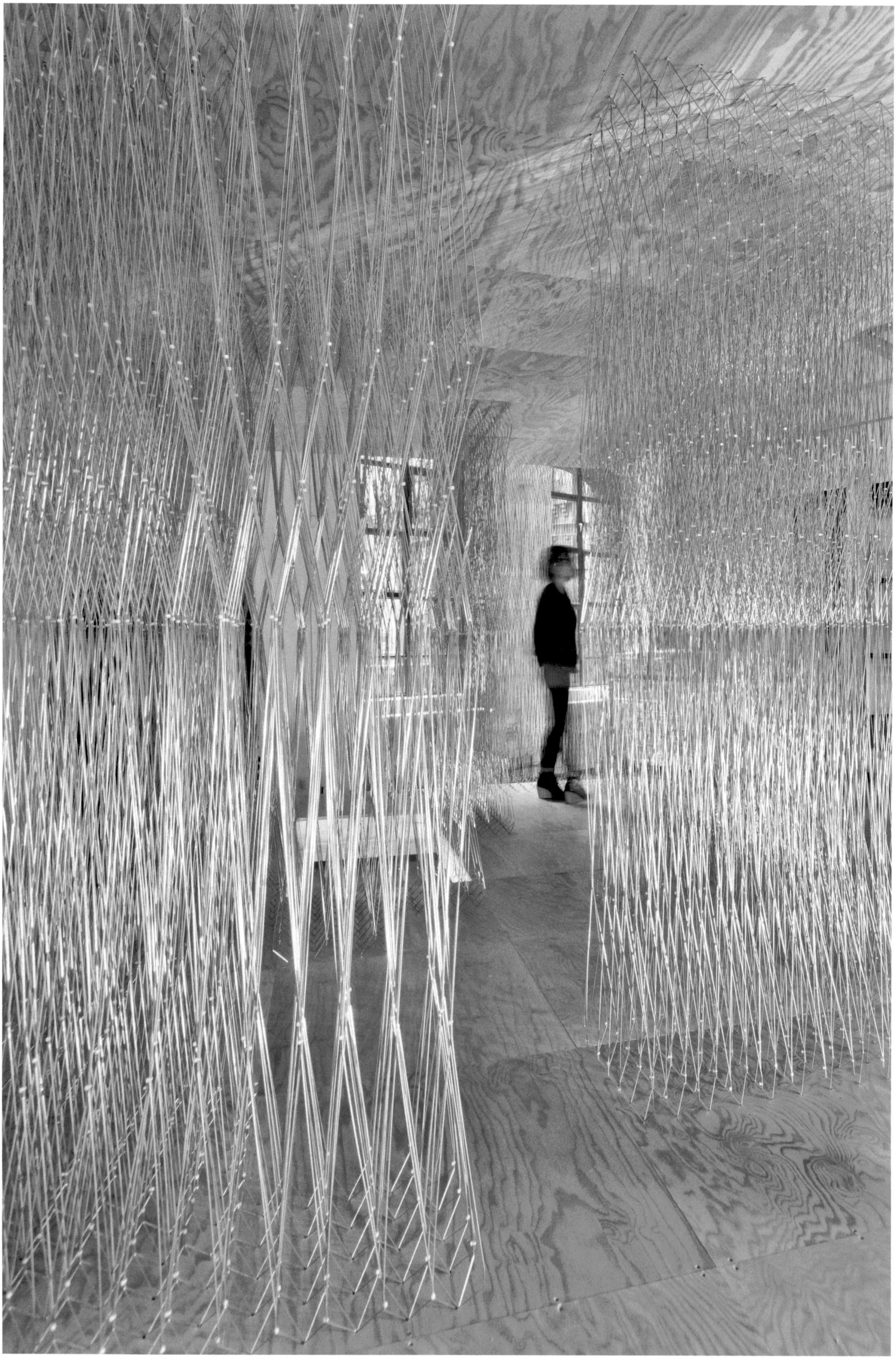

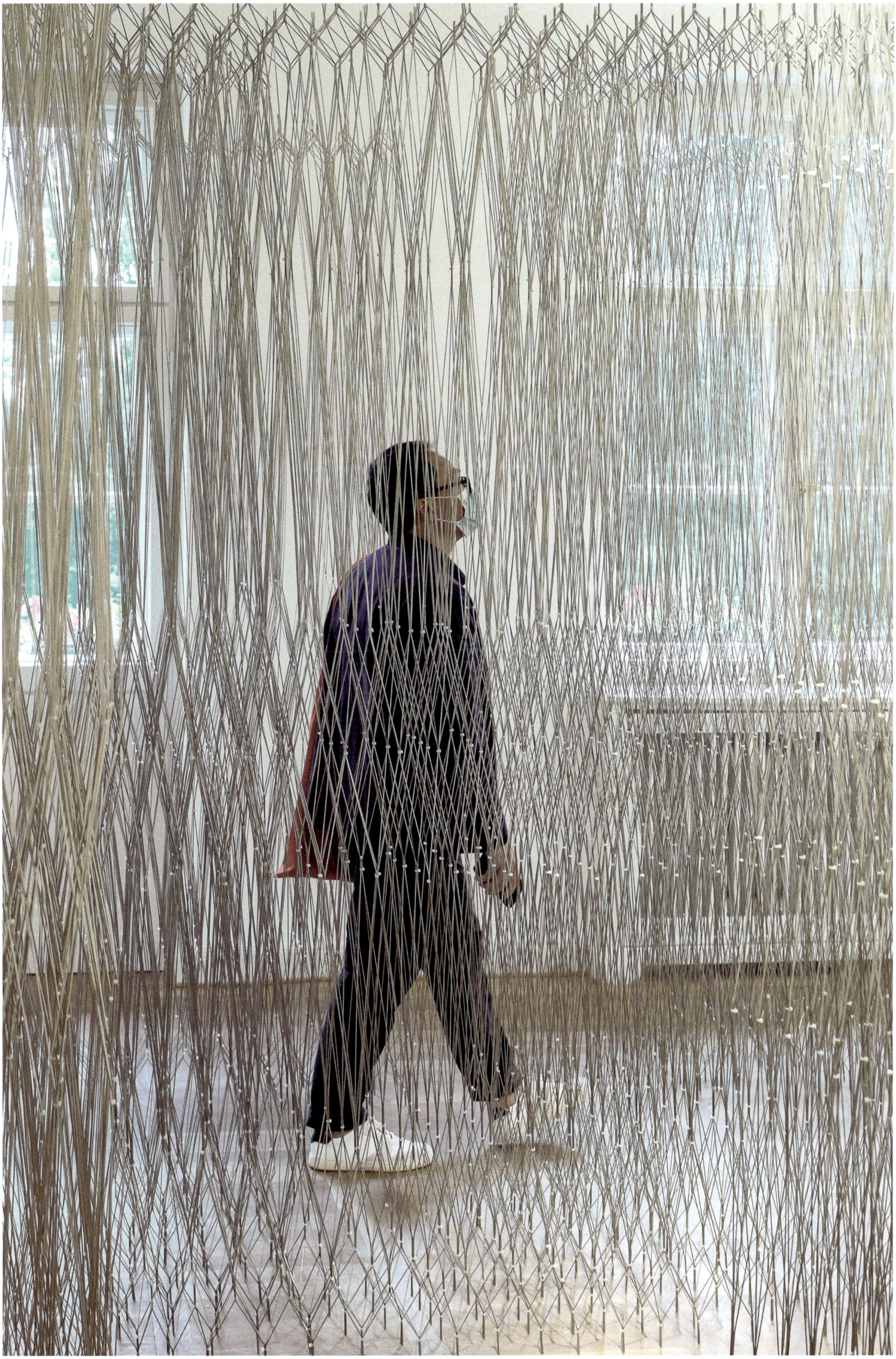

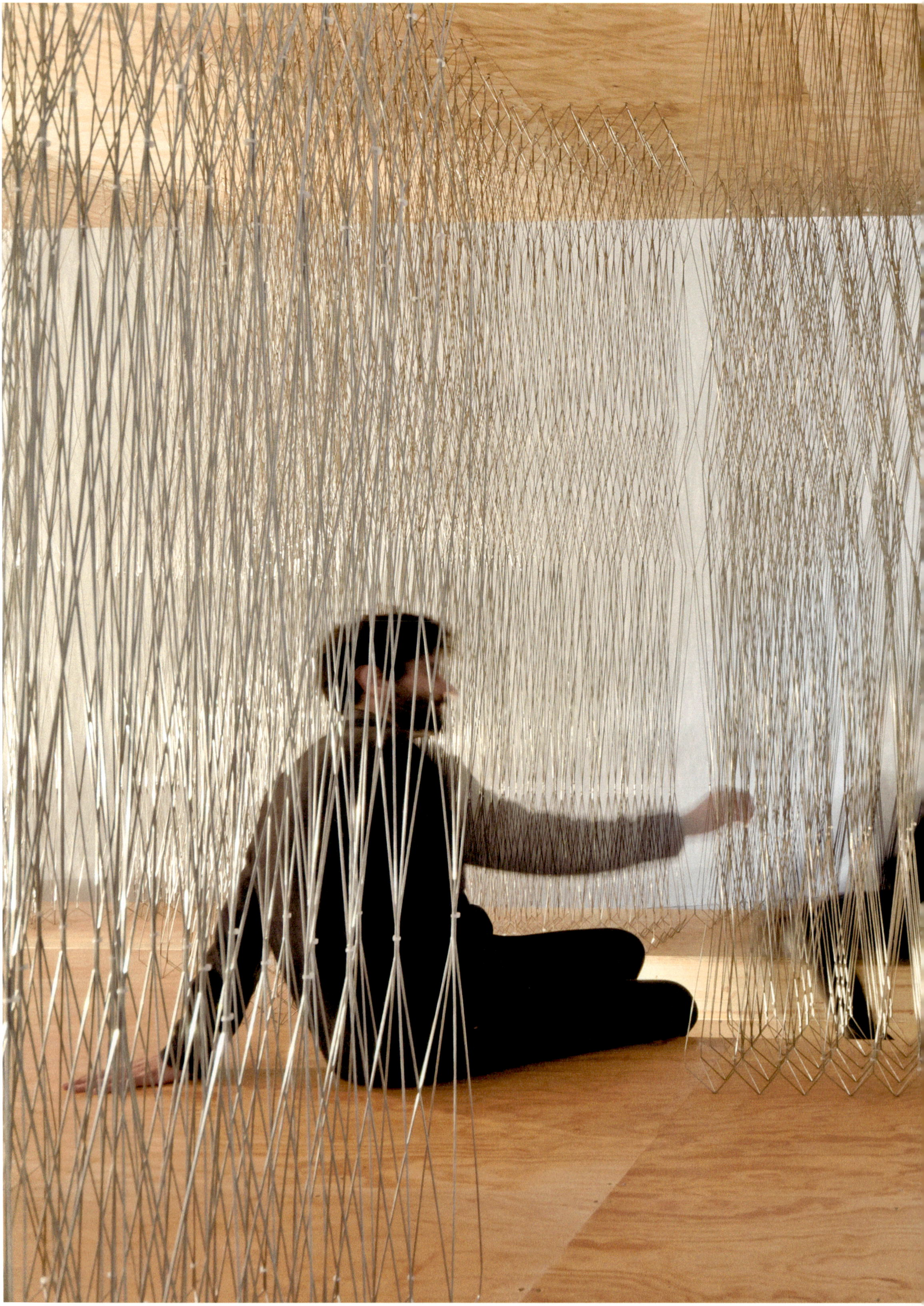

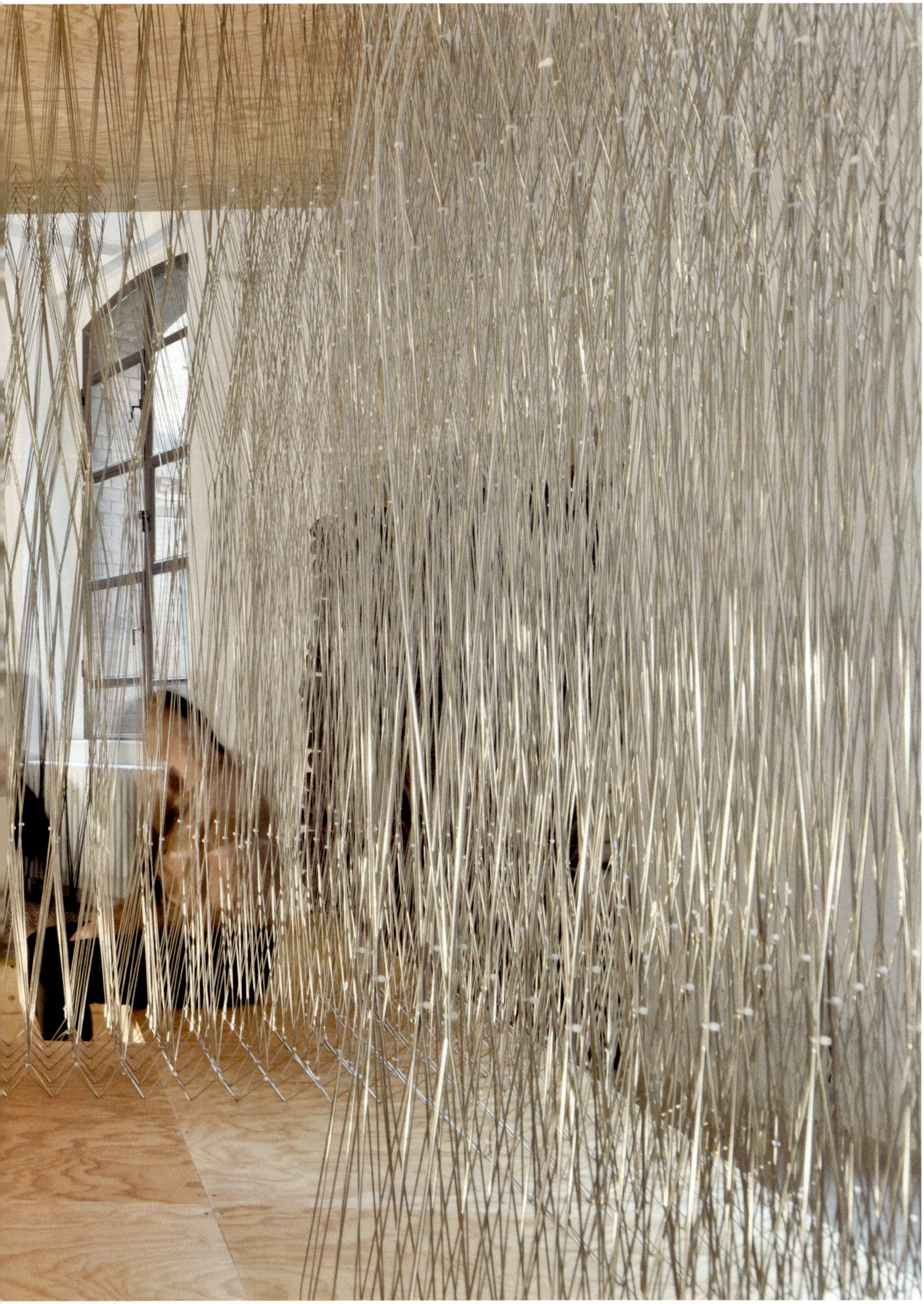

Casting

79

Casting: an age-old technique, whereby a liquid material is poured into a mold, which then hardens and can be removed as a solid positive. Casting can be done with pourable materials like concrete, plaster, terracotta, earth, clay, or polymers.

80

The formwork can be made of earth, sand, stone, wood, rubber, foam, or plastic. The process must be undertaken in such a fashion that the positive can be slipped from the mold intact. Casting can be done on-site or off-site; the product may be transported as a prefabricated element.

81

Digital fabrication techniques allow for differentiated molding or molds that can change from cast to cast. And so we updated an ancient art form and archaic technique for incredibly modern ends.

Zoning regulations demanded that a cafeteria/workspace in an industrial campus be located underground. According to the logic of sand casting, two construction possibilities existed for us: a large roof could either be poured in place (excavating the earth below to create useable spaces) or cast in the ground and lifted into place.

Casting in the ground or lifting into place allows a variety of forms and surface textures – from rough and organic to more geometrically precise. For the sake of ease, we chose to cast in place concrete that was formed with acoustic panels.

The effect was that of occupying a space under an elevated ground. We further punctuated the roof by adding openings into courtyards to light the spaces below. We planted a garden on top of the roof, replacing the disturbed ground.

85

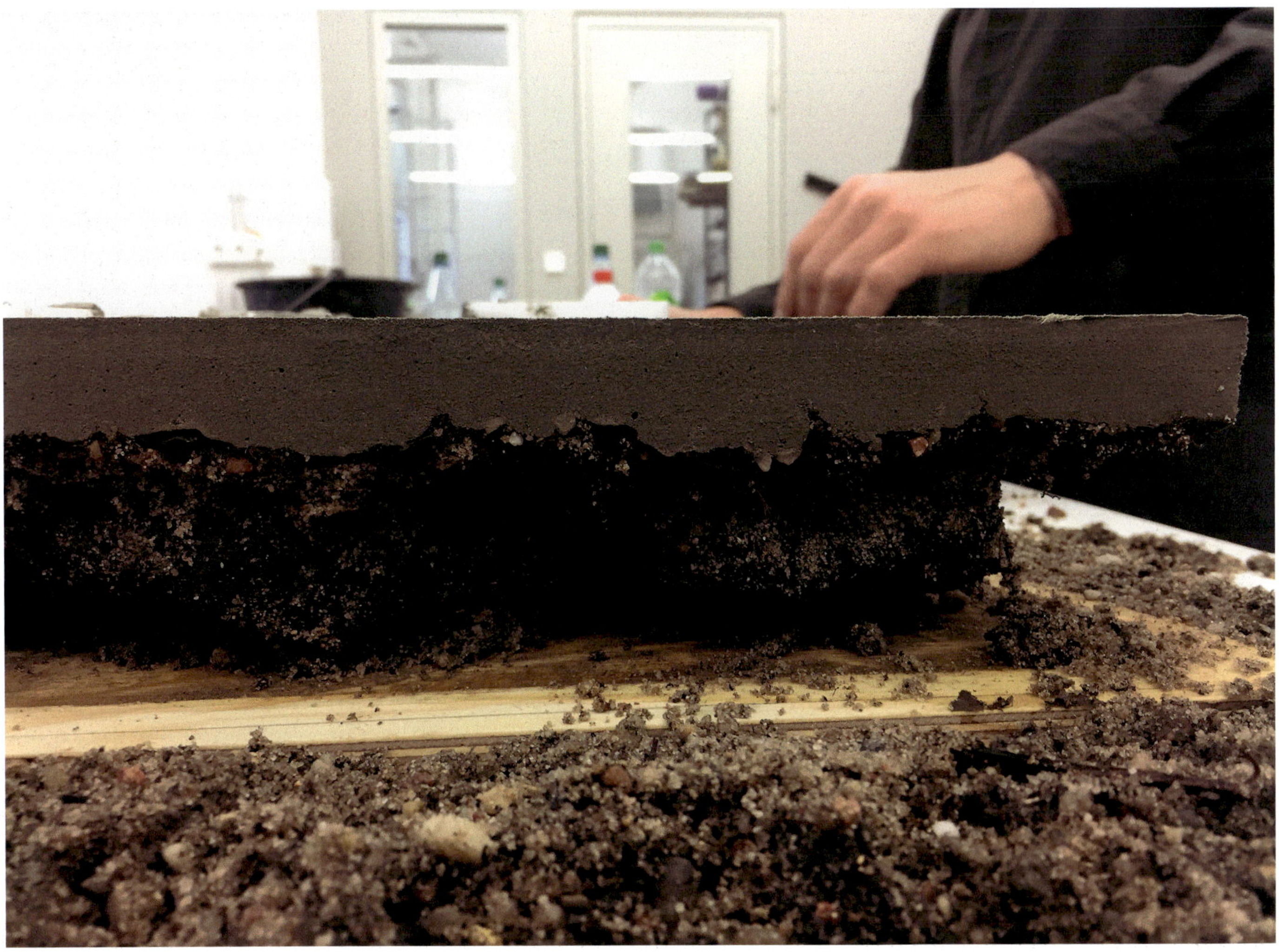

86

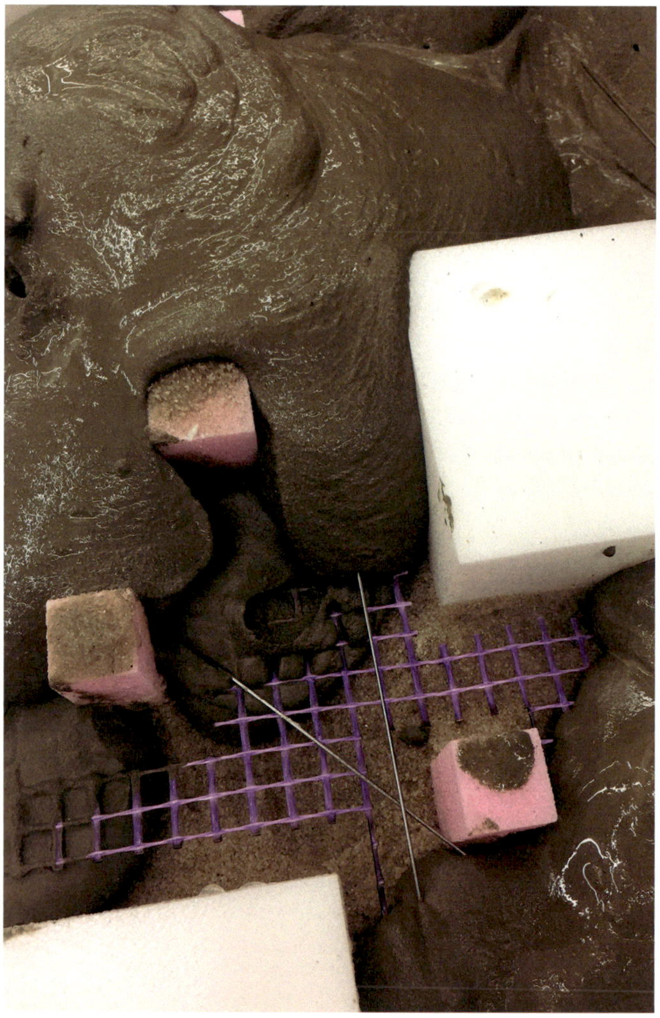

87

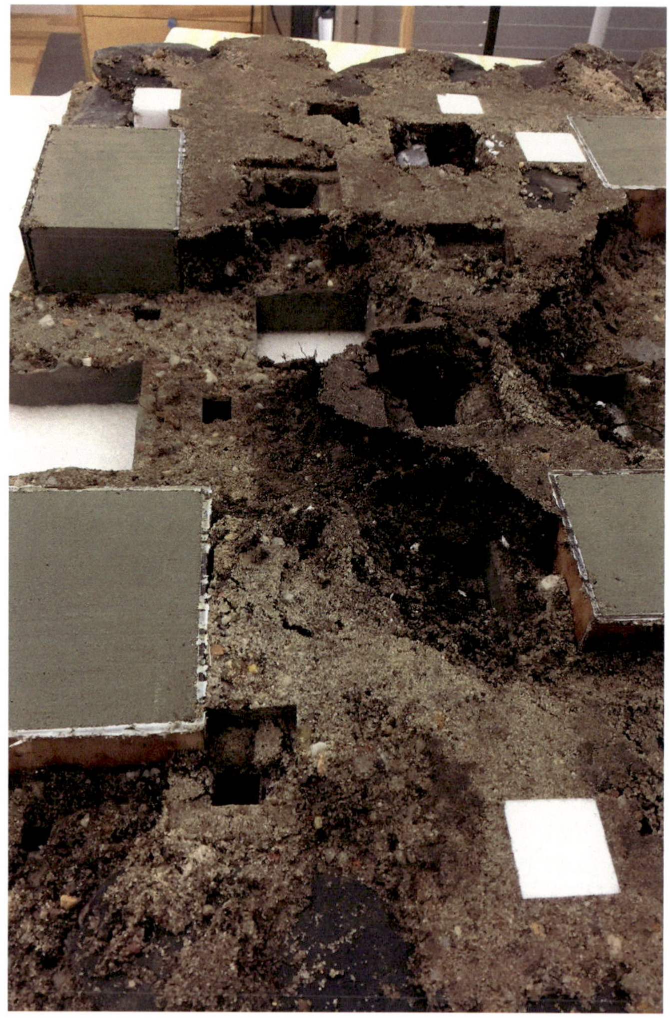

88

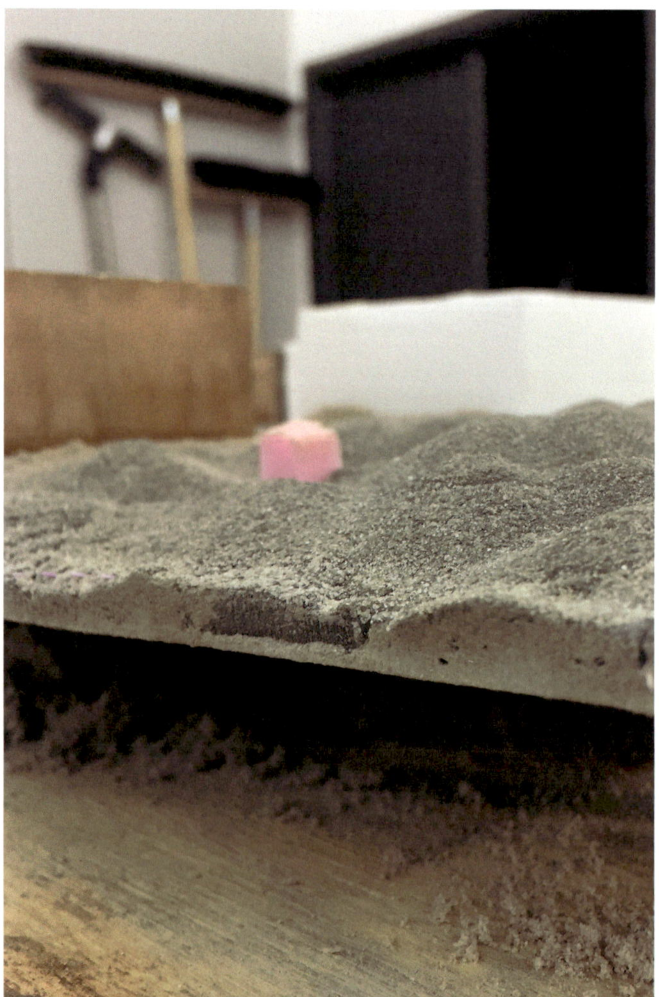

89

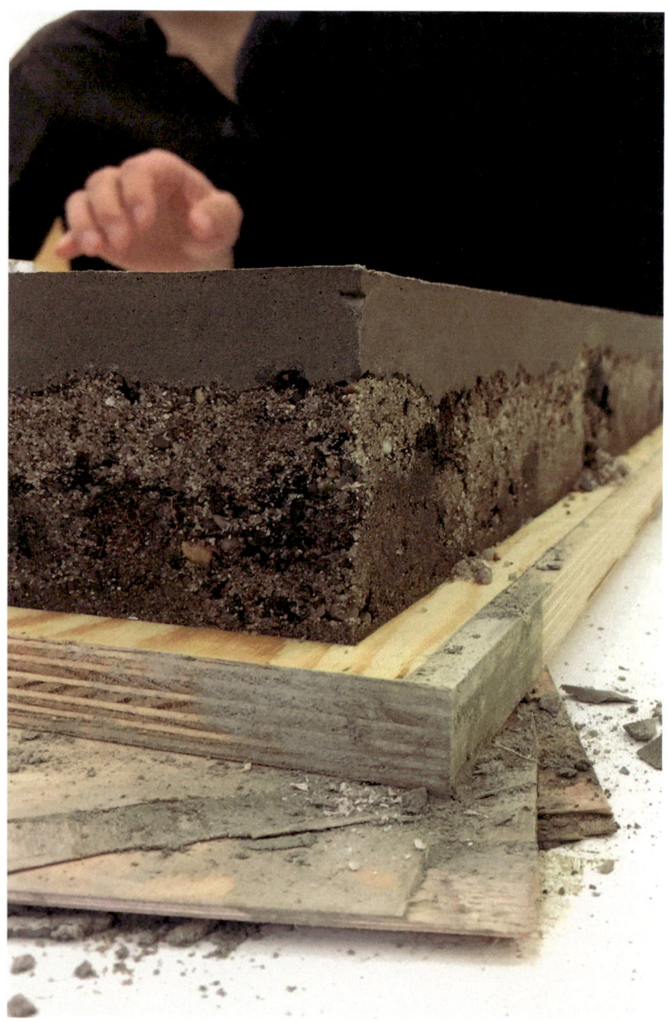

90

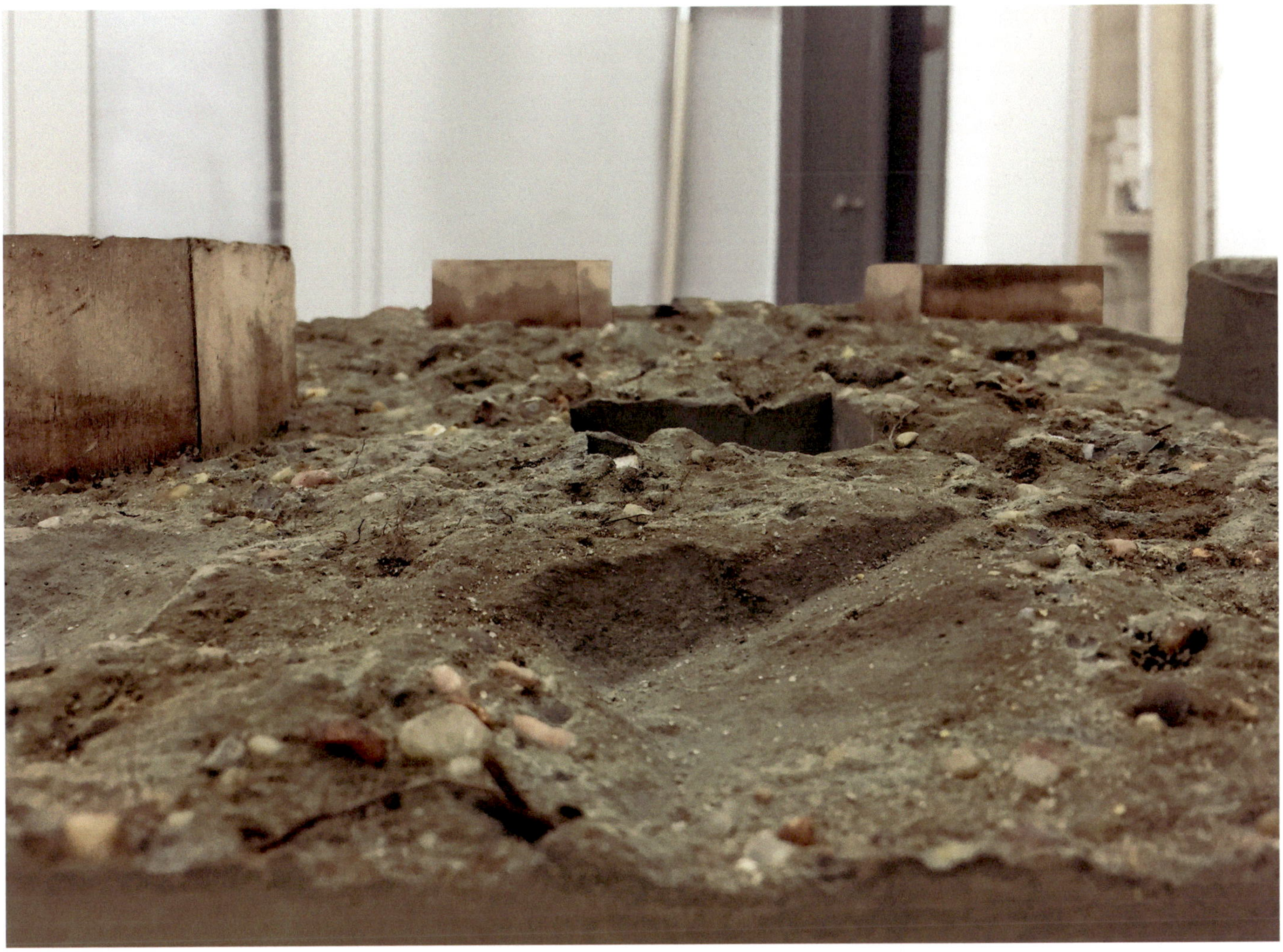

91

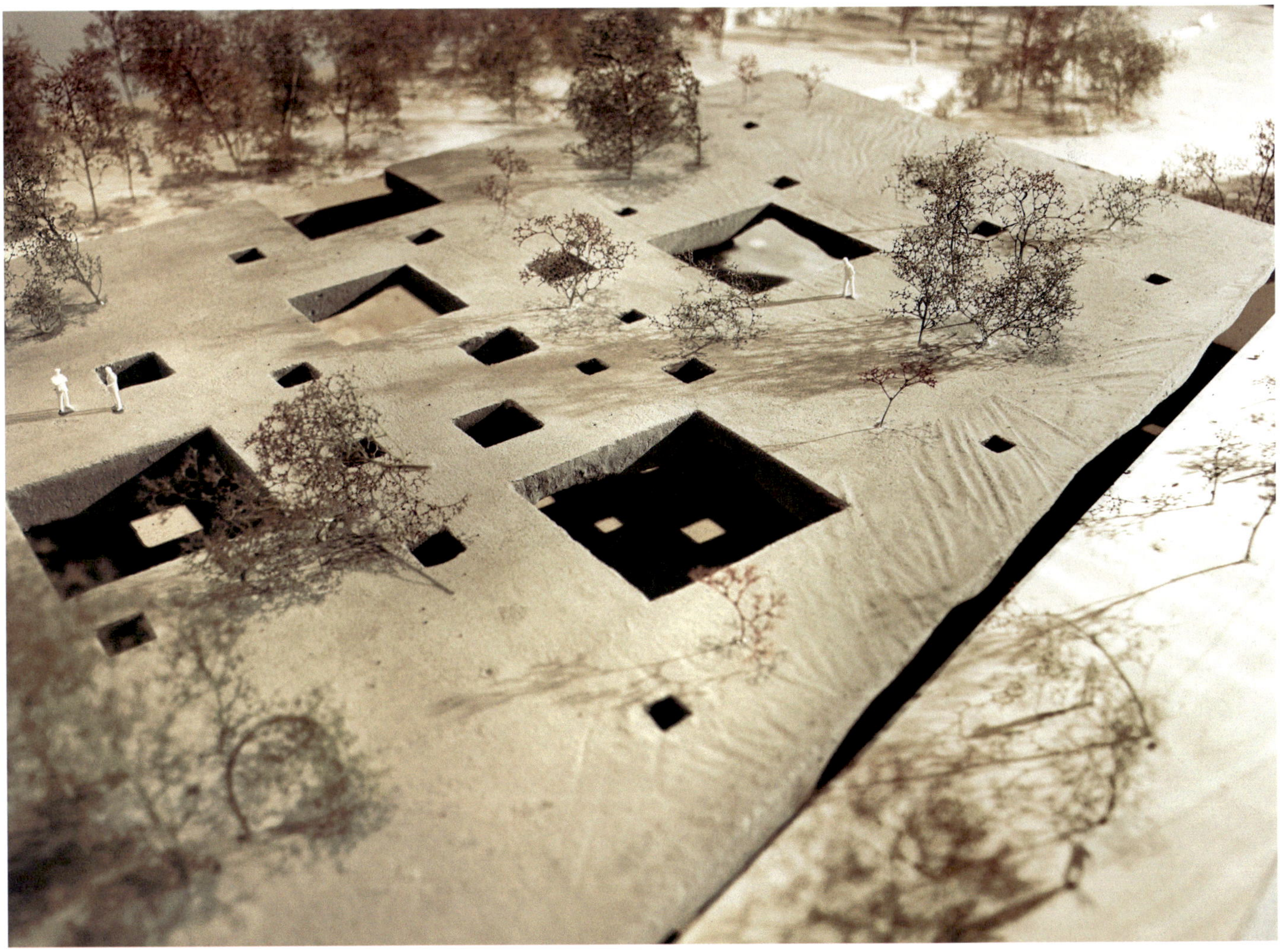

92

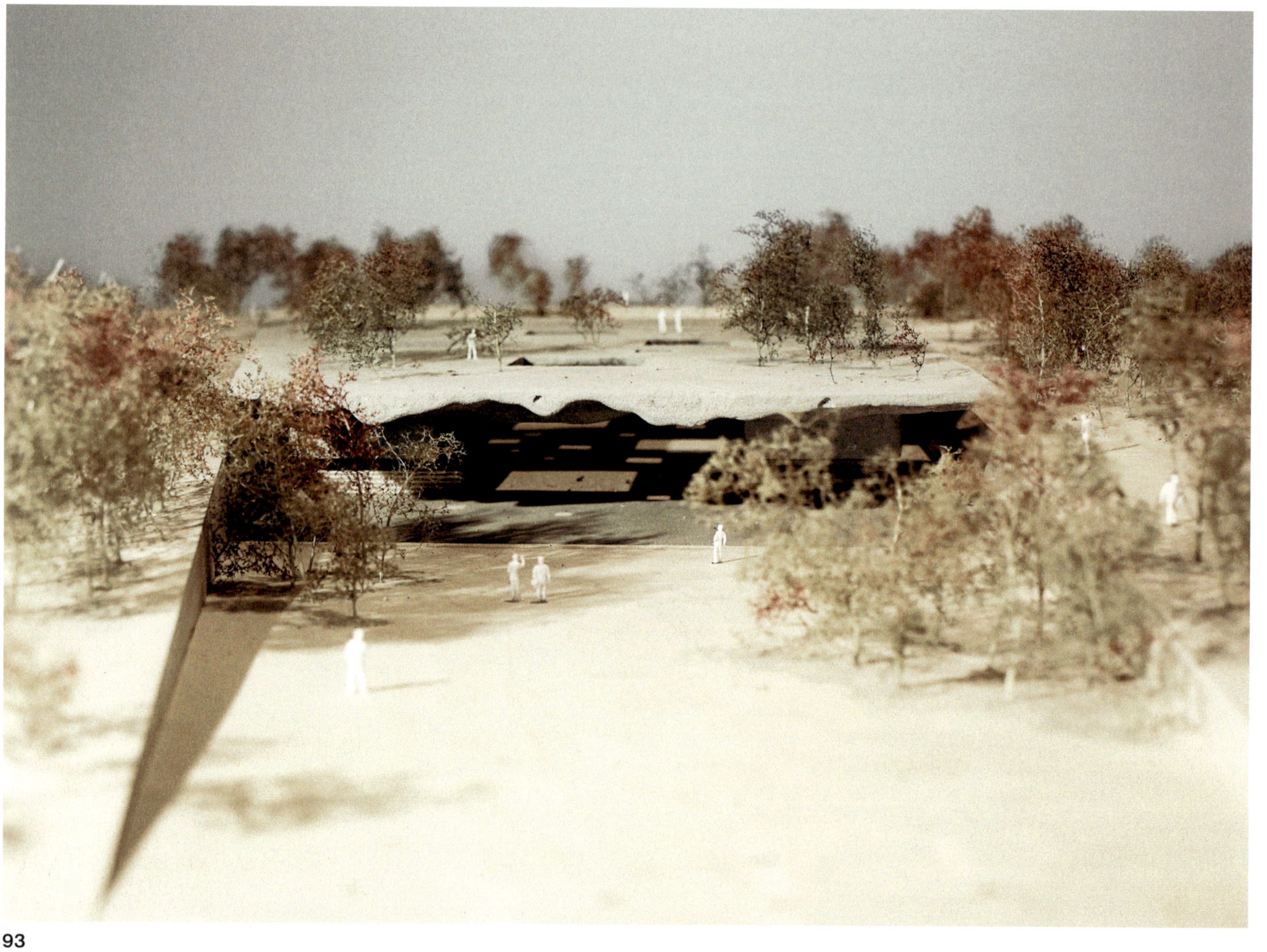

93

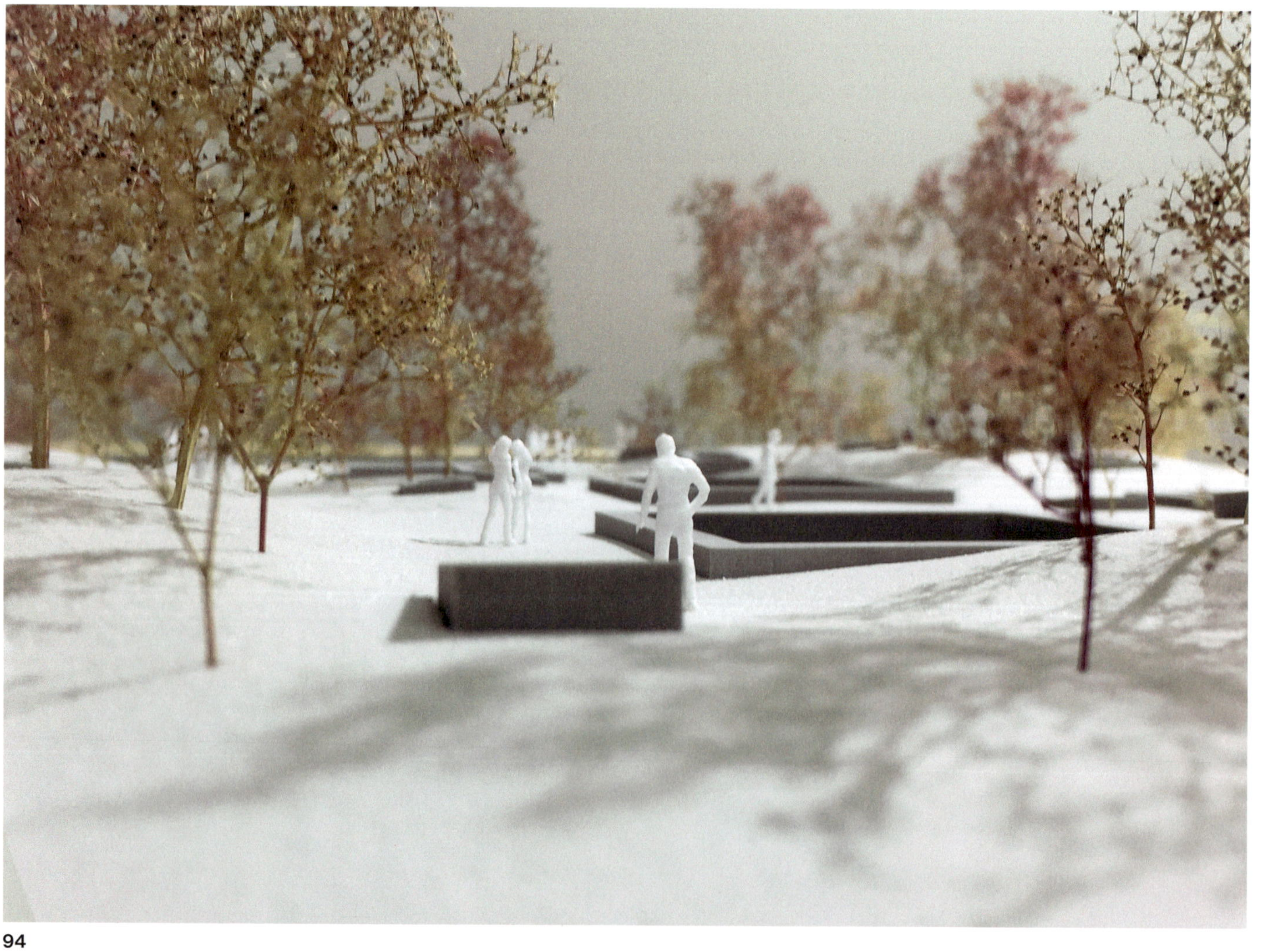

94

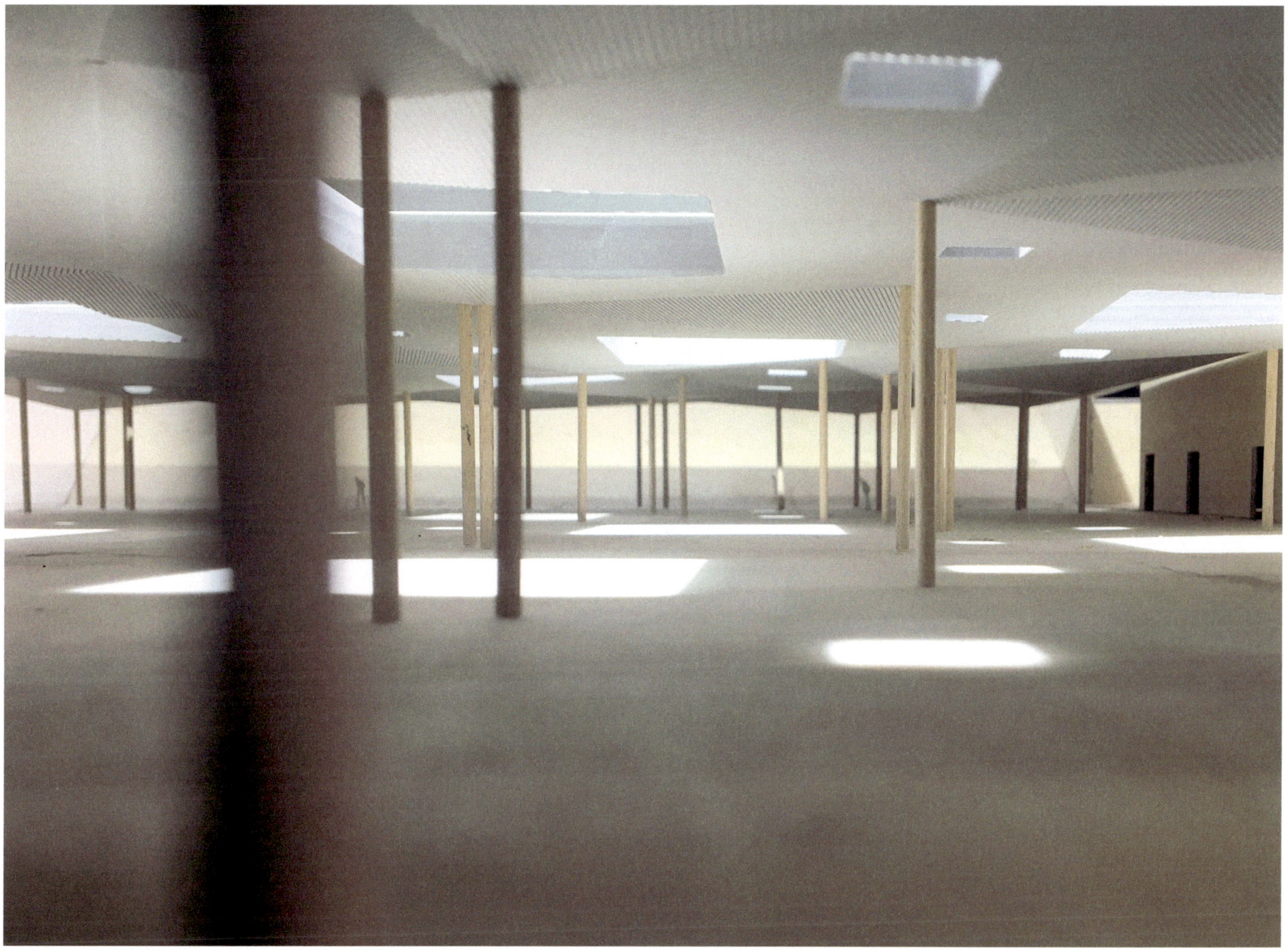

95

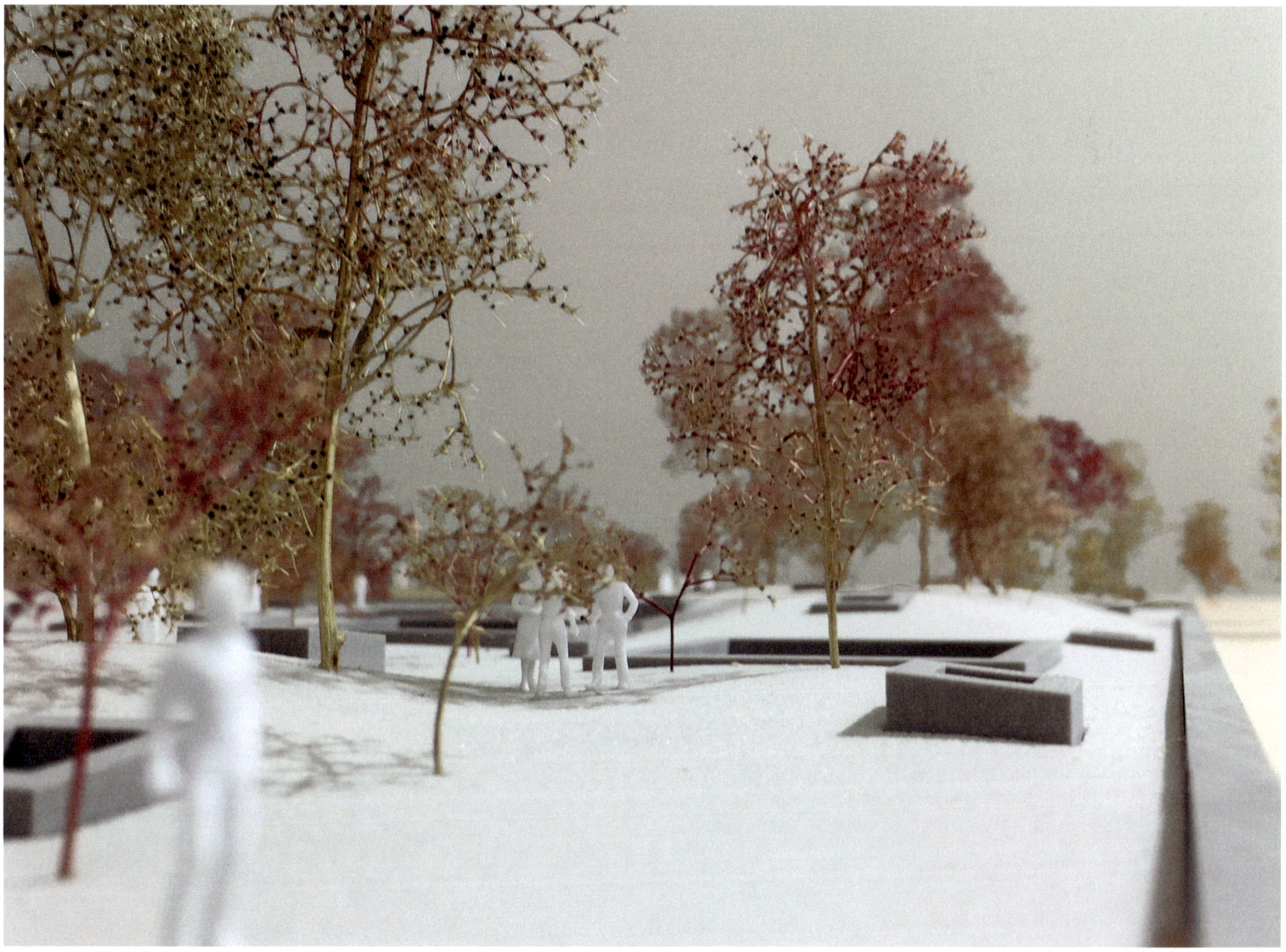

96

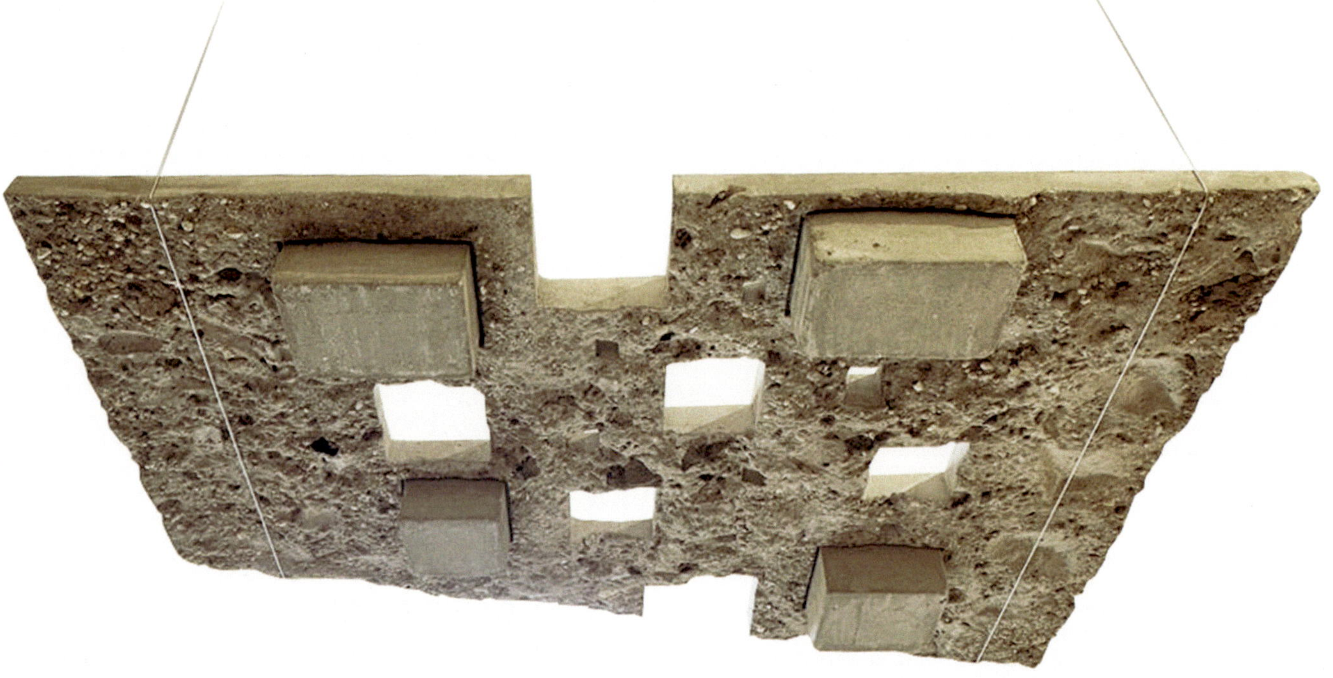

97

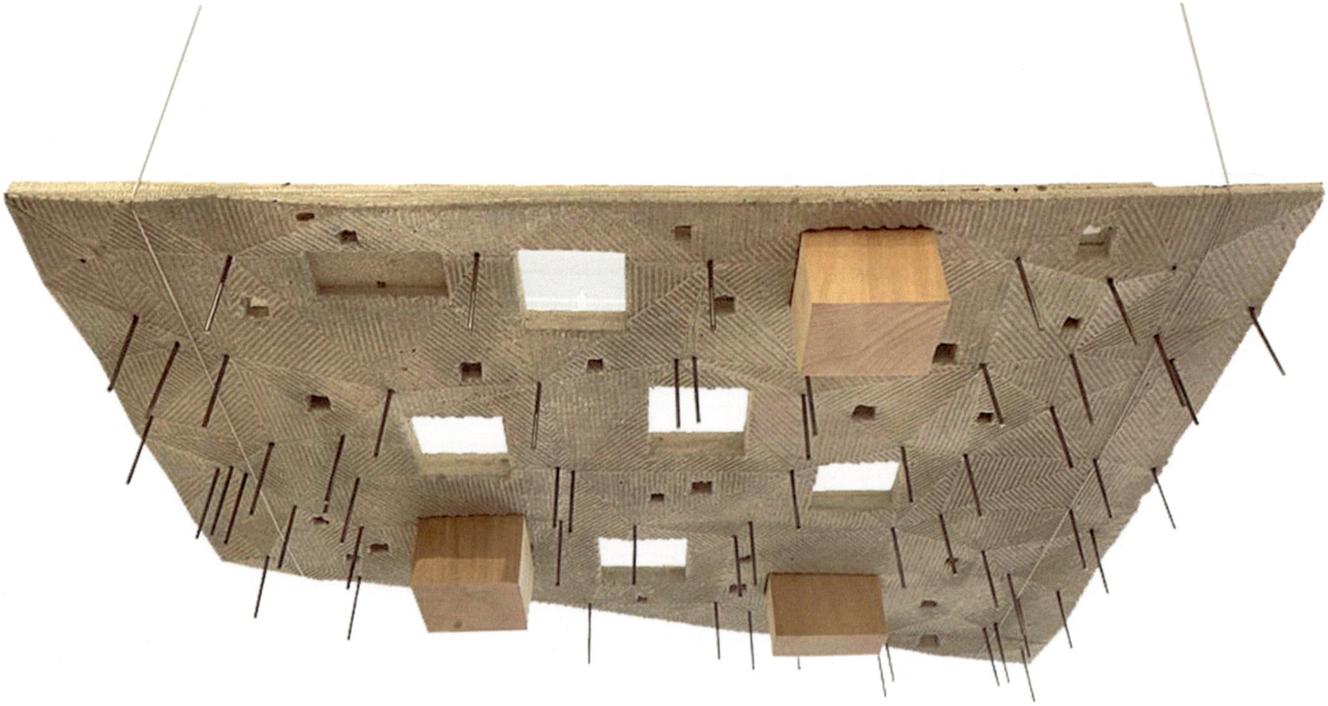

98

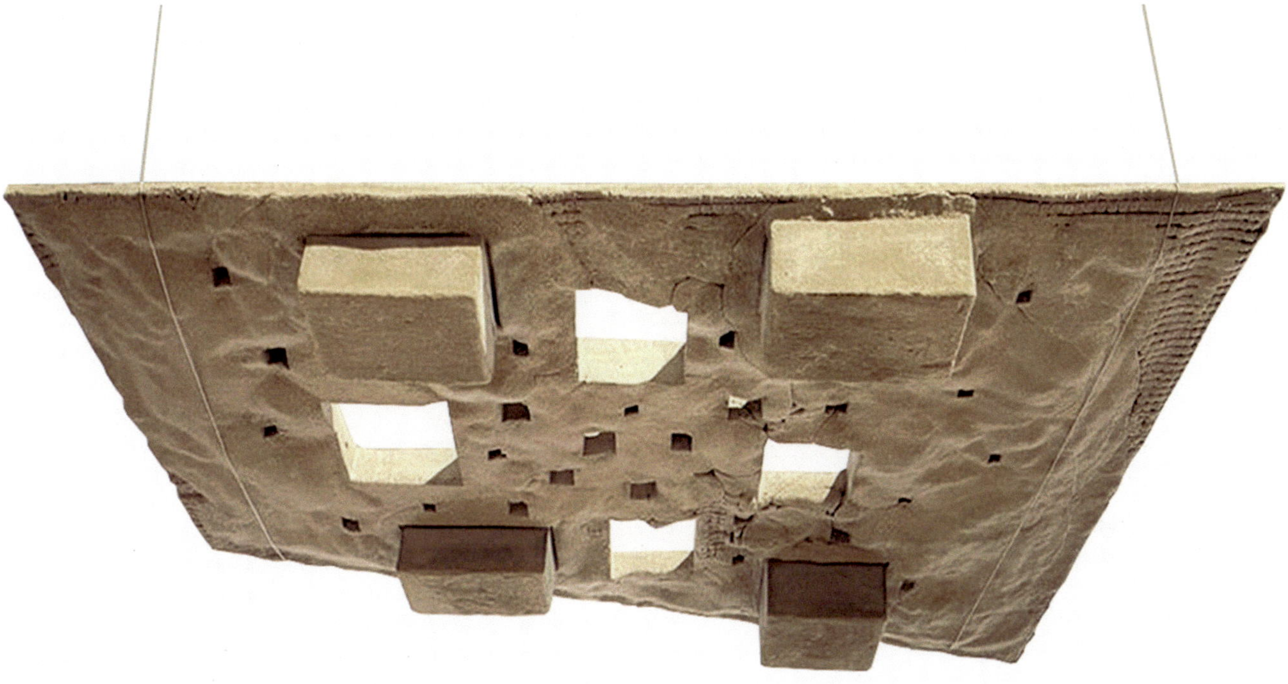

99

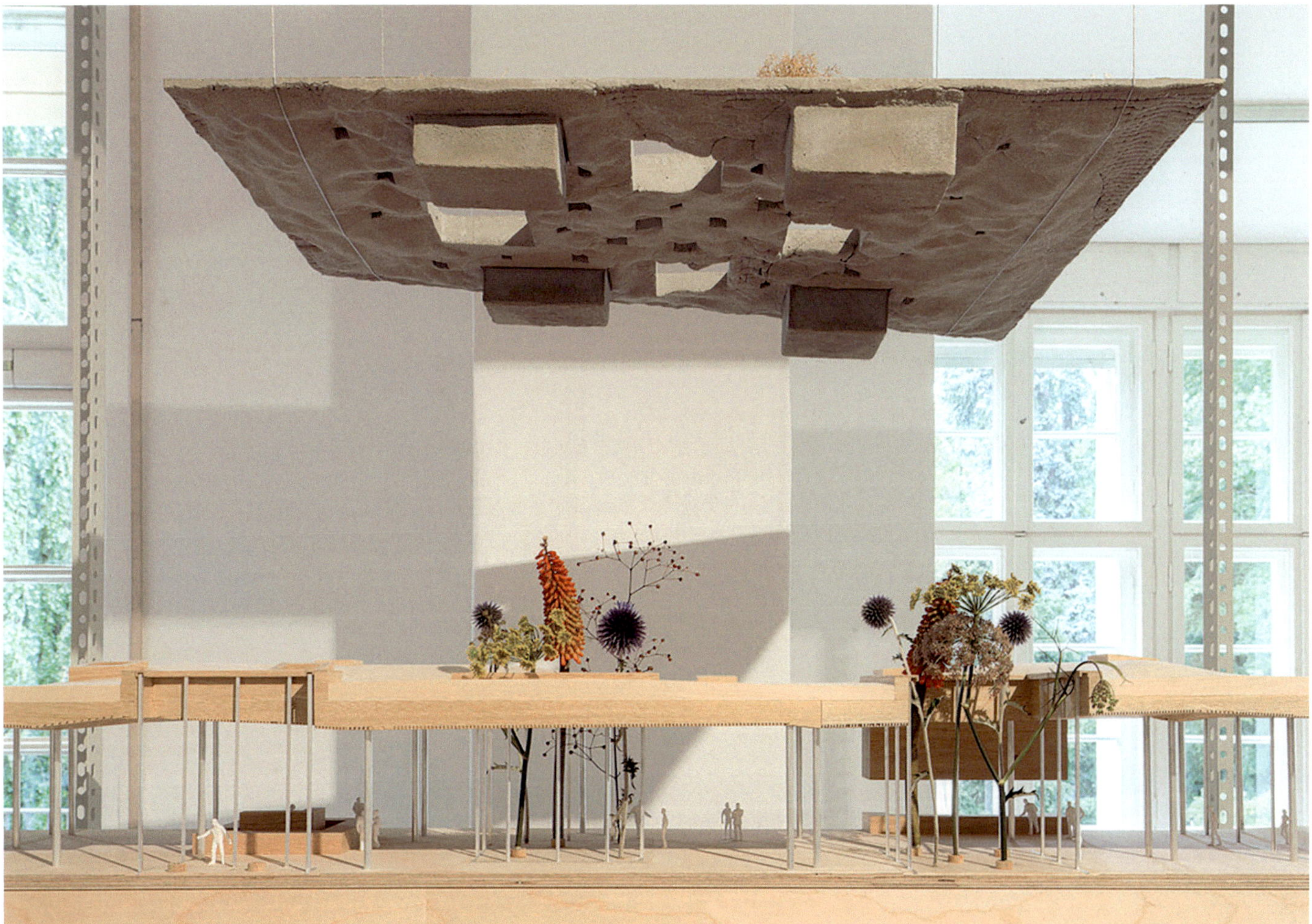

100

For years, we have been fascinated by clay and its potentials. During a series of ceramic workshops, held between 2016 and 2017, we investigated clay as both a generator of form as well as a building element – in particular for facades.

Techniques and what they produced varied: casting, clay thrown on a potter's wheel, stacked coil structures, slab clay, extruded pieces – the material lends itself to a wealth of forms.

The value of these exercises resided in the use of a specific material (clay) that can be transformed by different tools and techniques to produce formal experiments. The work we produced was stored and can be retrieved or added to as an active archive.

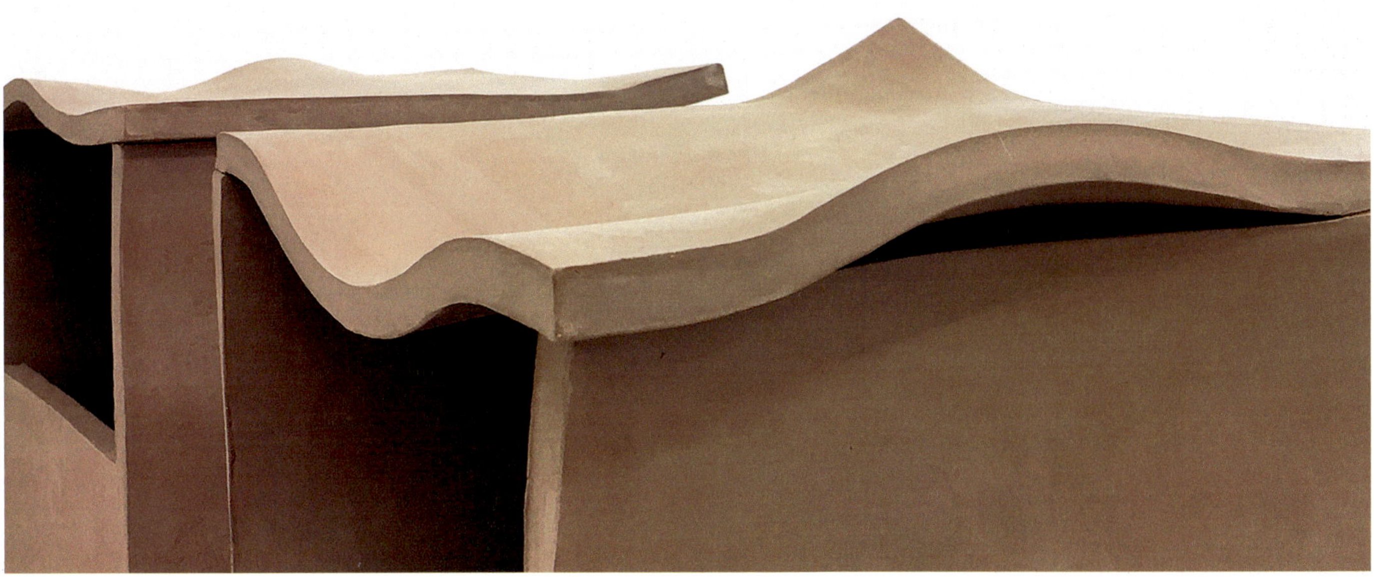

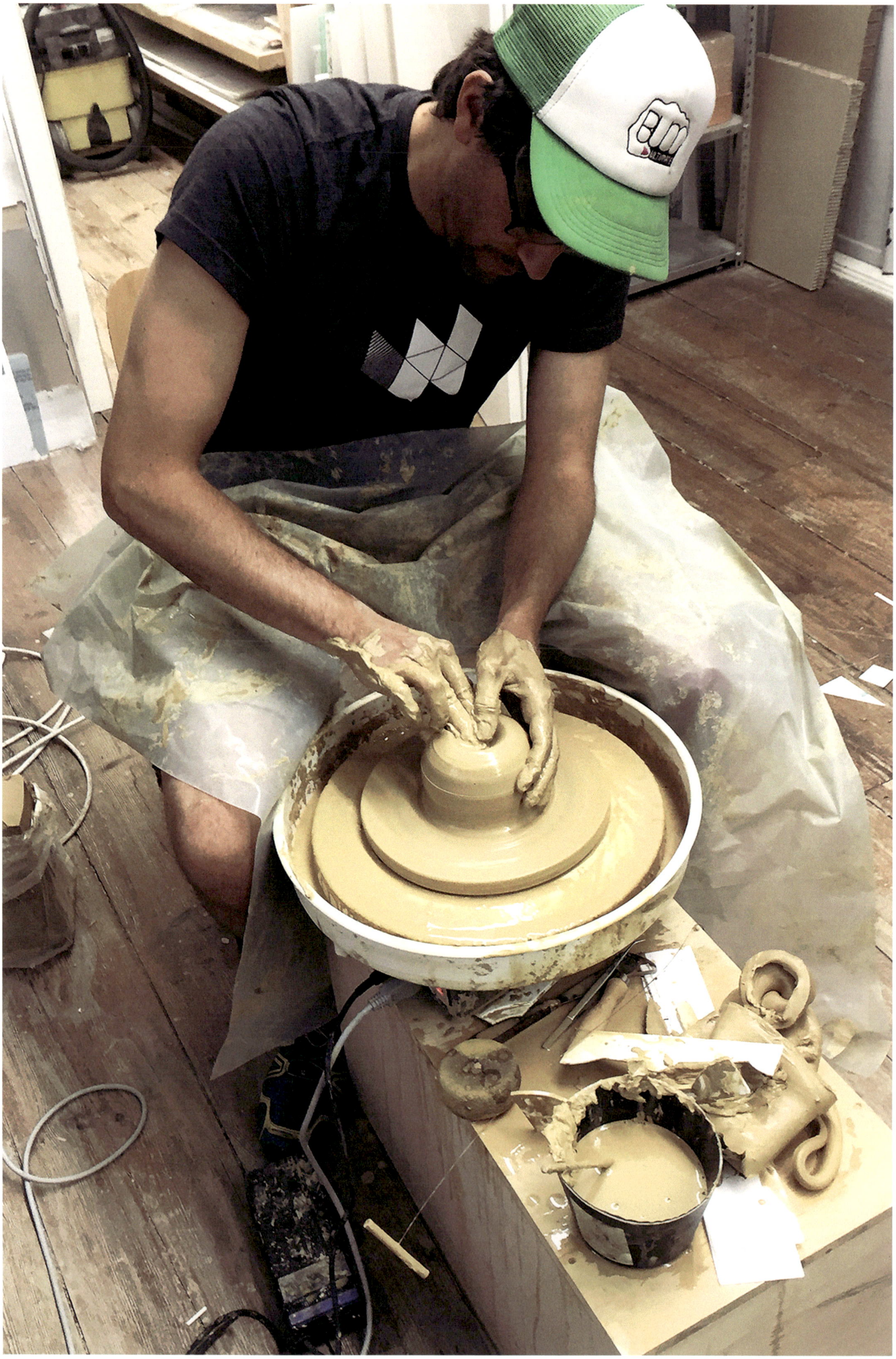

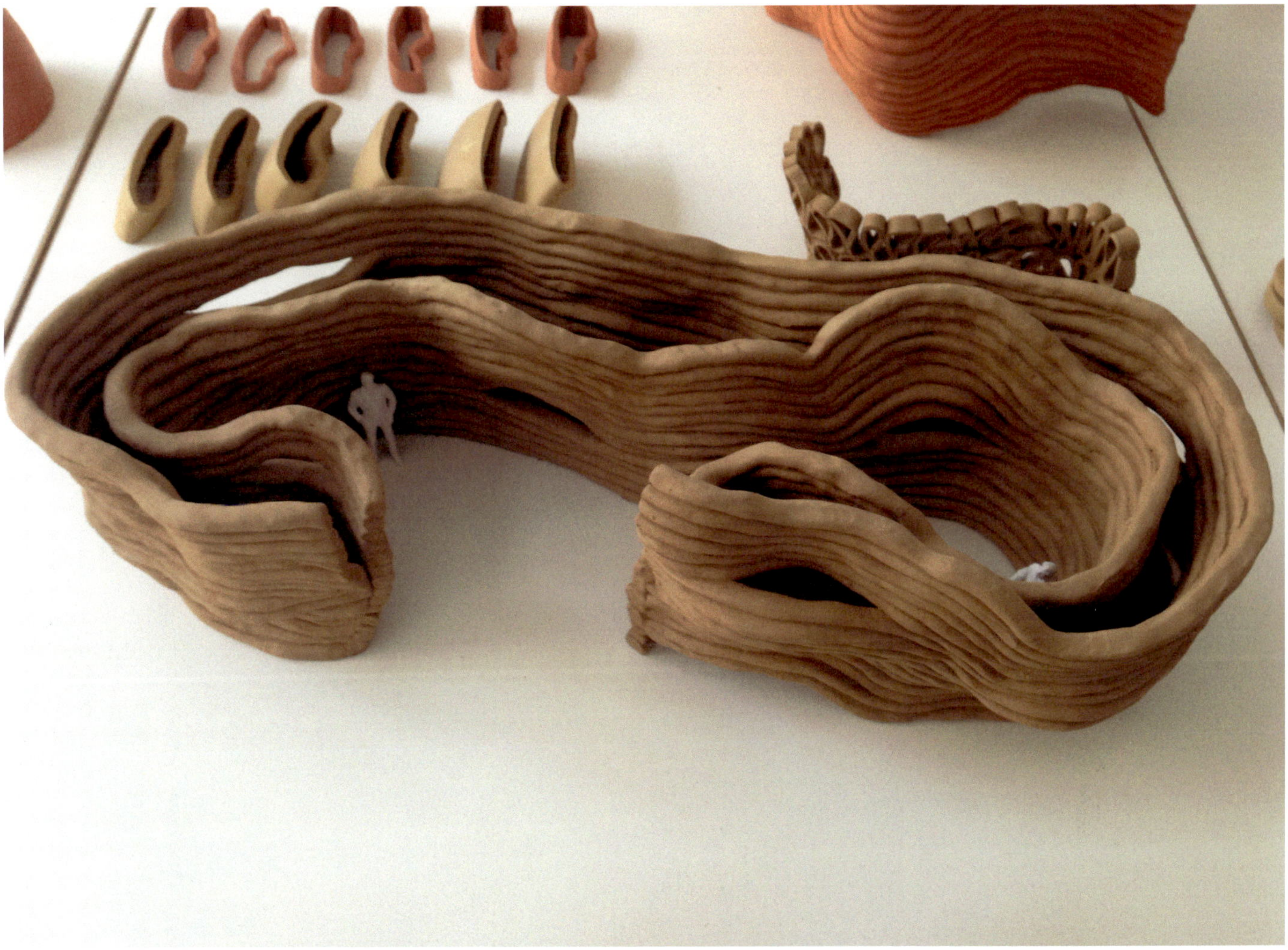
106

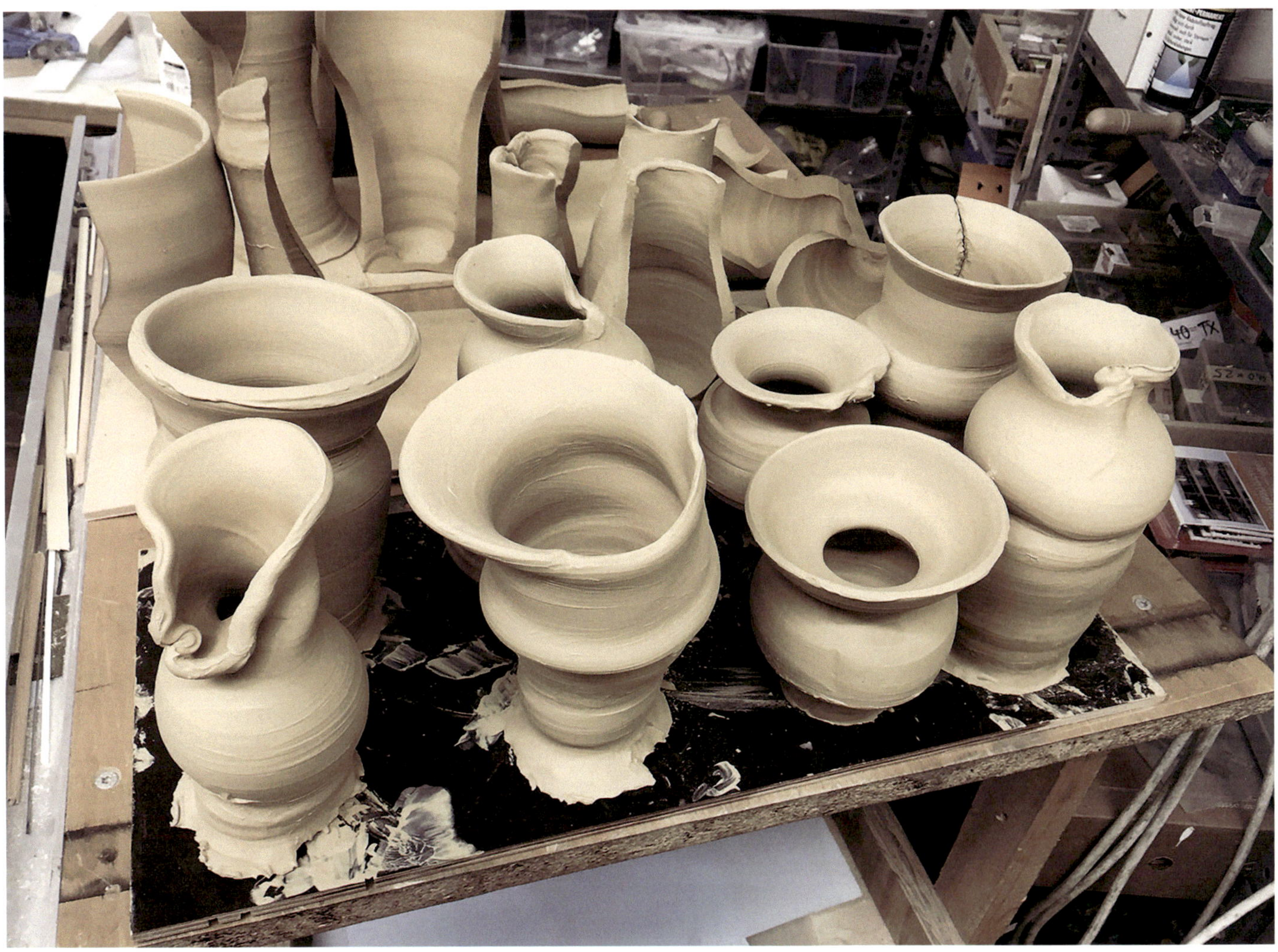

107

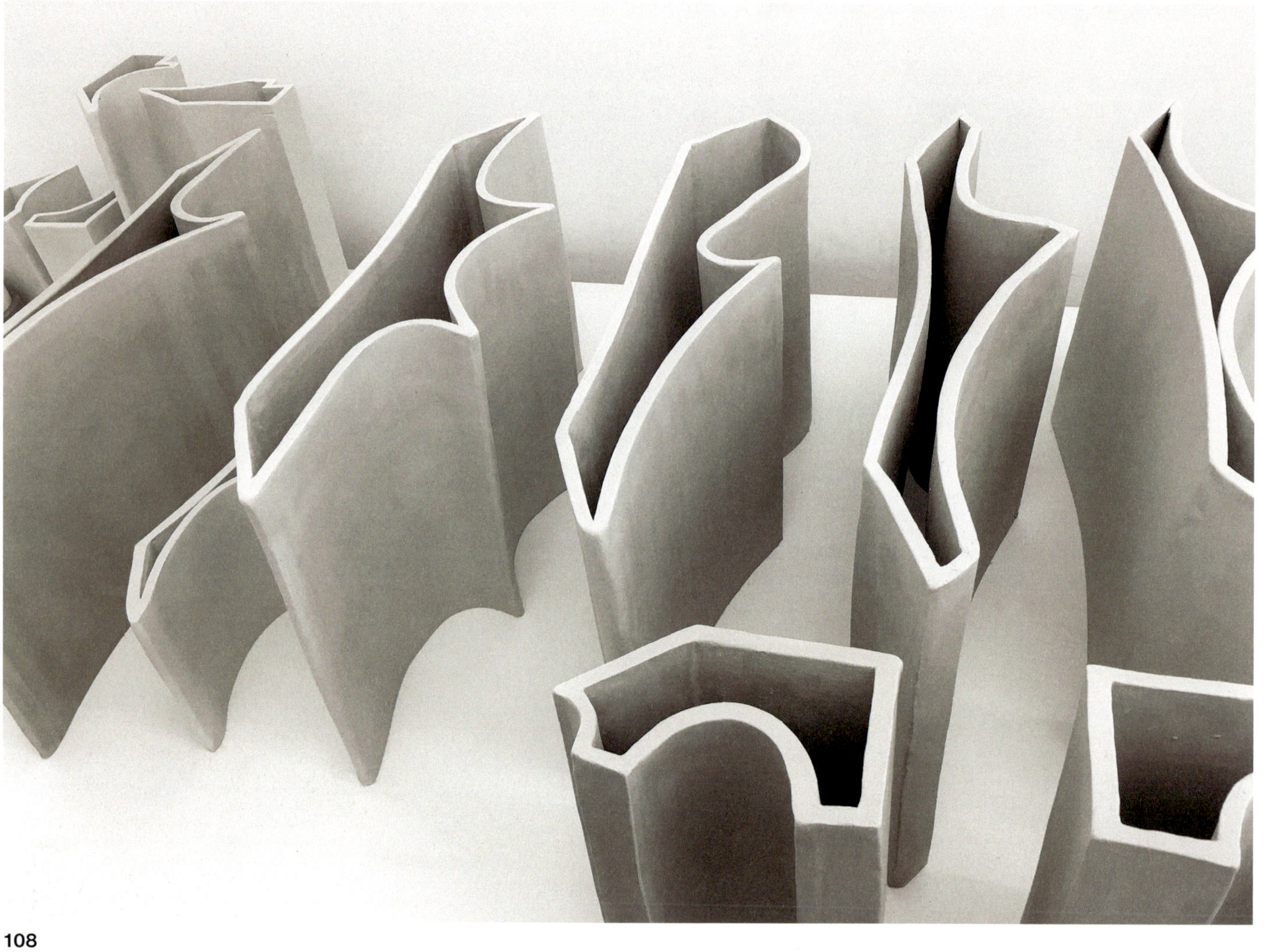

108

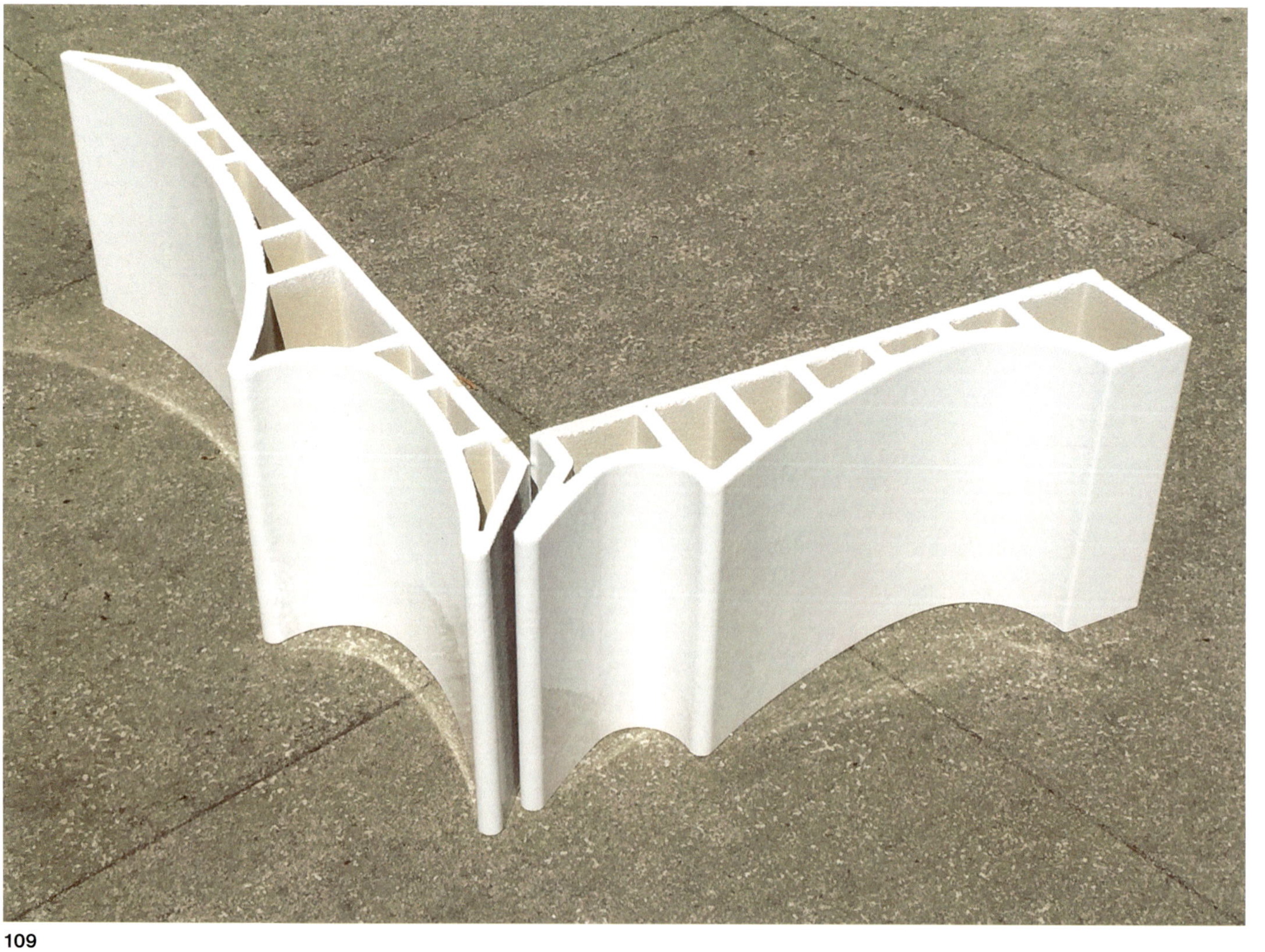

109

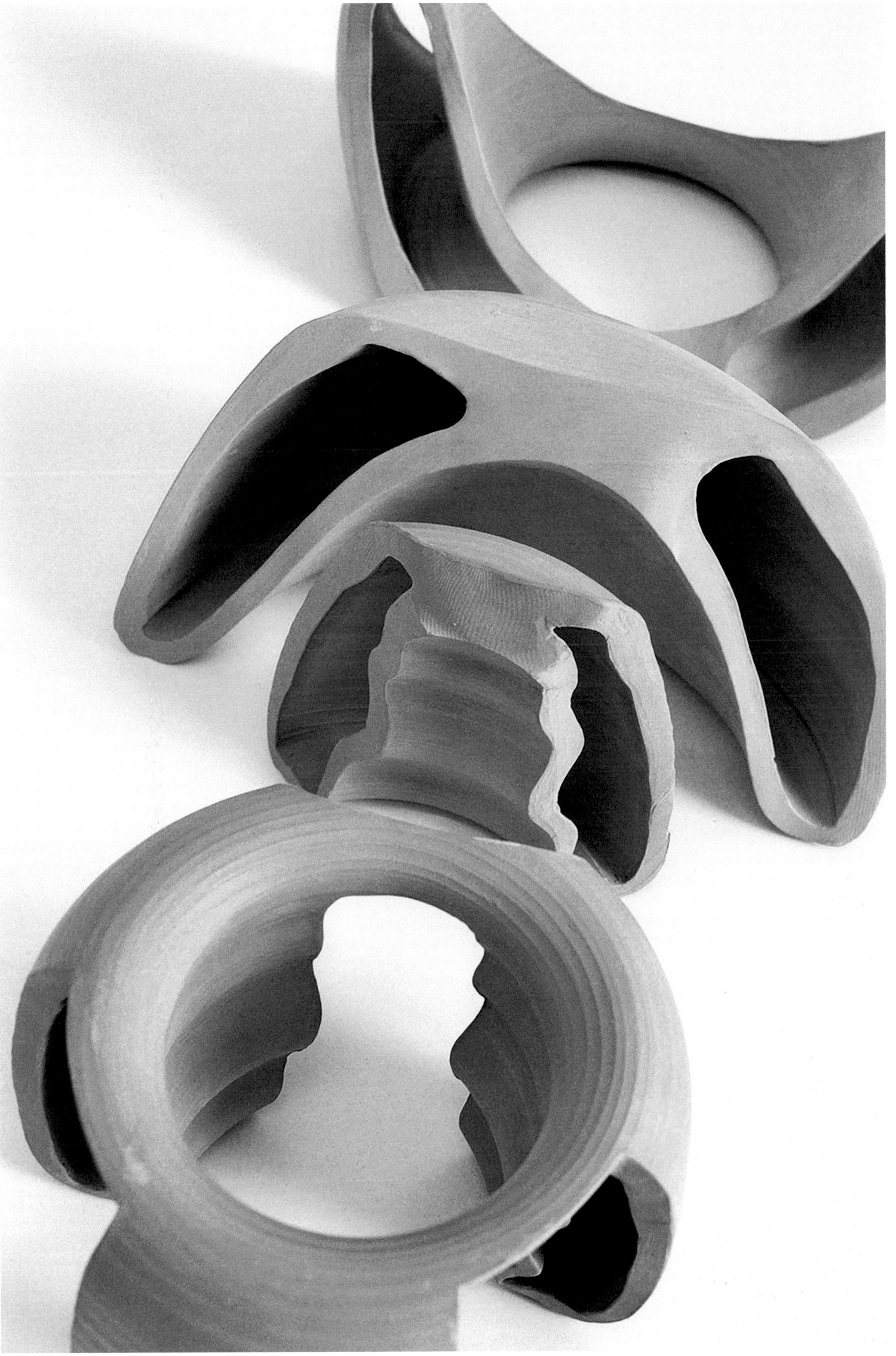

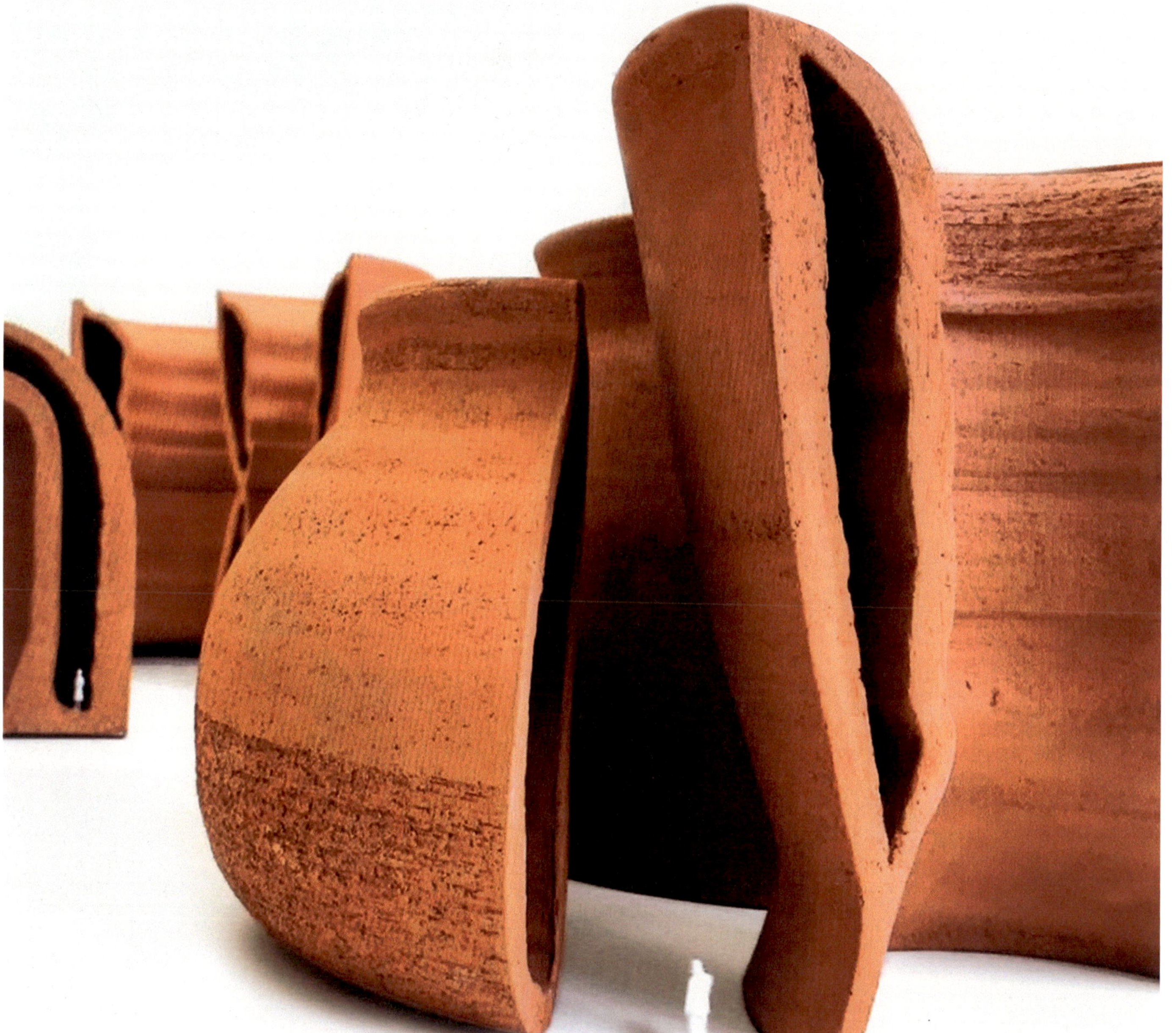

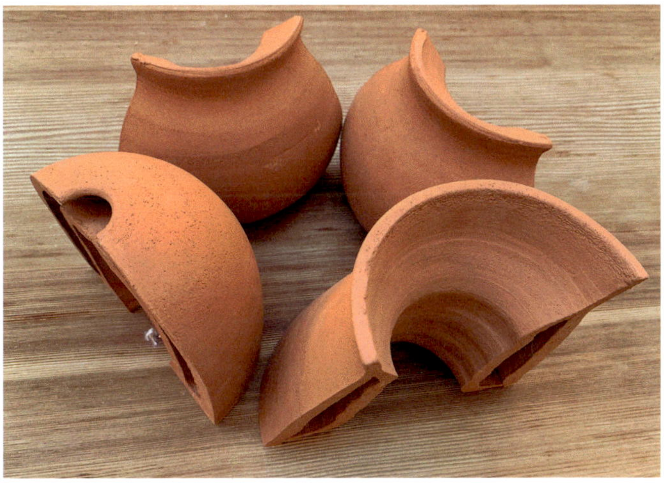
112

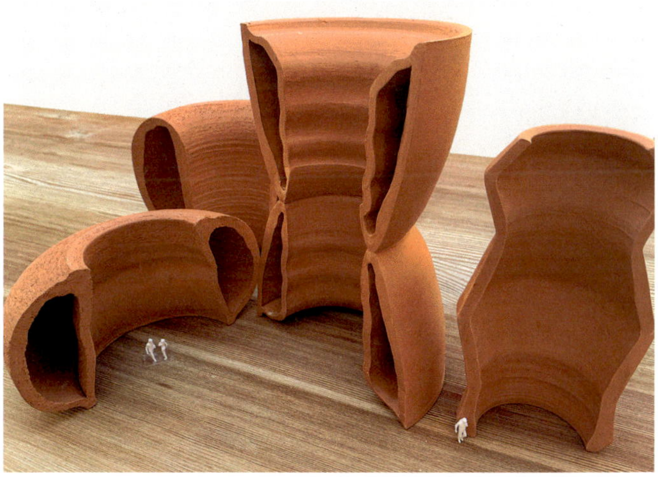
113

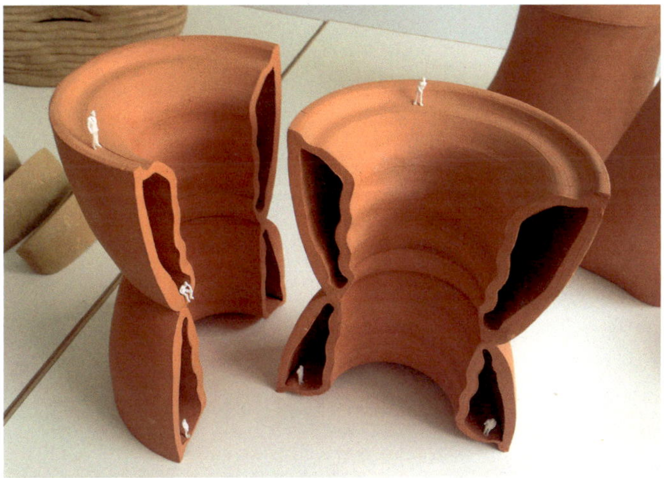
114

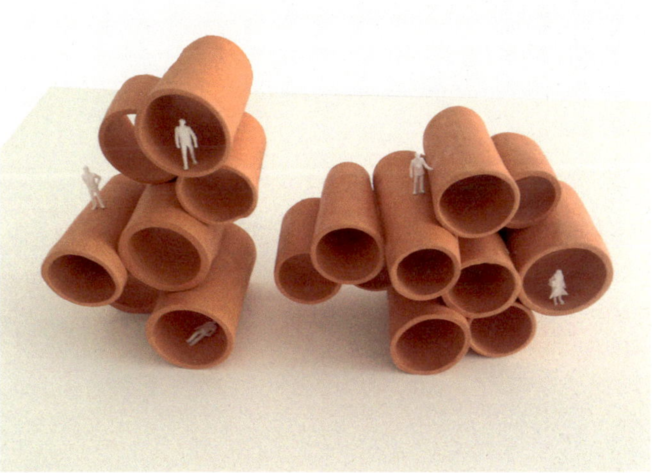
115

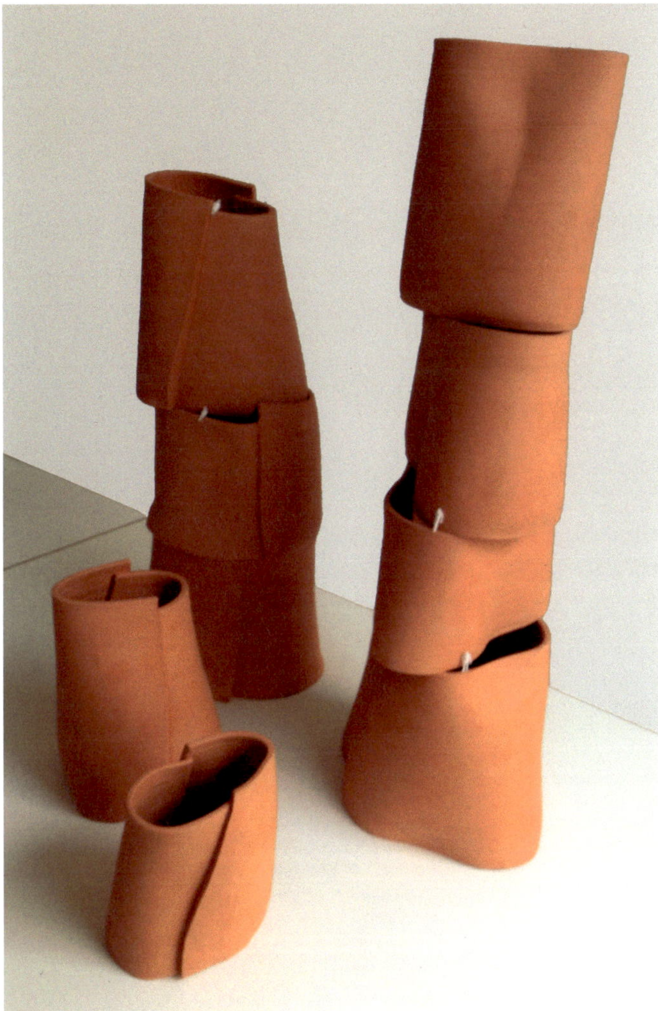
116

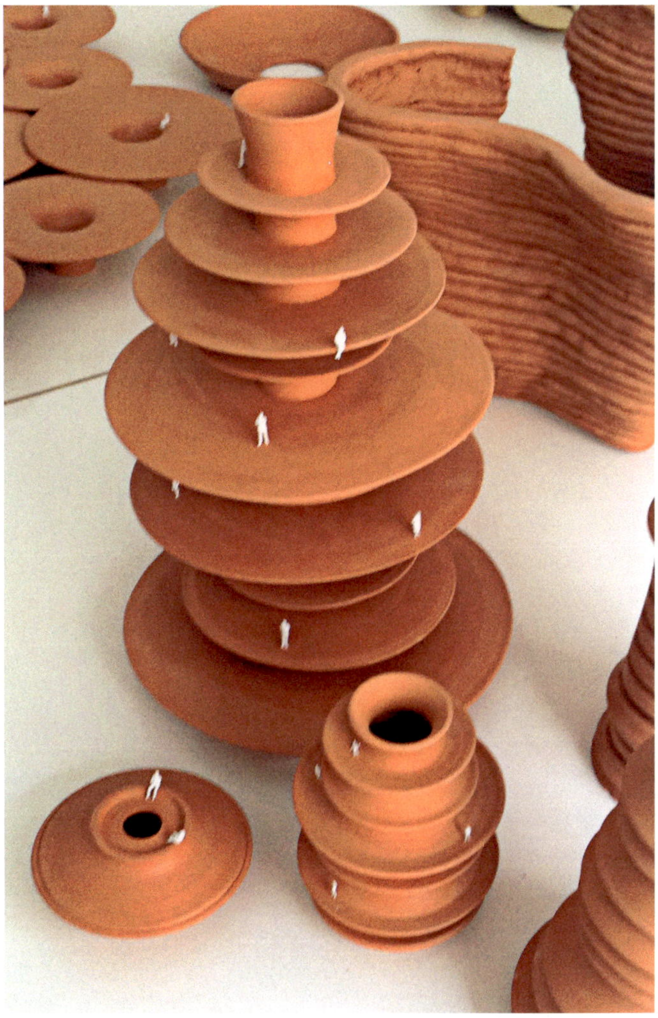
117

Our experiments with casting derive from – and continue to deliver us back to – the question: How does one make space itself something tangible and experiential? We produced two walls using sand and stone casting techniques to form a gateway entrance to the Haus am Waldsee exhibition.

We infilled large plywood boxes with a CNC-cut ridged foam topographical positive, which was then covered in sand and stone. Infralight concrete, a self-insulated lightweight concrete, was then poured.

Much of the debris used to form a surface texture was comprised of stone, brick, and tile – and much of it was found in Berlin's archeologically and historically potent ground.

The two walls – highly textured on one side and smooth on the other – faced each other, creating a dynamic canyon-like threshold to the exhibition spaces inside the historical villa museum. Large cranes were used to lift the walls onto steel foundations.

How, then, do the walls make the space itself tangible? They defied pragmatism as normal walls, and nevertheless transcended the merely conceptual. Together, they were legible as forming an entrance – their solidity, their assertive stance, demarcated and reorganized their context.

The walls split the existing area and generated another. They formed a passageway. They enclosed the visitor – compelling awareness of a momentary experience of interstitiality. The walls will be reused within the gardens of a local artist. What kind of space will they generate there?

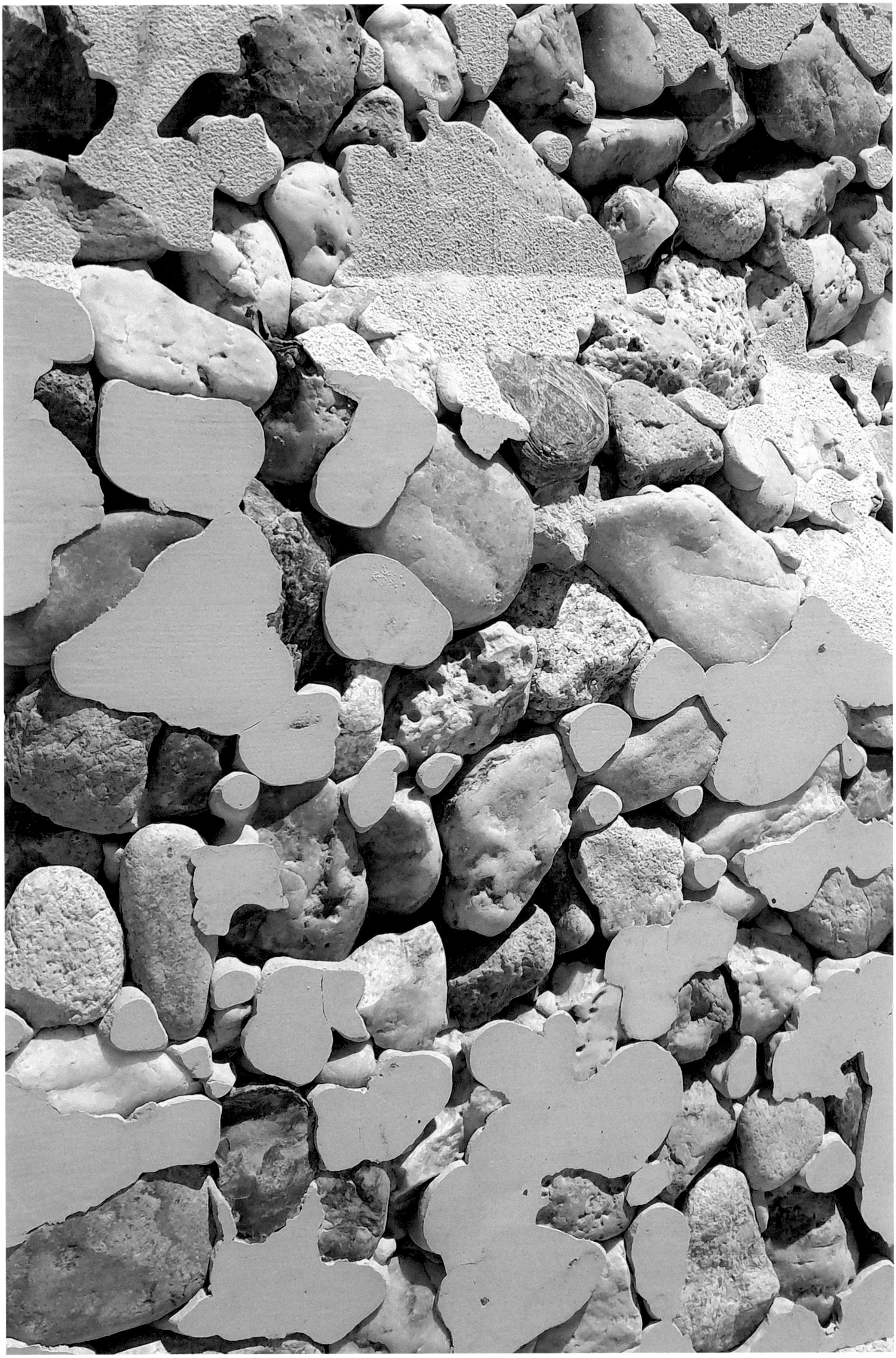

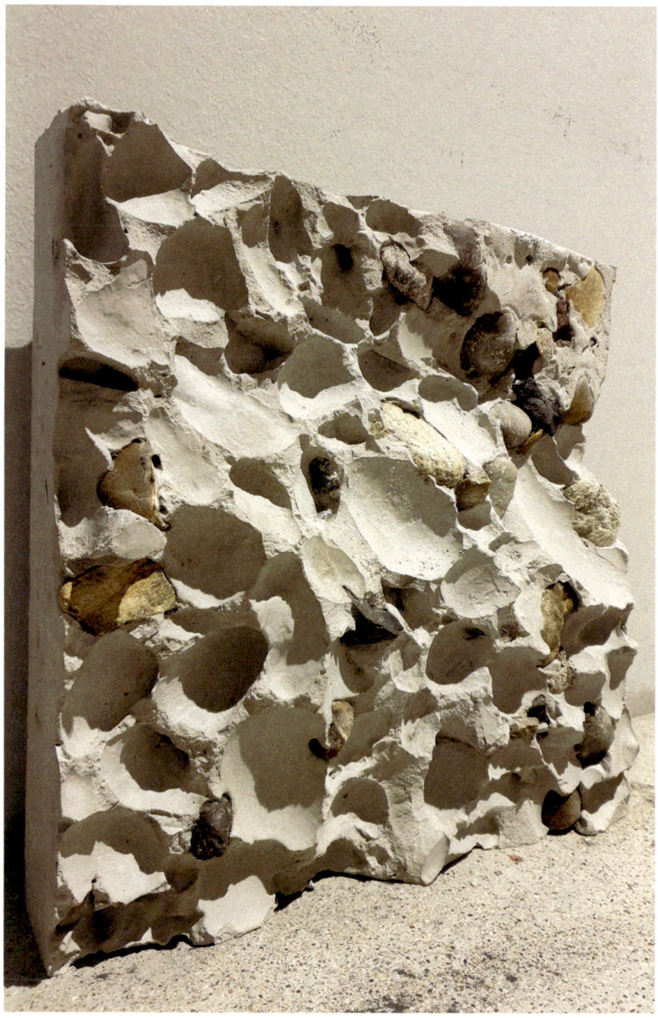
125

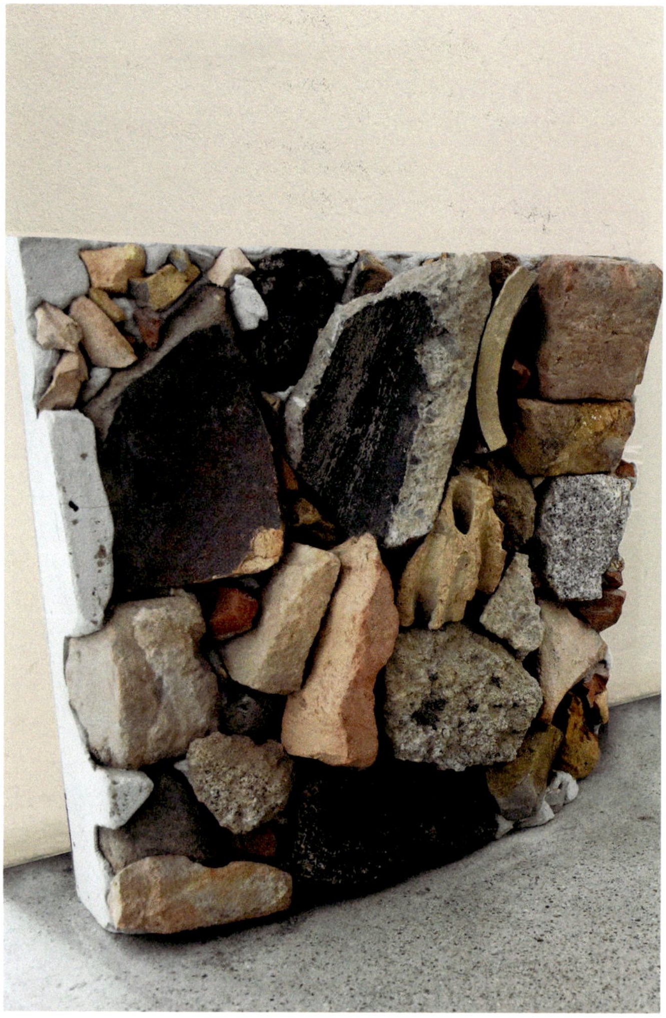
126

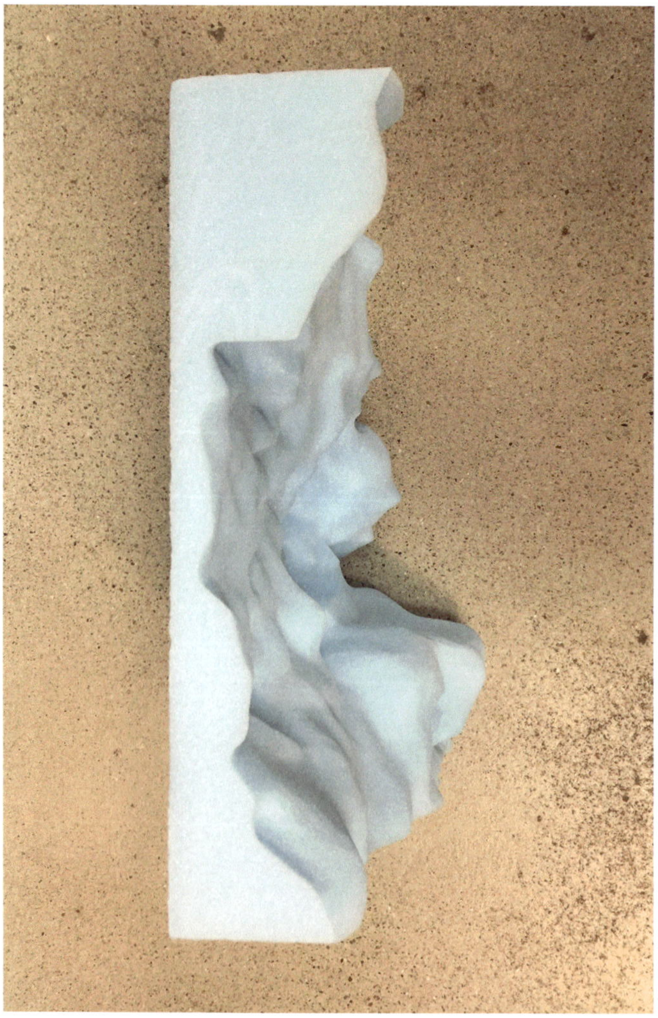
127

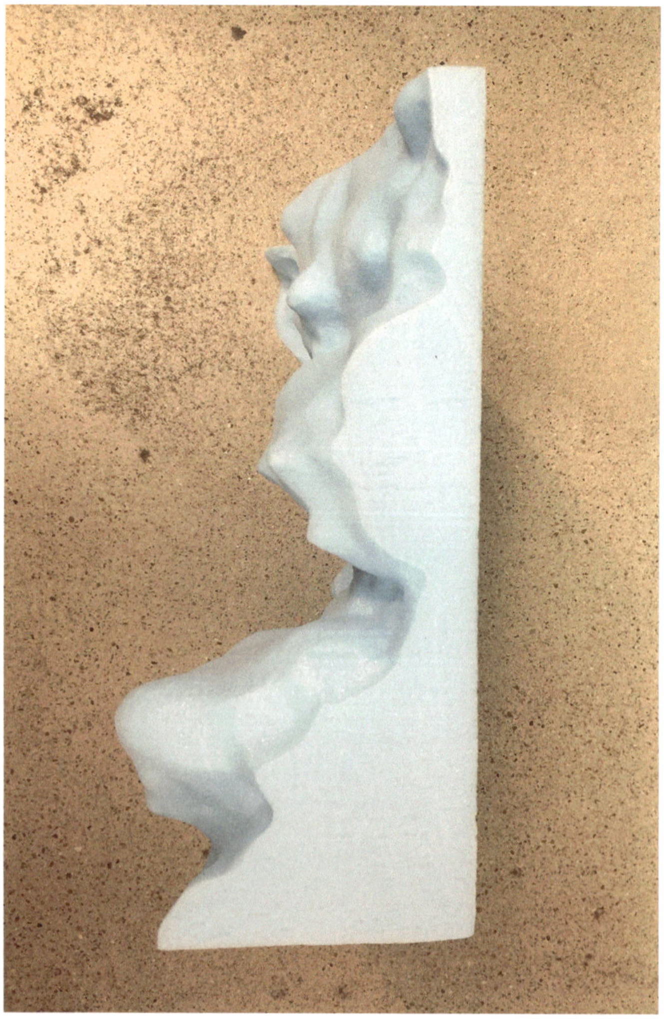
128

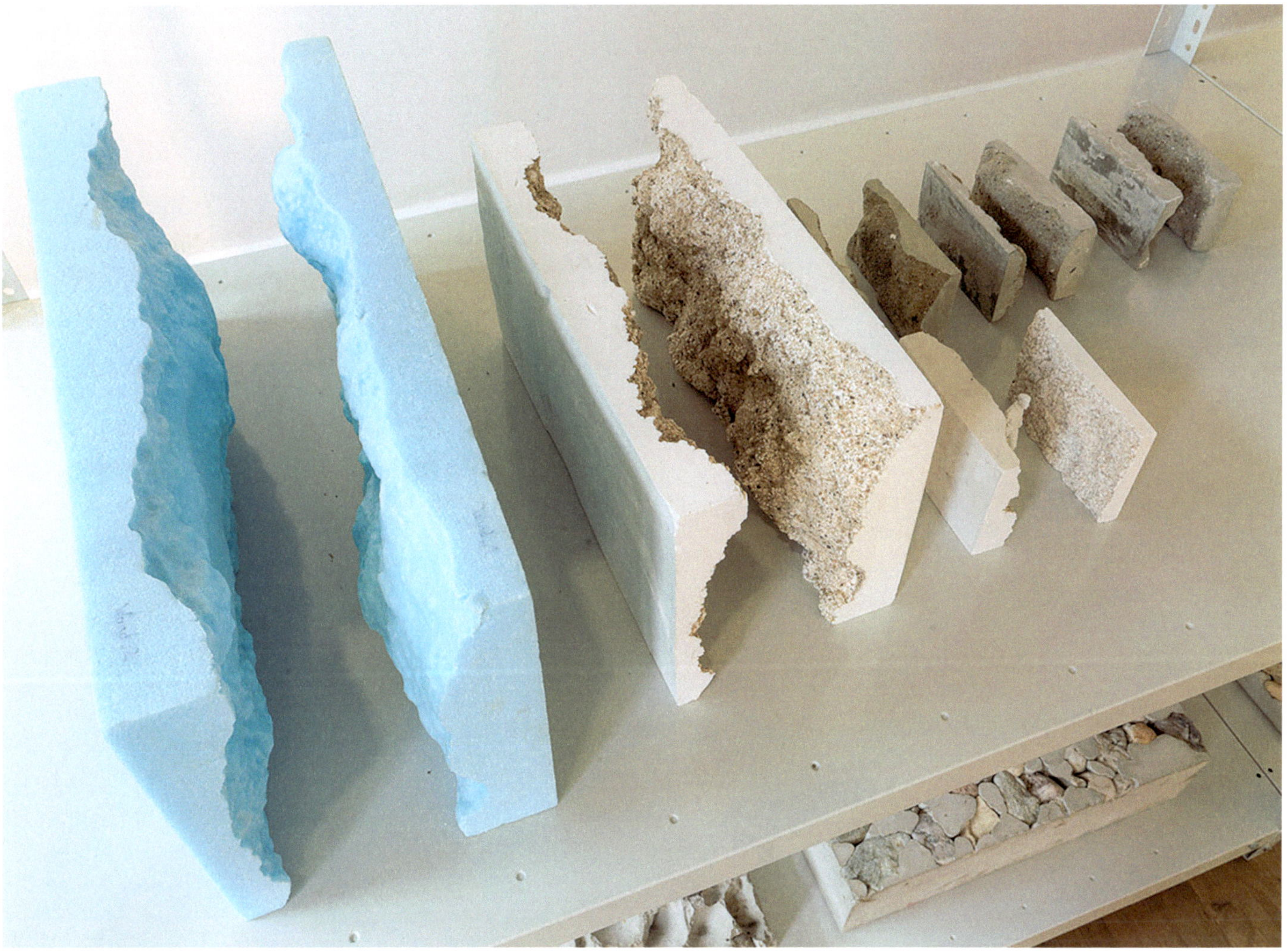

129

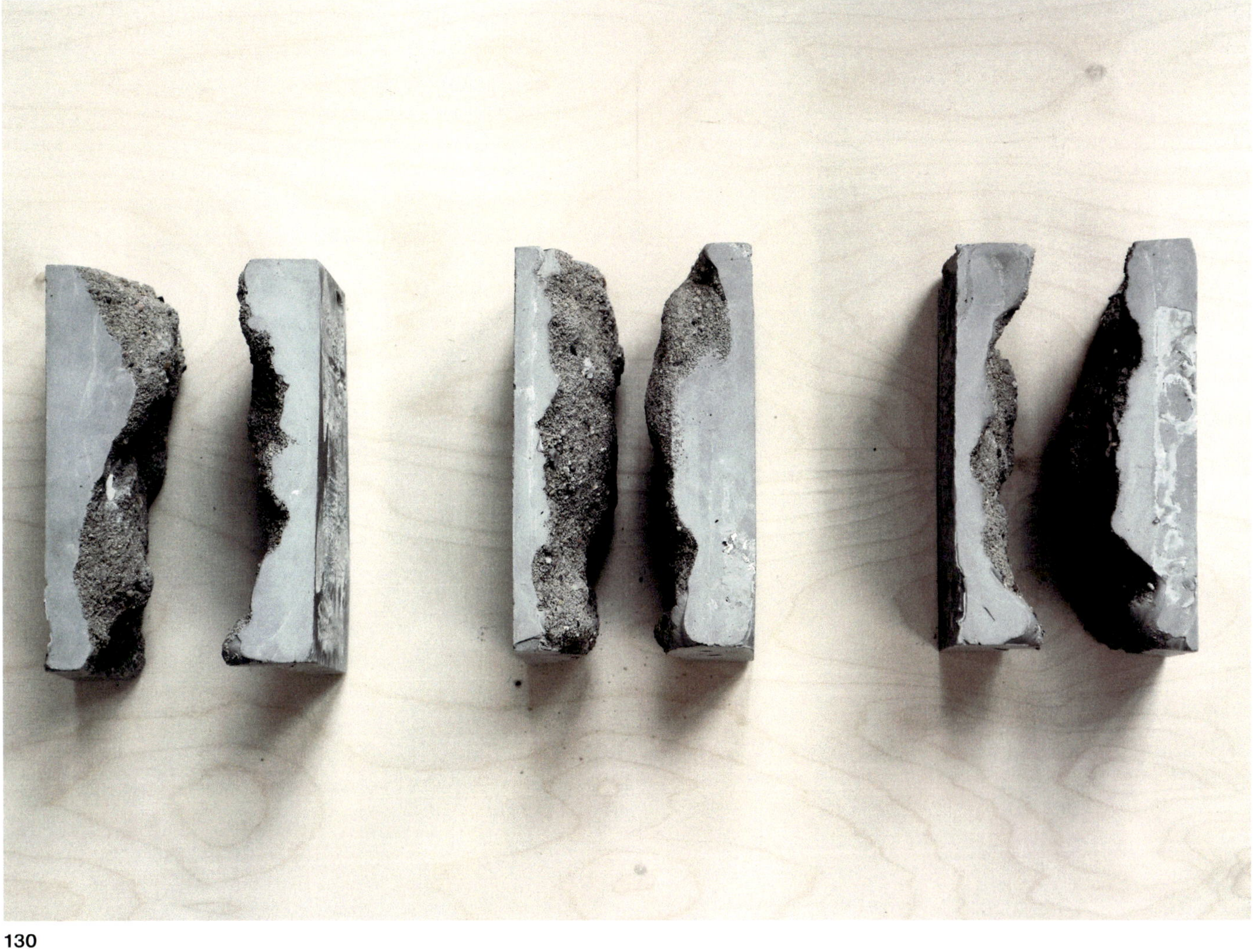

130

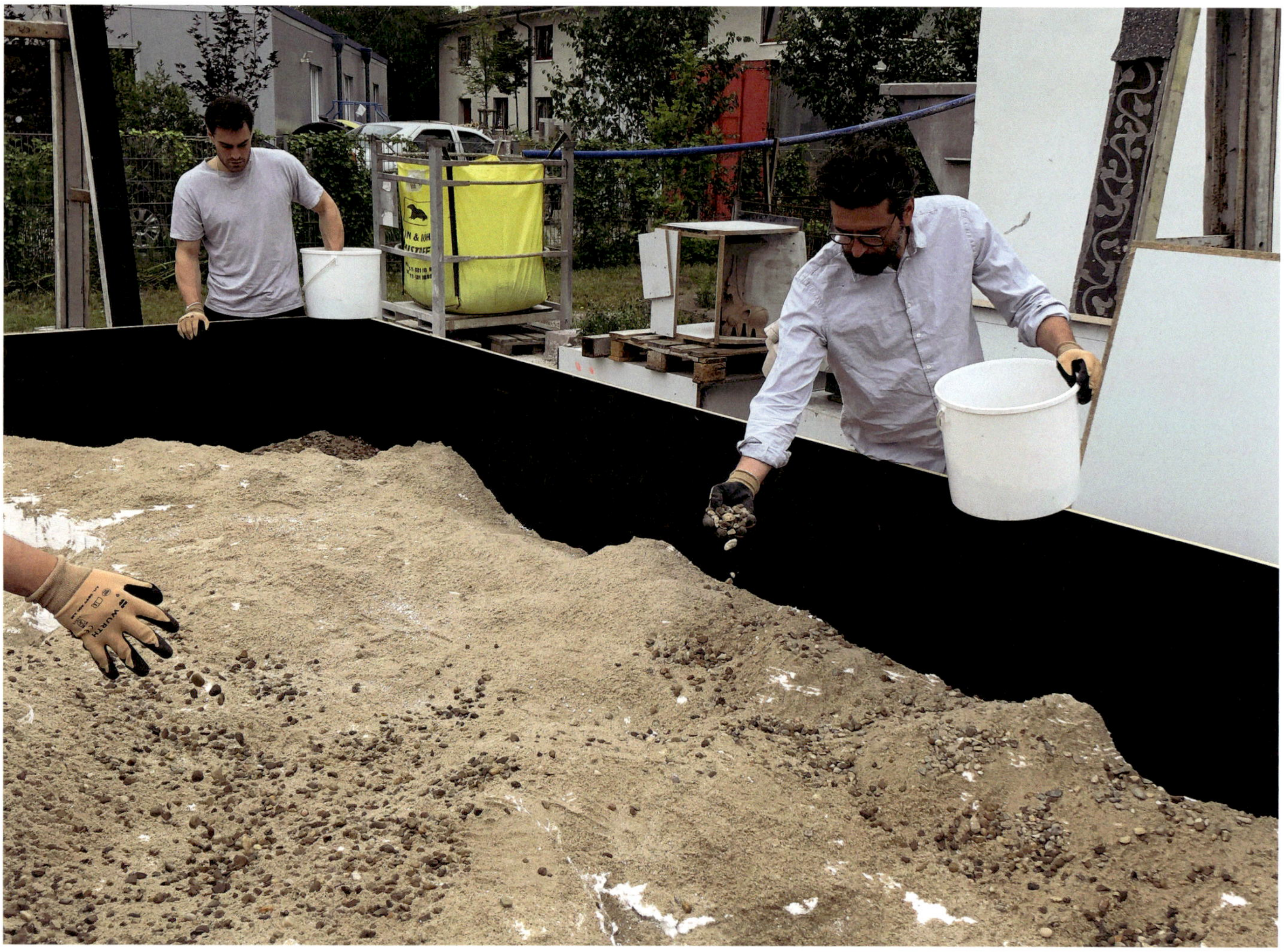

131

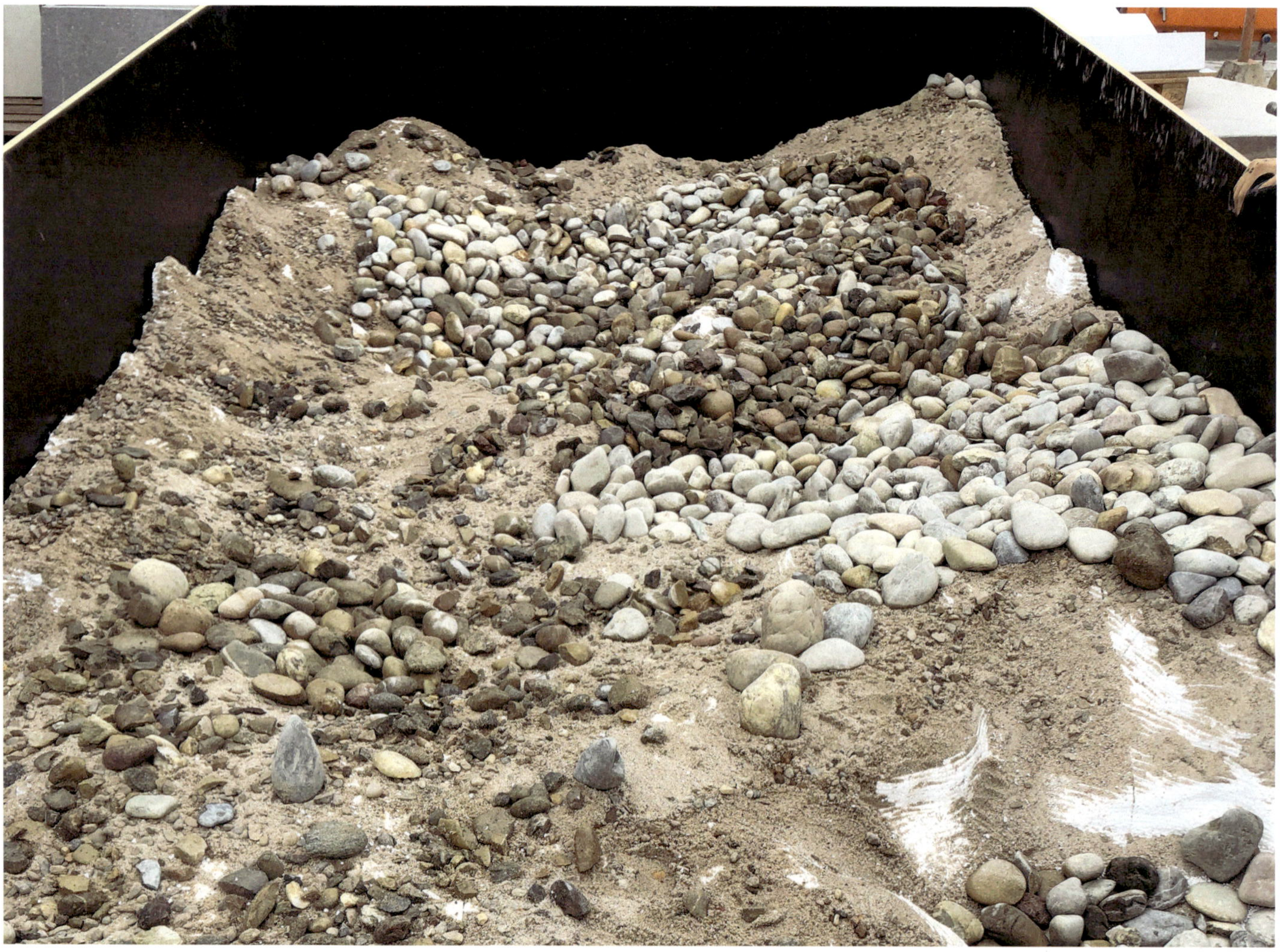

132

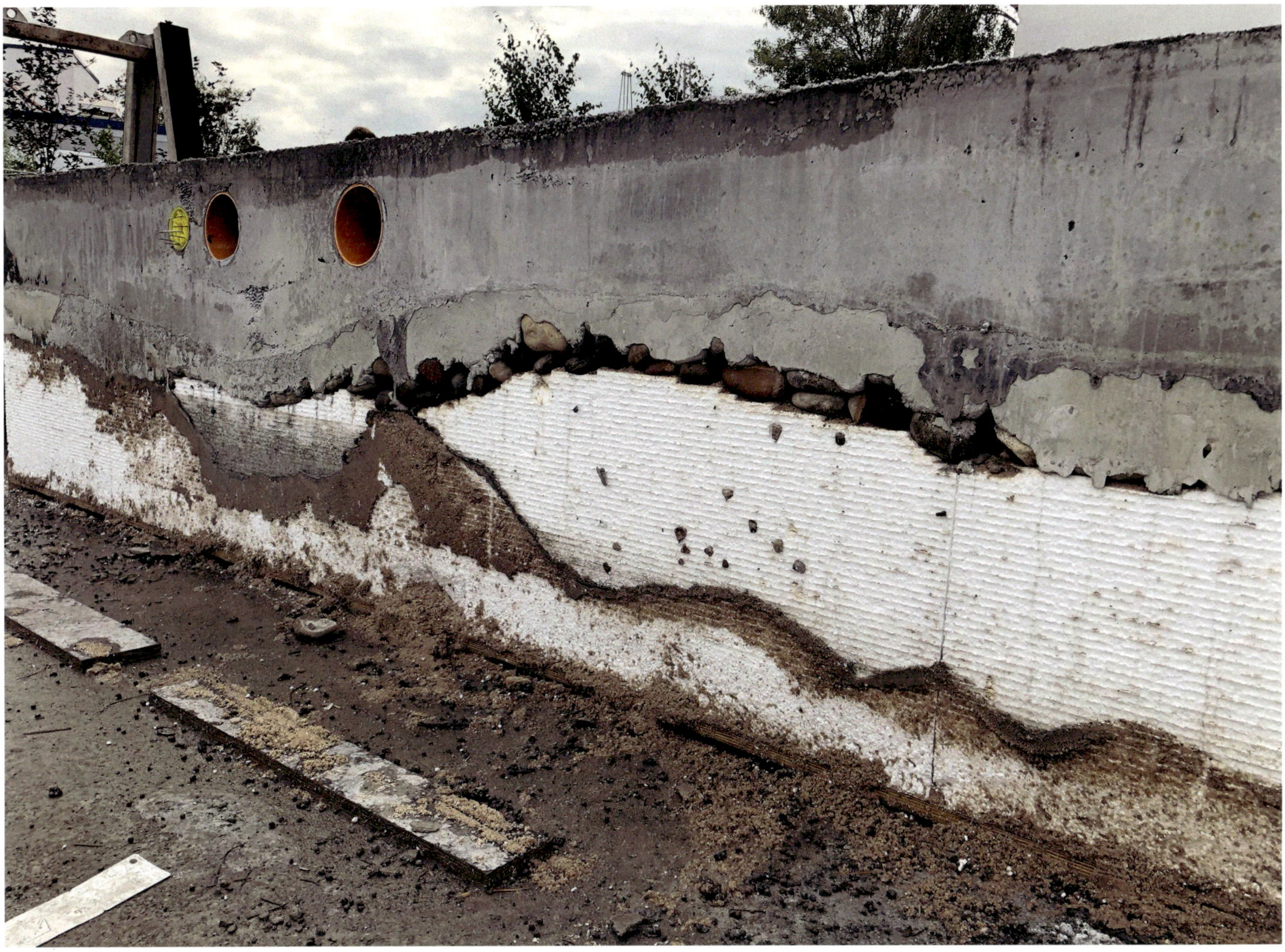

133

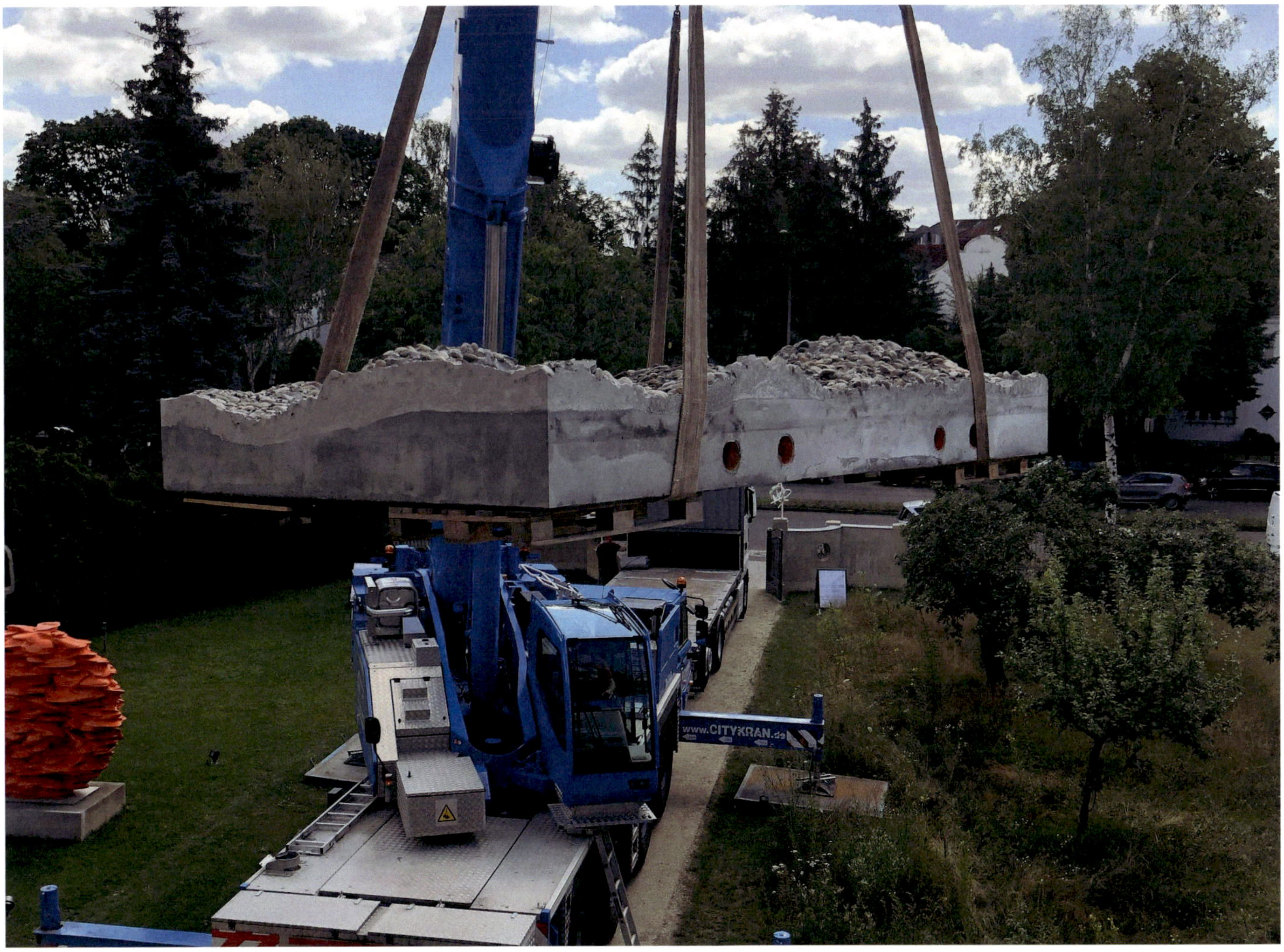

134

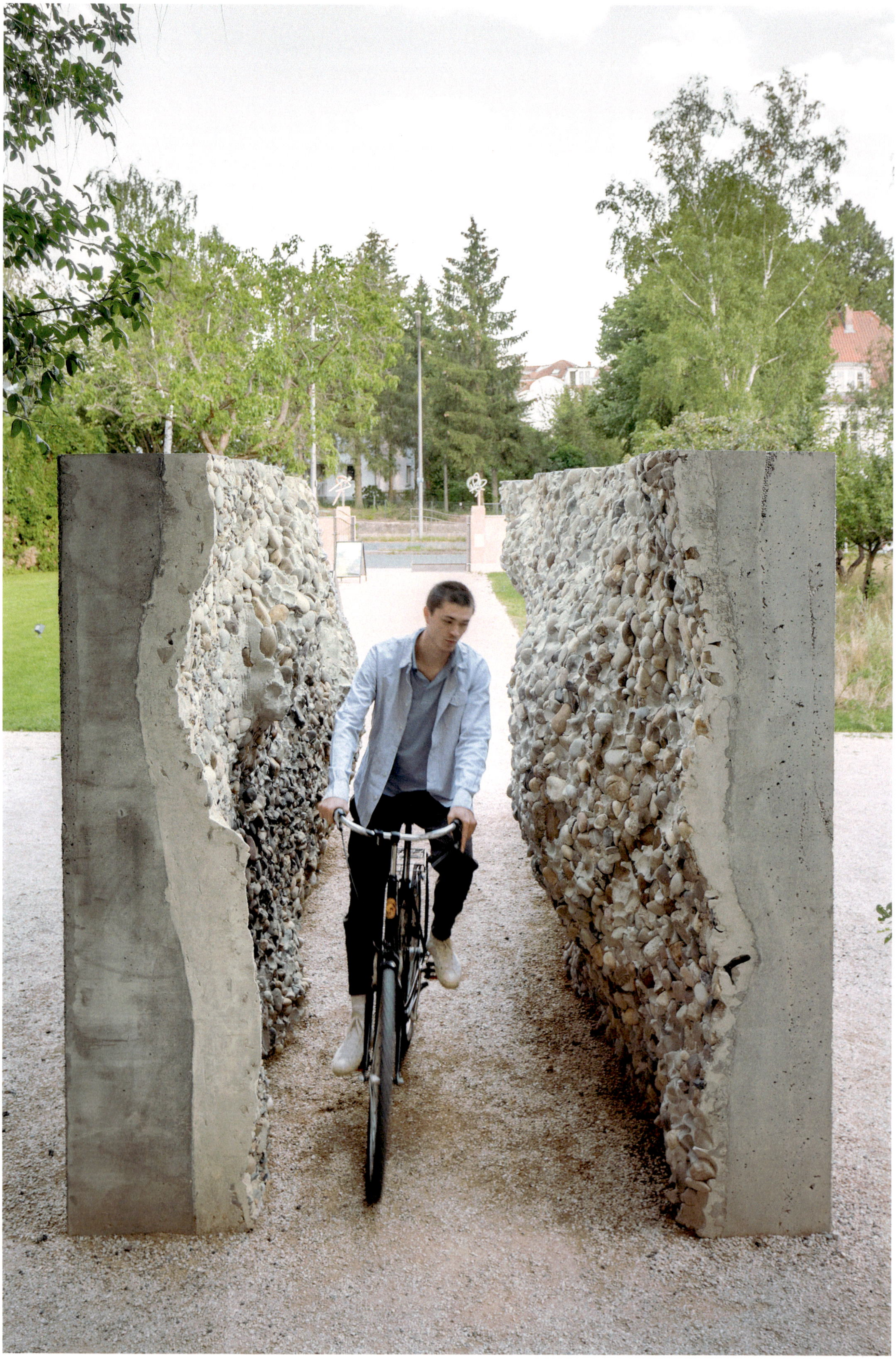

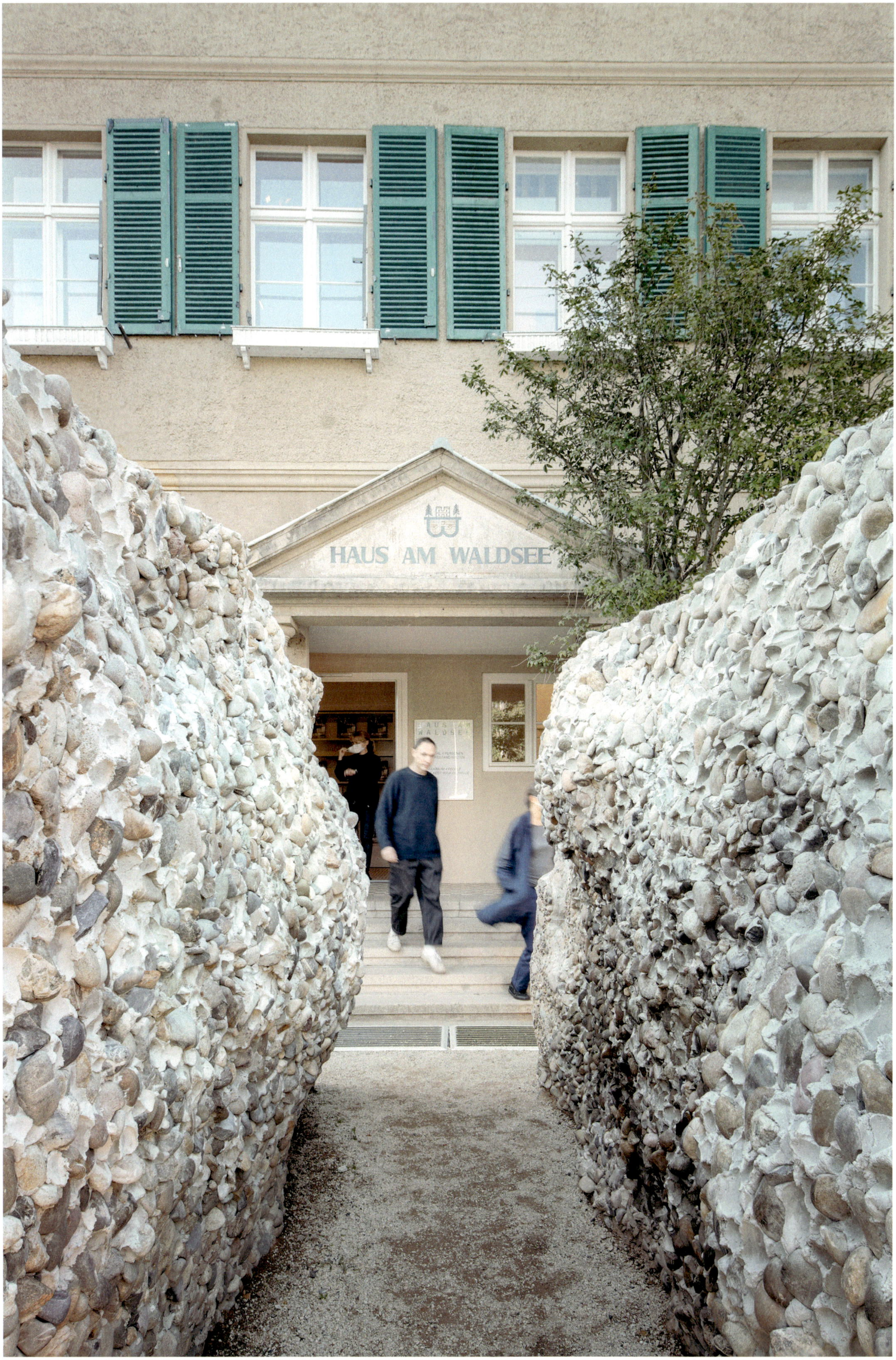
HAUS AM WALDSEE

Topographical

137

A primary interest in Land artists and Process artists (Robert Smithson, Michael Heizer, Richard Serra, etc.) during graduate school engendered a life-long engagement with topography as an integrated and formative dimension of architecture.

138

In other words, we developed a devotion to the site itself as something to be constructed or built: the site as a vital grammar without which the language of any structure cannot be articulated. We are interested in an architecture that is situated upon, into, or embedded within the ground.

Topographical describes architectural form-making, which resembles, indeed responds to, natural ground variation configured either horizontally or vertically.

Fidelio, Beethoven's sole opera, was performed at Theater an der Wien (where it originally premiered in 1805), in March 2020, under the direction of Christoph Waltz. The opera was set in a prison. The vertically disposed set was a double helix, constructed of timber.

The structure allowed the performers to ascend and descend, to move forward and backward, across its stepped terraced surfaces. The set was prefabricated off-site in Poland and moved into the theater piecemeal (like a ship-in-a-bottle) using digital cutting technologies.

As a spatial and three-dimensional volume, the set complimented and extended the curvilinear spaces of the historical baroque theater, and acted as an acoustical instrument projecting the sound of the musicians and singers.

The set was an abstract and topographical extension of the tiered seating of the theater, both extending it and completing it through a spatial dialogue.

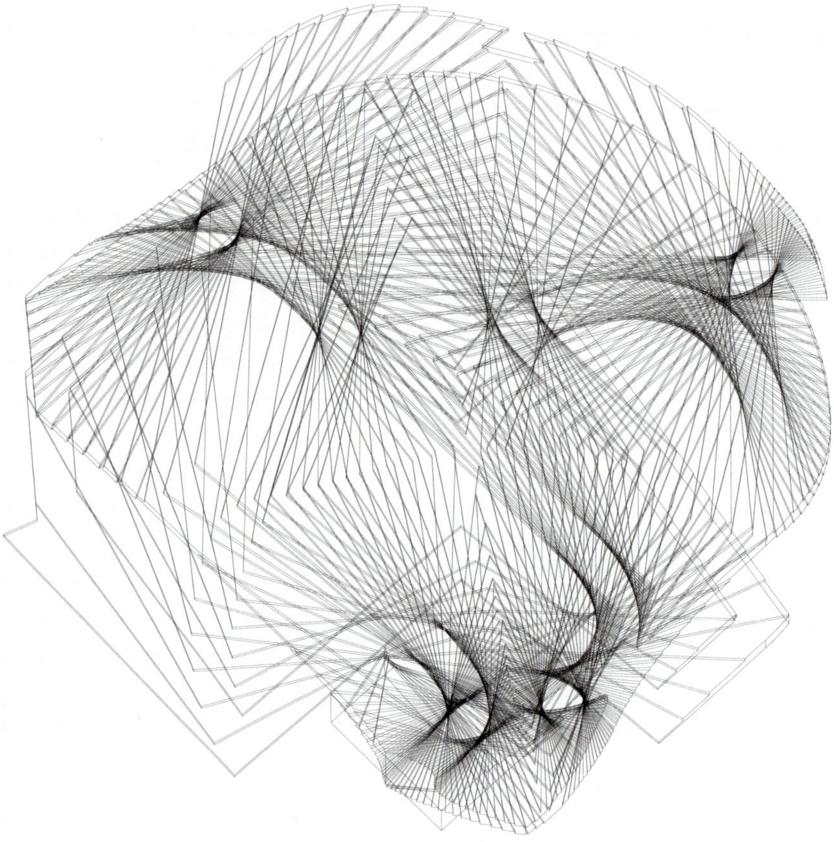

145

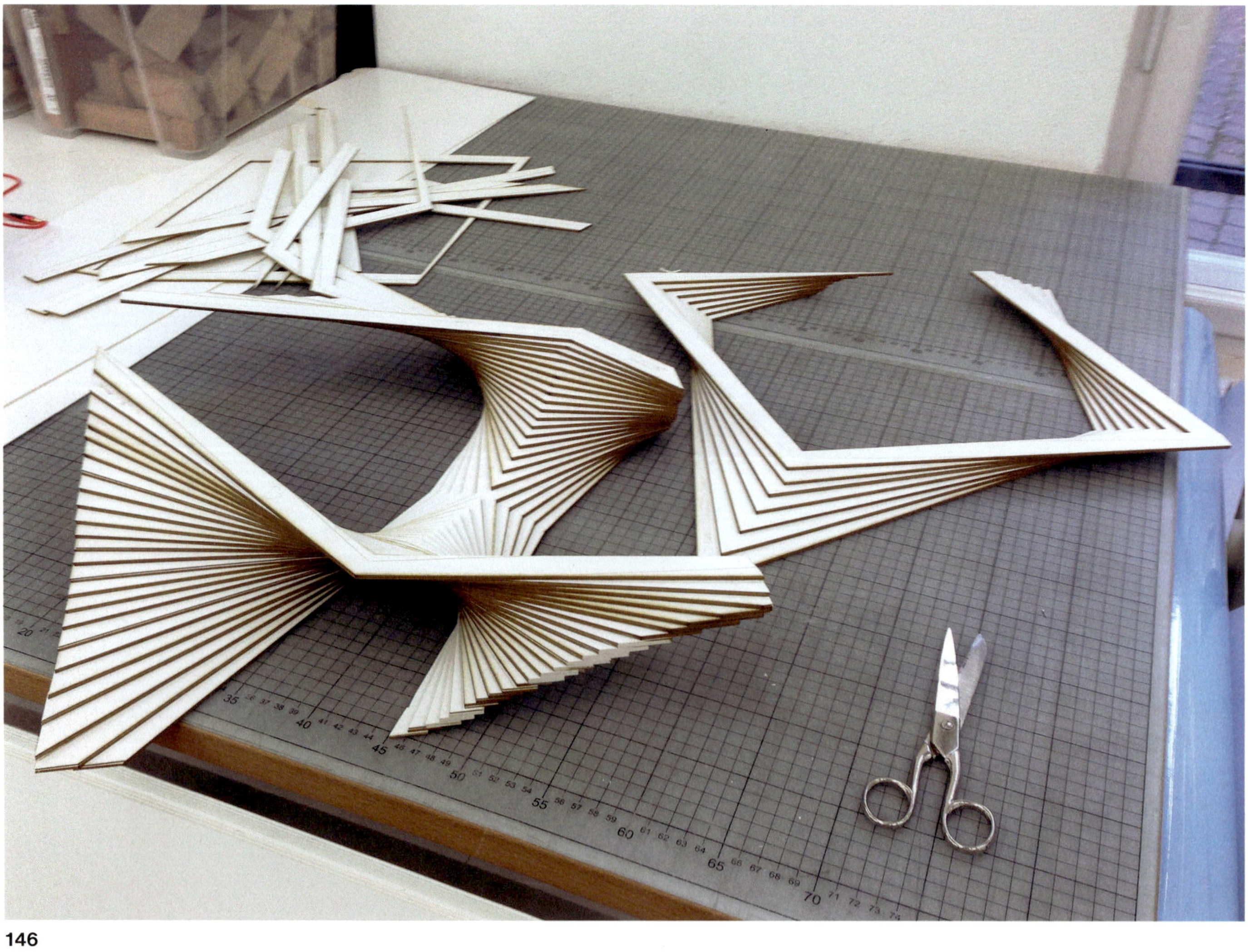

146

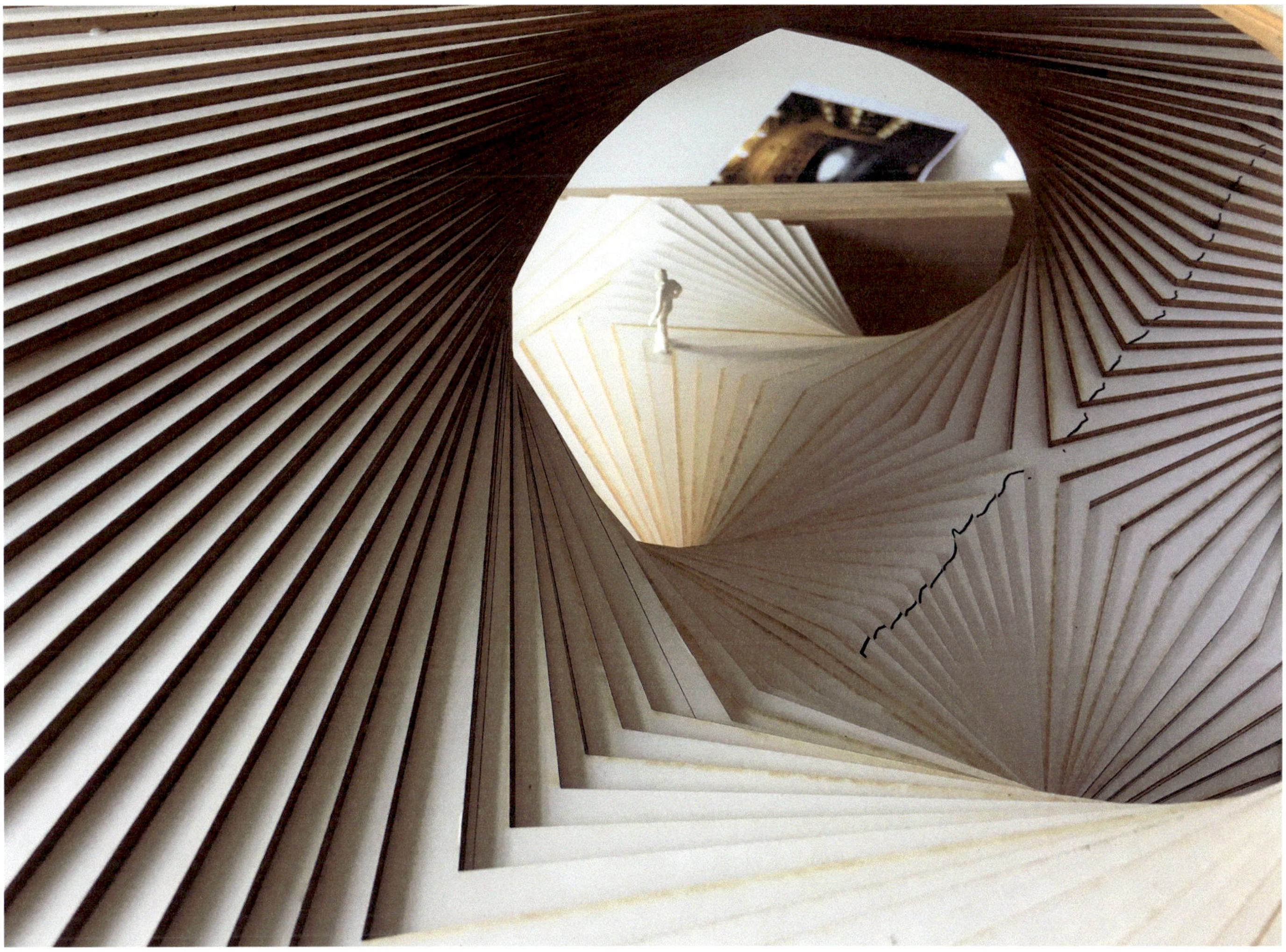

147

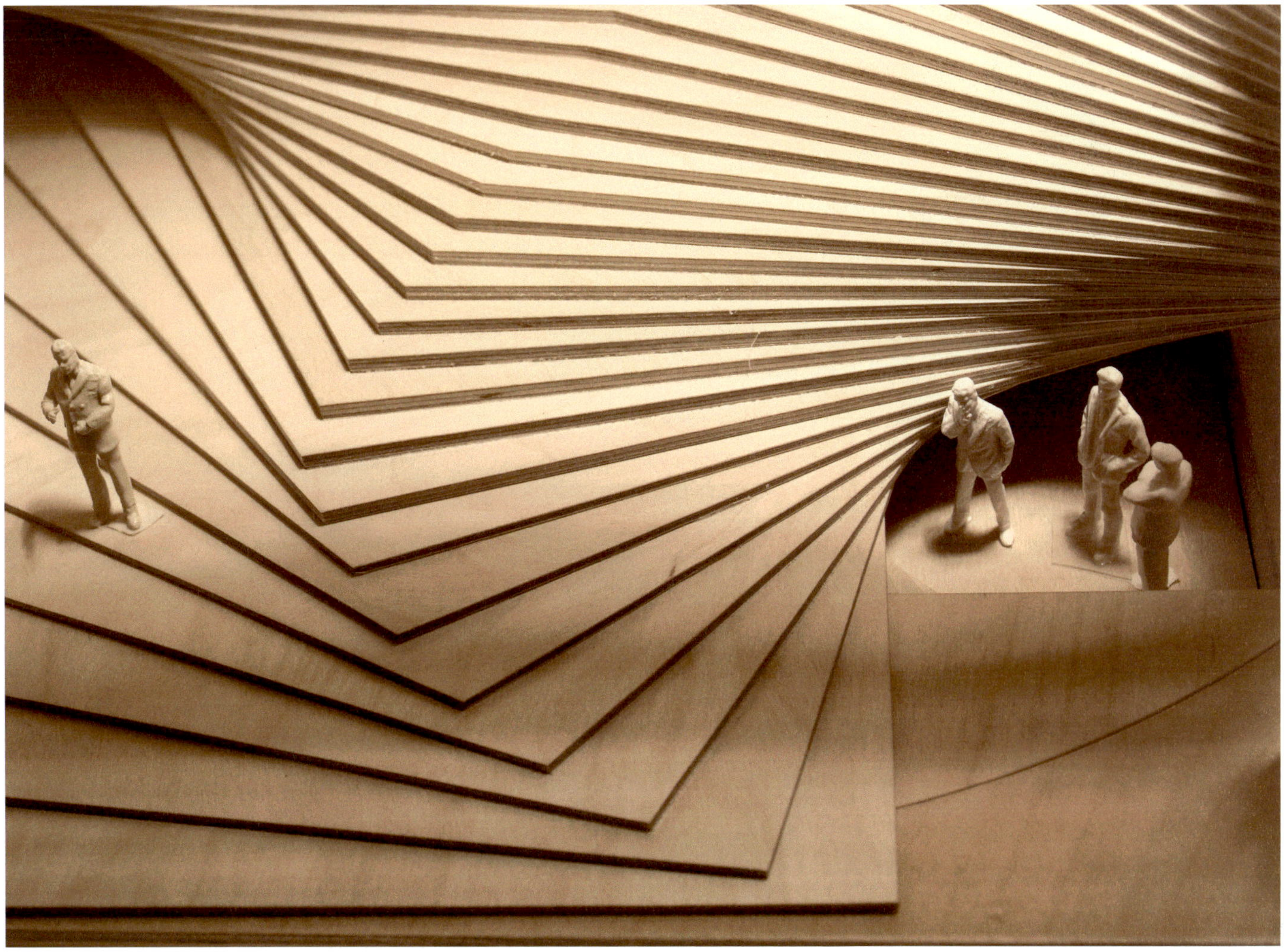

148

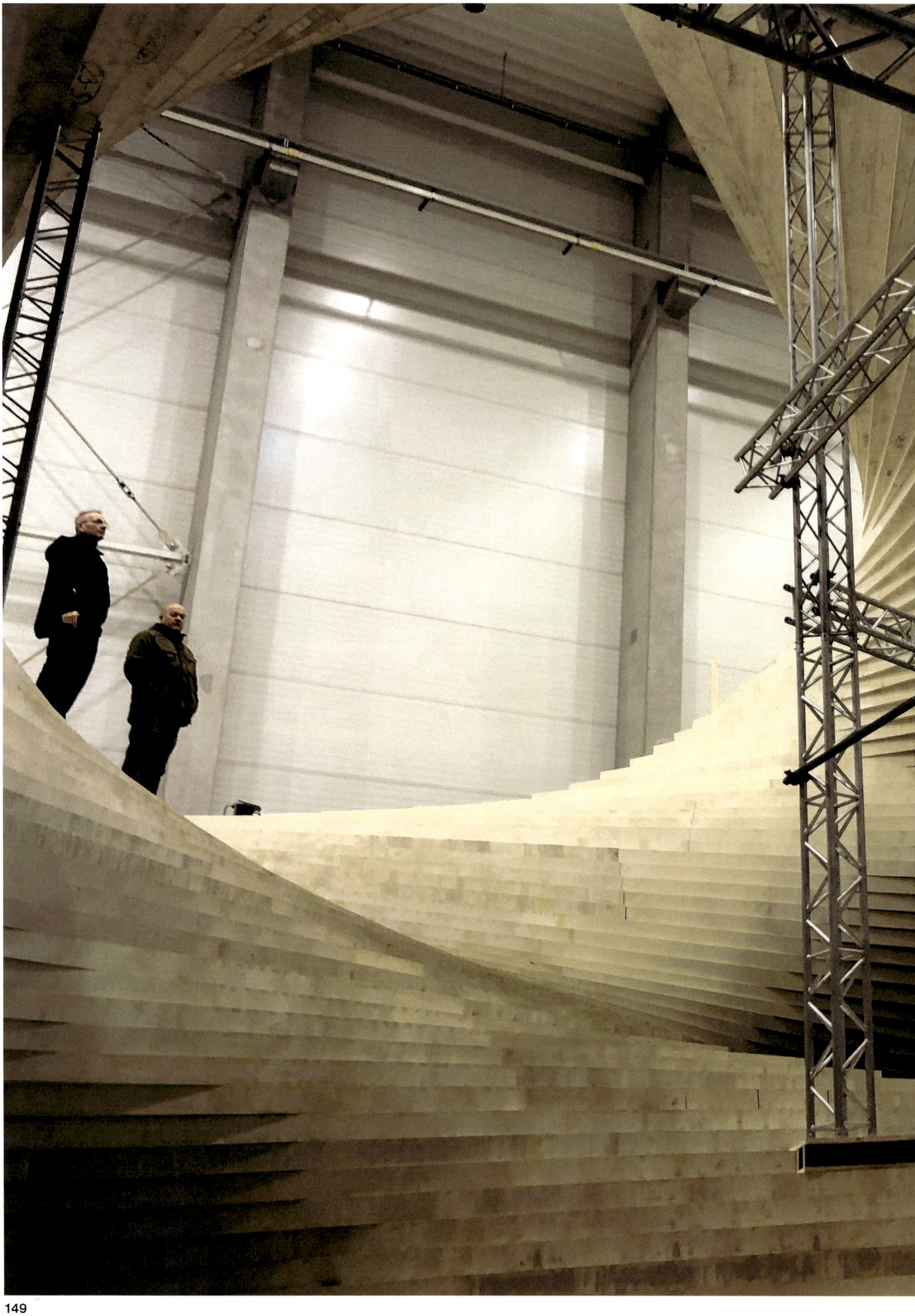

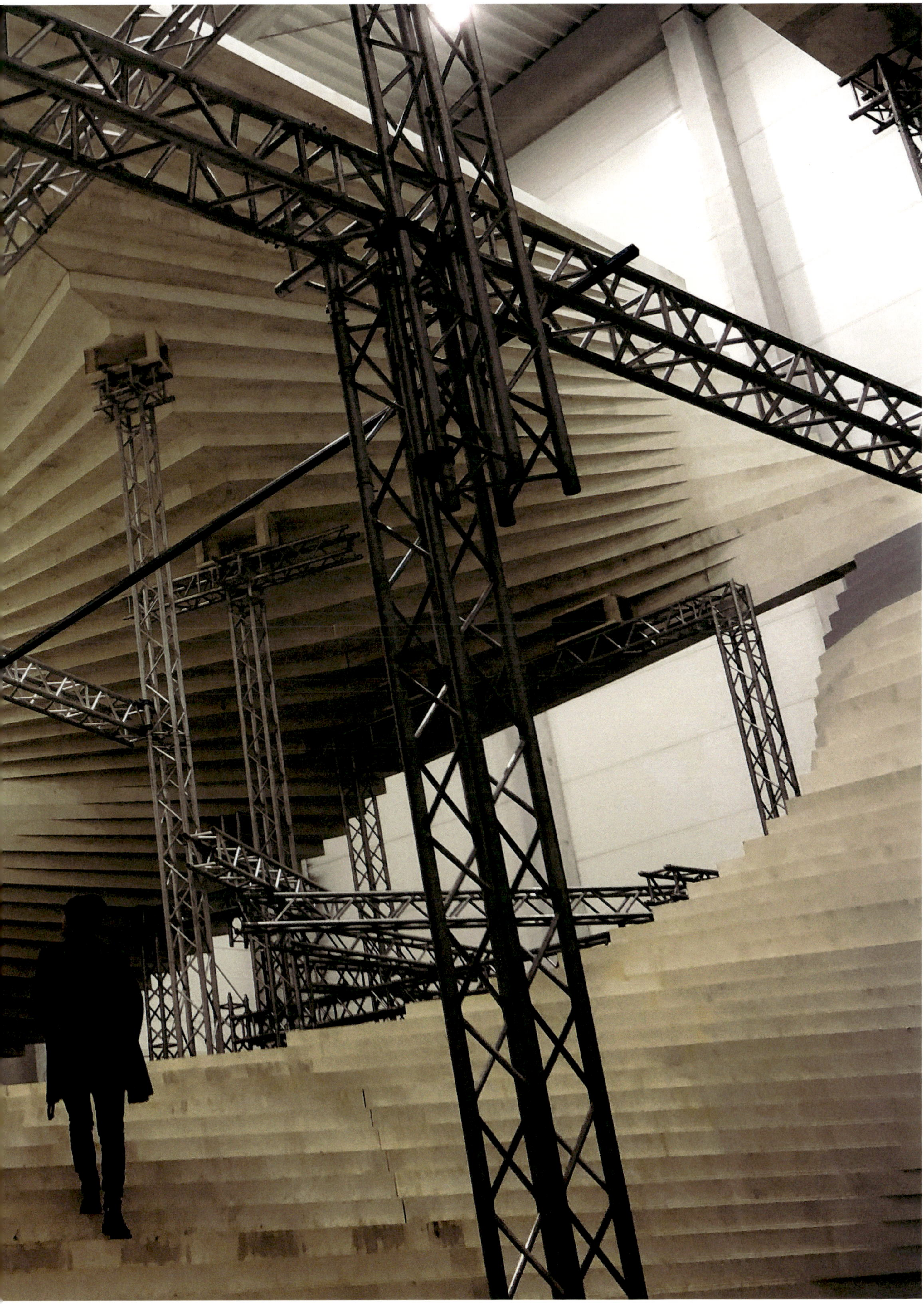

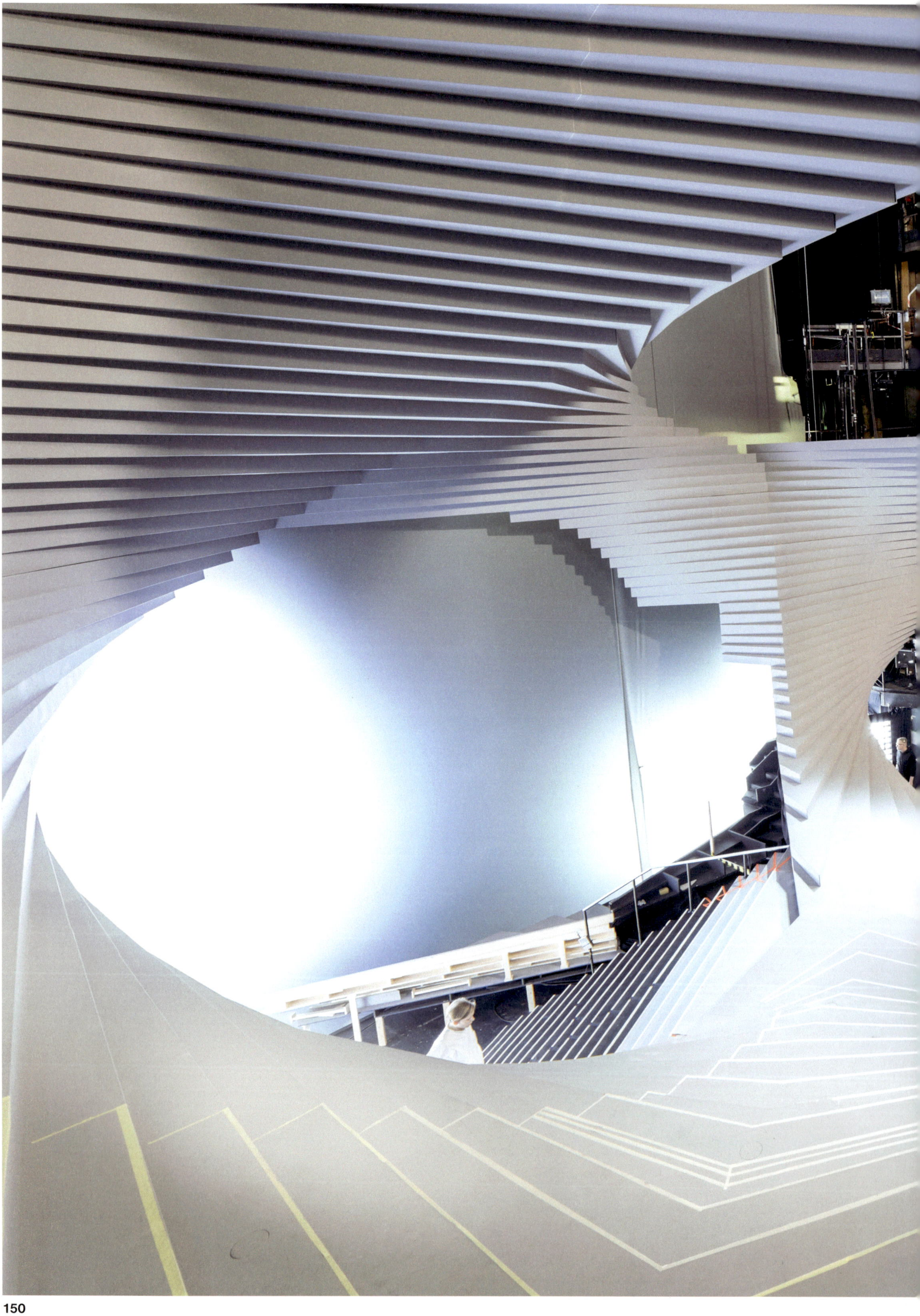

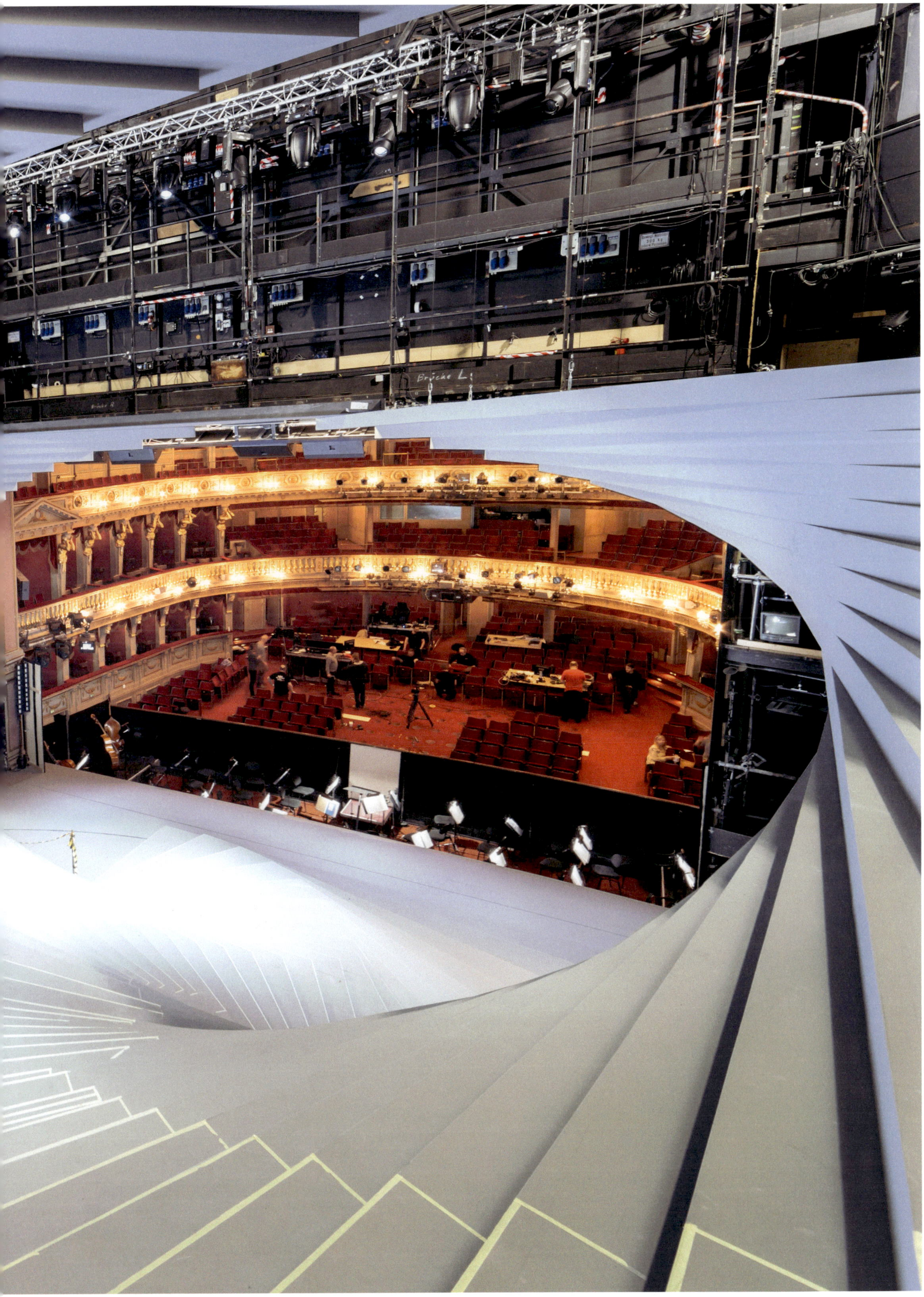

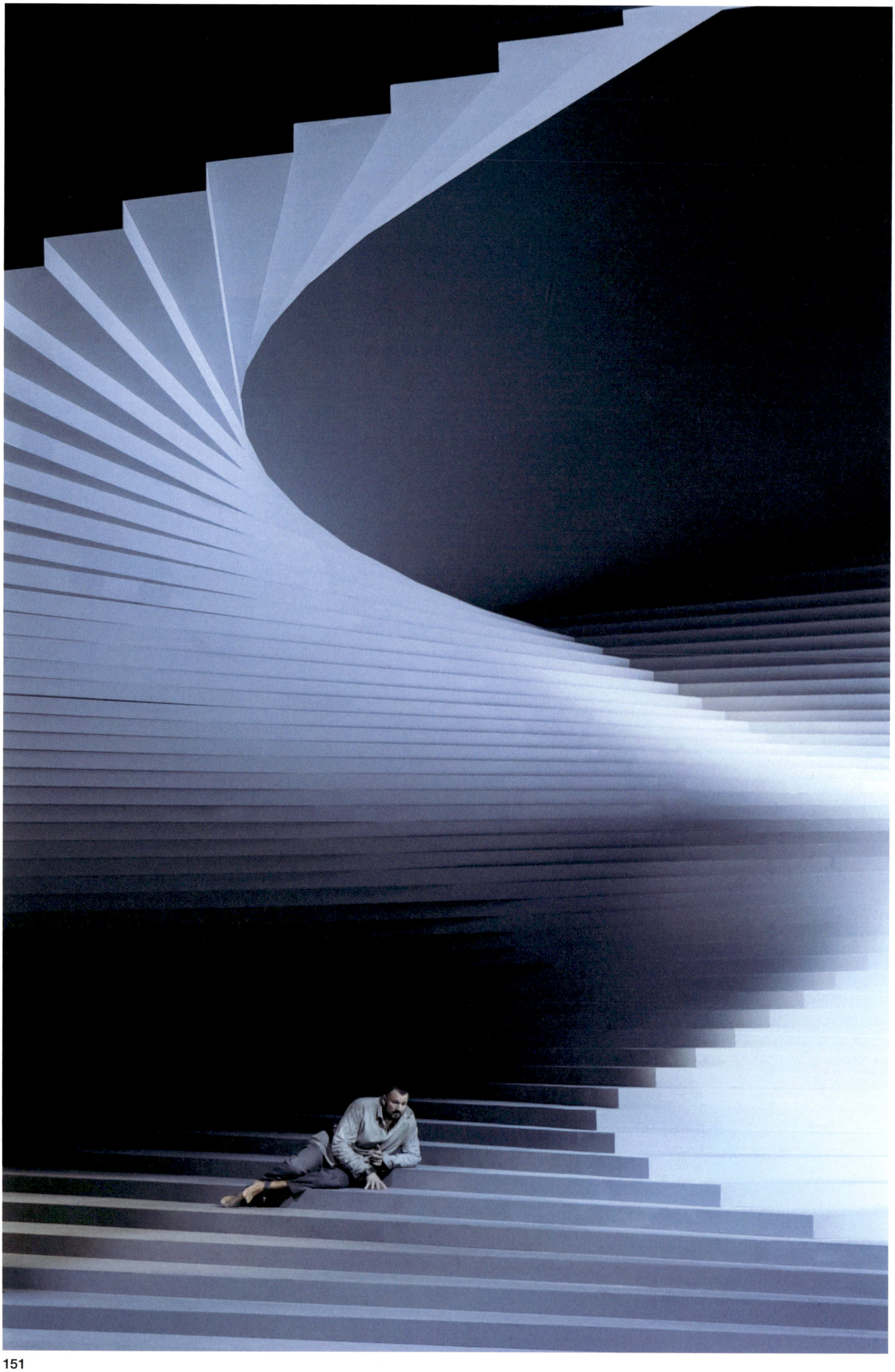

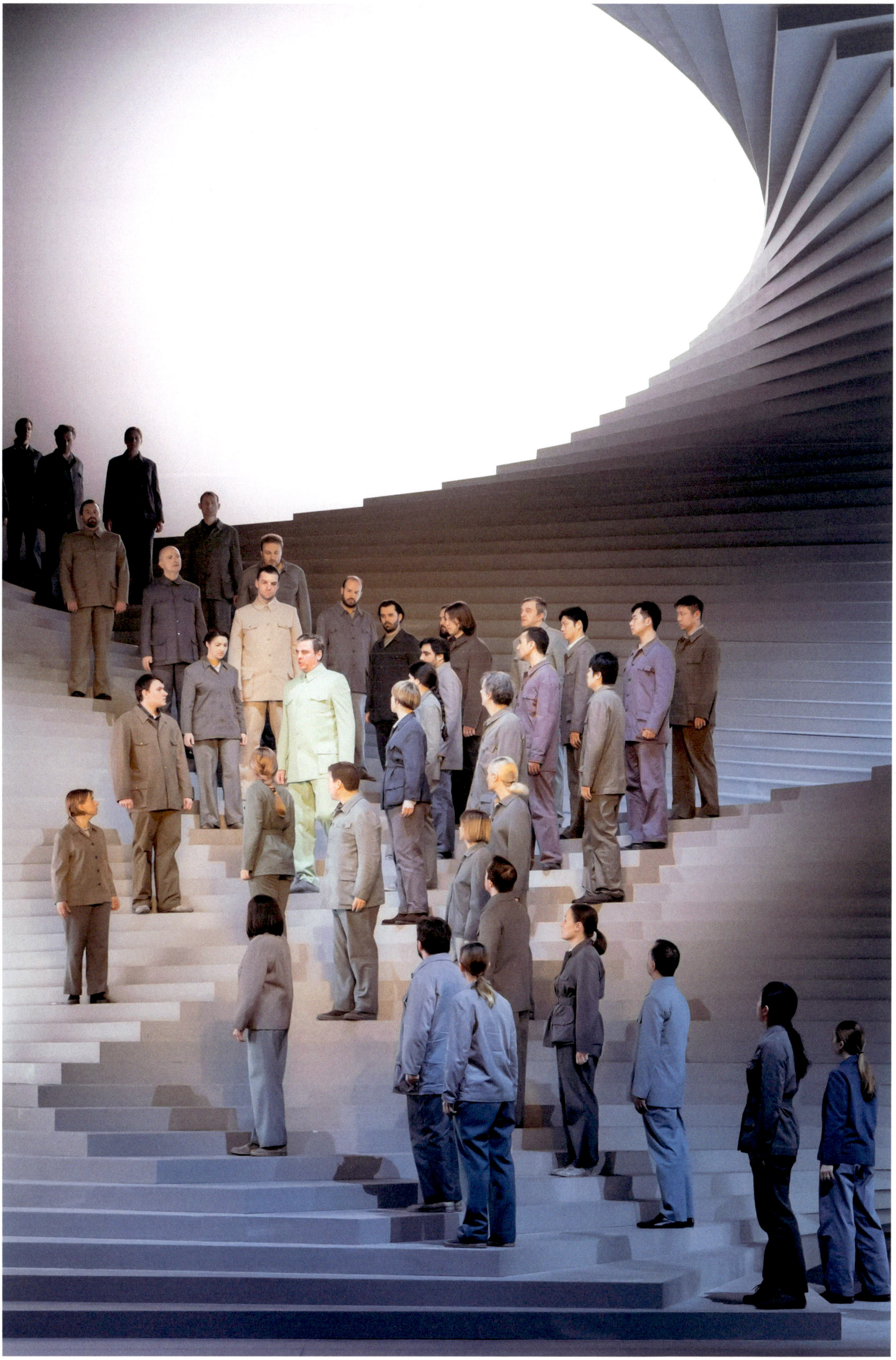

The Estrel Berlin is a high-rise complex organized as a topographically extruded ensemble of buildings. In flat Berlin, it assumes an island-like presence, and has a resonance as a gateway to the city.

153

The two towers (the existing and the new) inflect toward one another and frame this gateway. Inspired by the game Tangram, we combined triangular volumes, extruded to different heights, to create a family of forms culminating in this 175-m tower over Berlin.

154

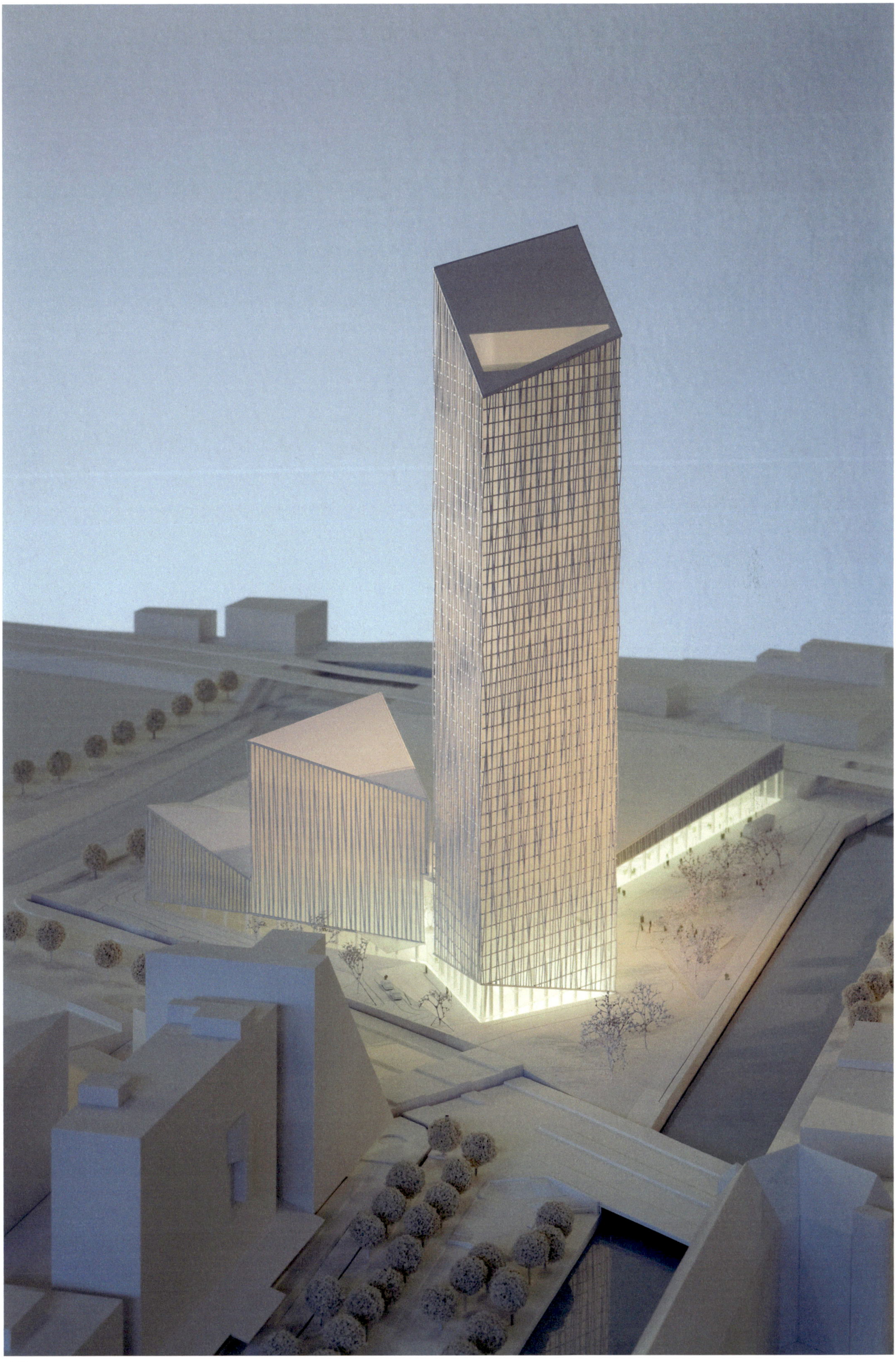

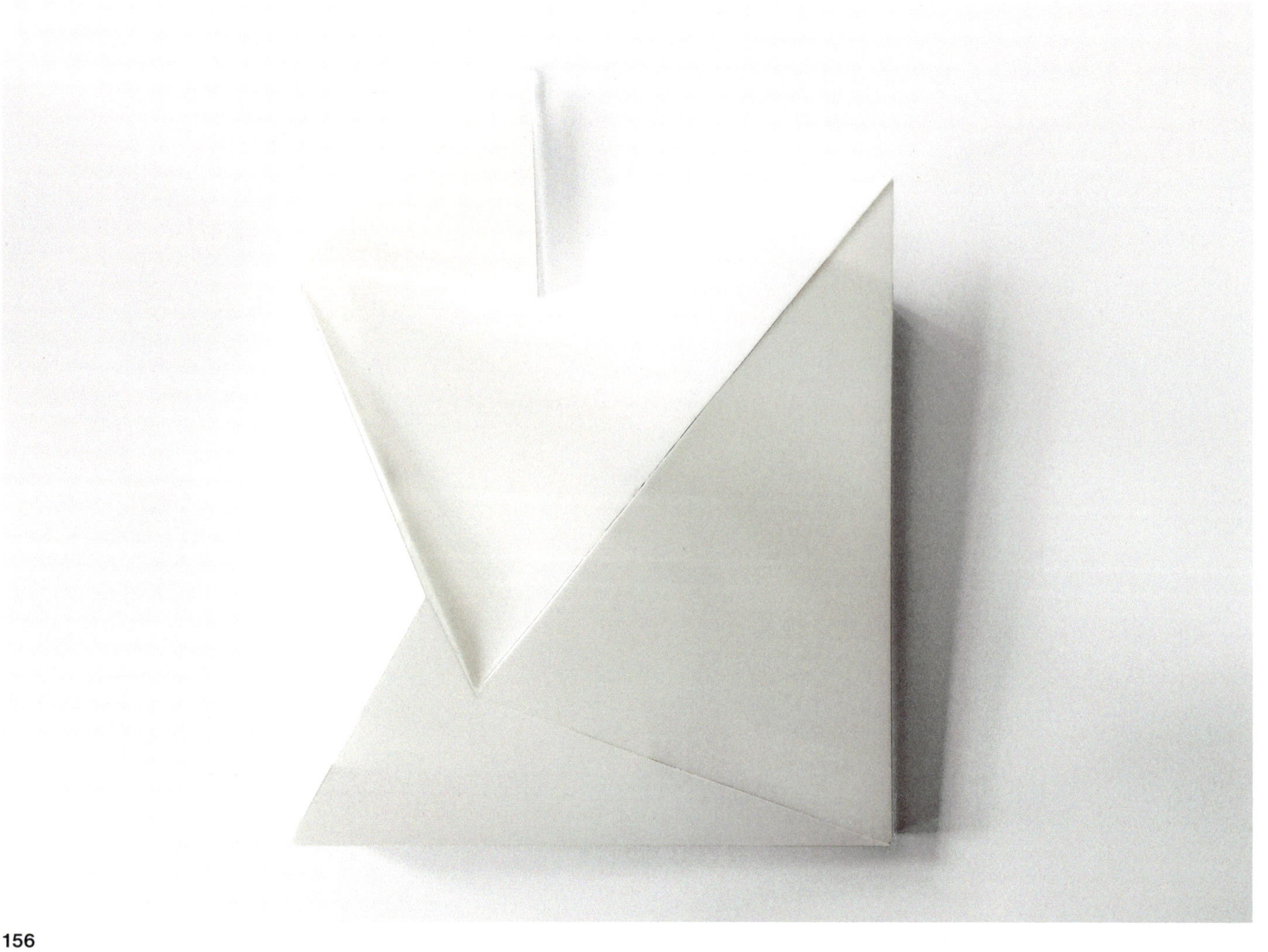

156

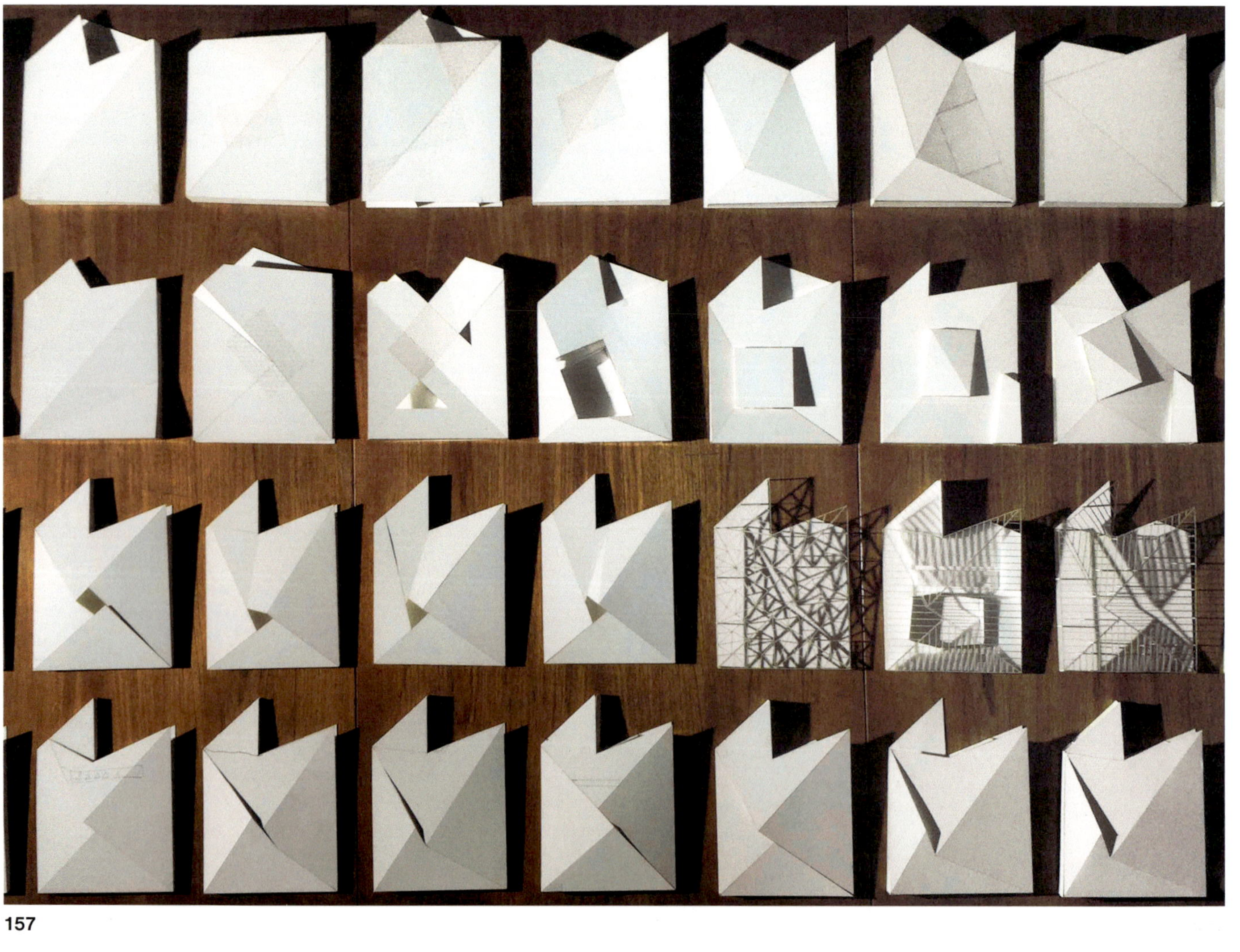

157

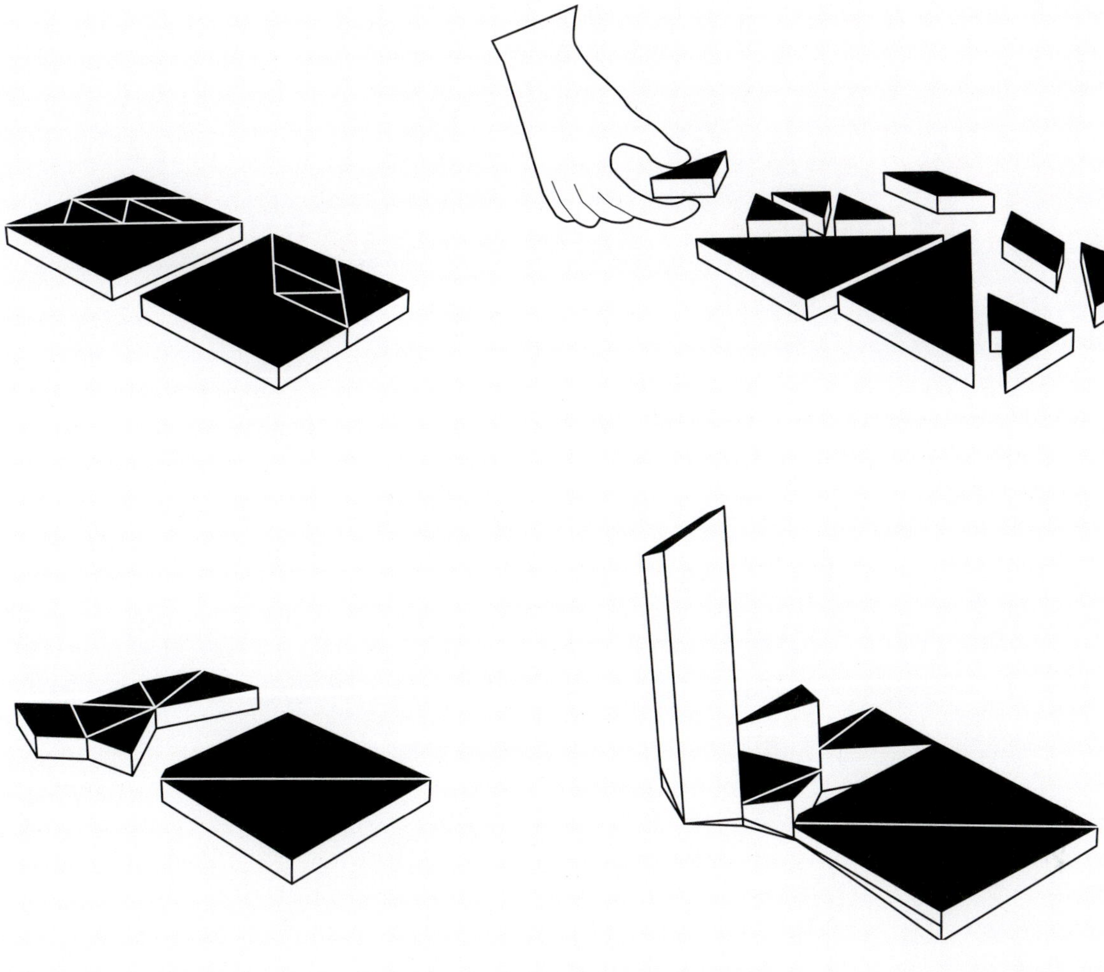

158

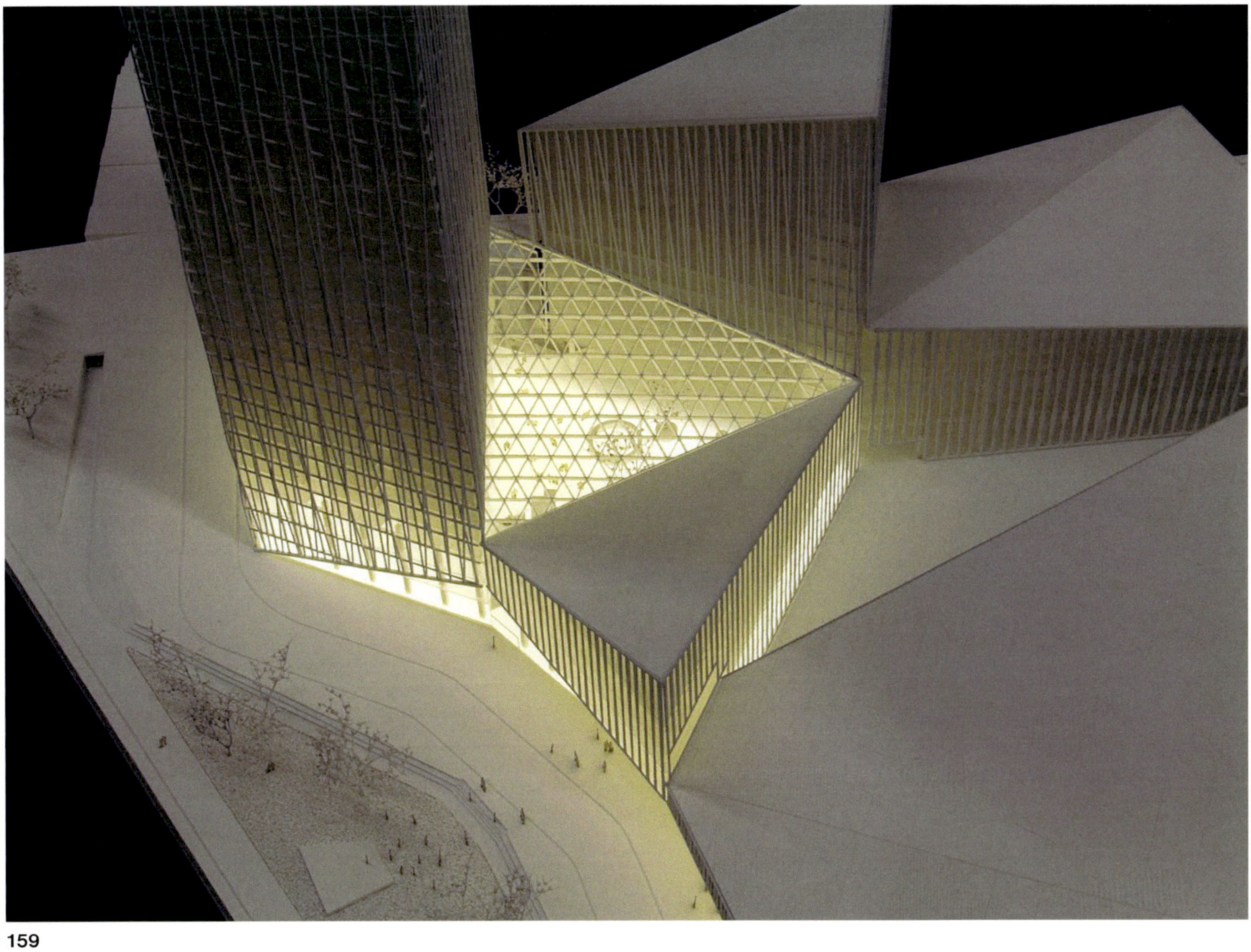

159

Challenging the idea of a standard high-rise, the Kurfürstenstrasse work/live building was massed as a topographical series of five stepped building volumes, ascending from low to high, each radially disposed to turn on an oblique corner site.

The volumes are clad in vertical bands of broken stone with a deep-set glazing, giving the impression of a series of stone cliffs. Like the ground that heaves upward along fault lines, the building extrudes vertically and unevenly, resulting in a stepped profile. Here, mimicry really is the highest form of flattery.

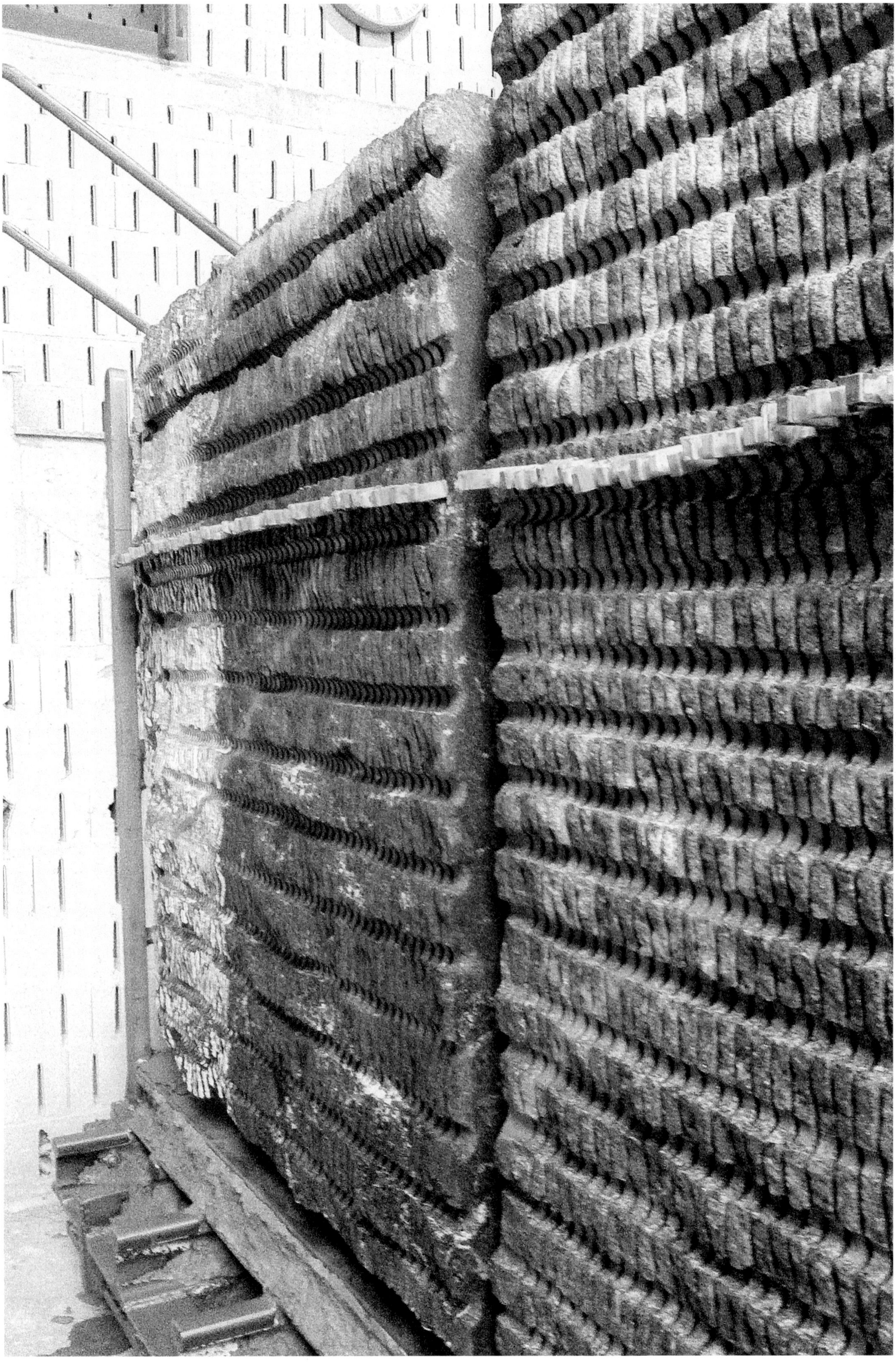

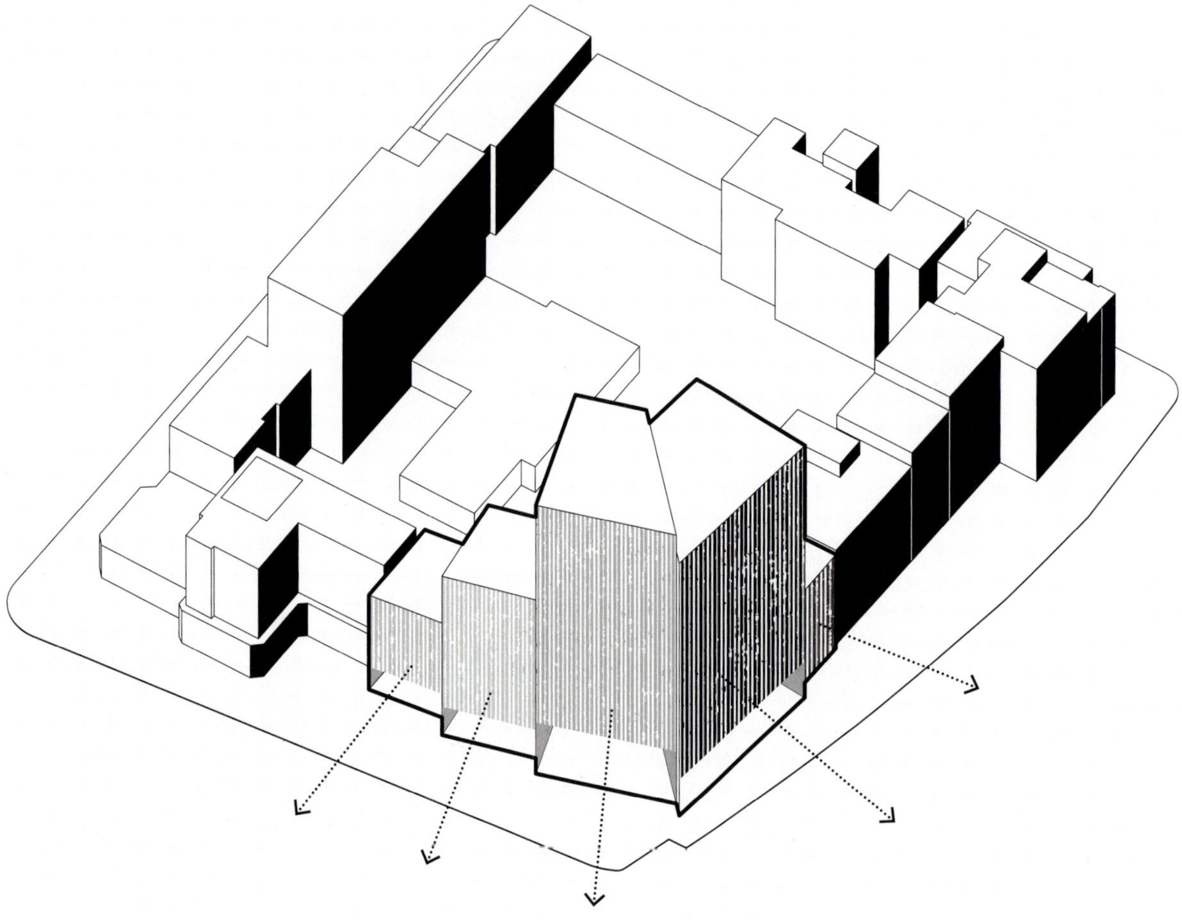

163

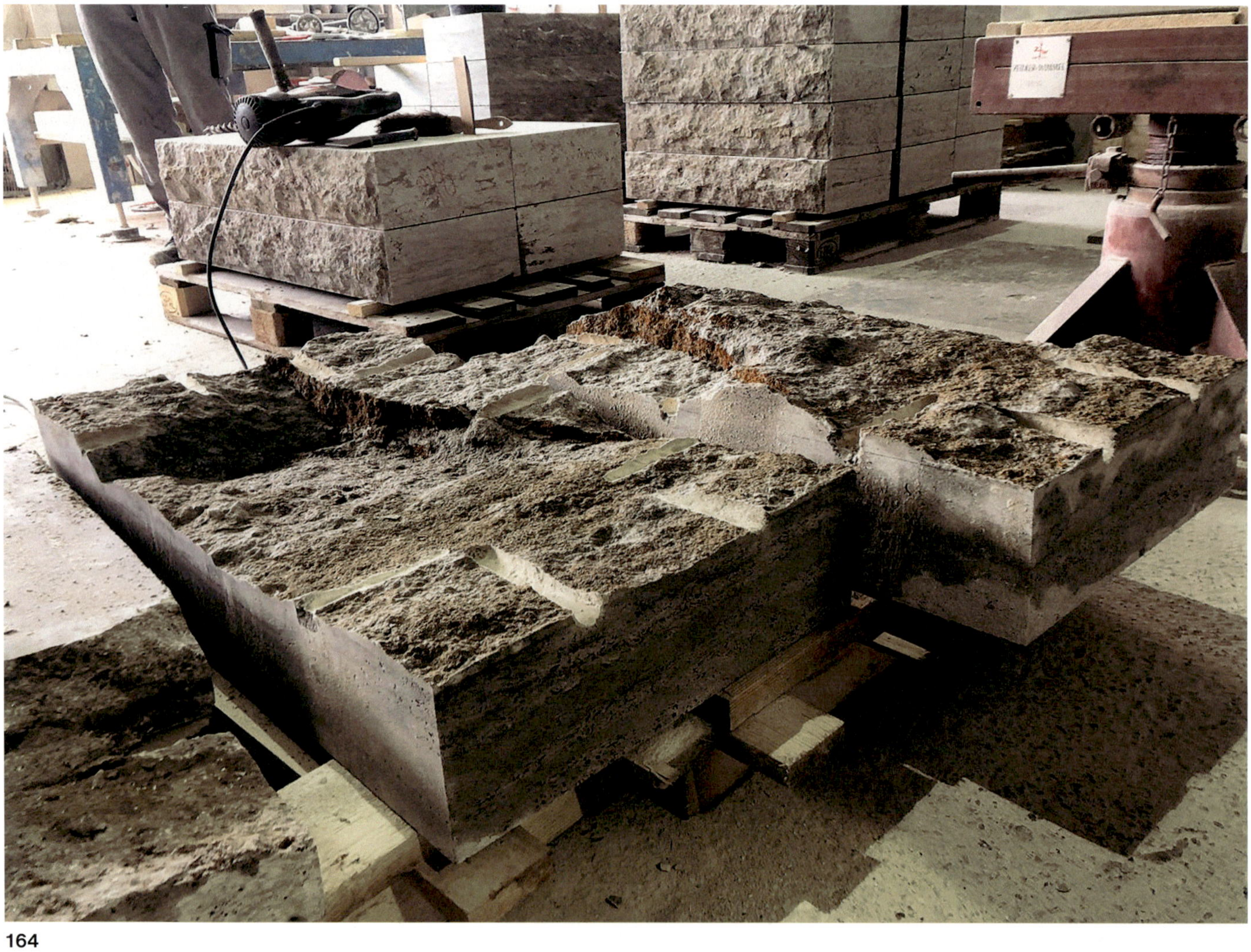

164

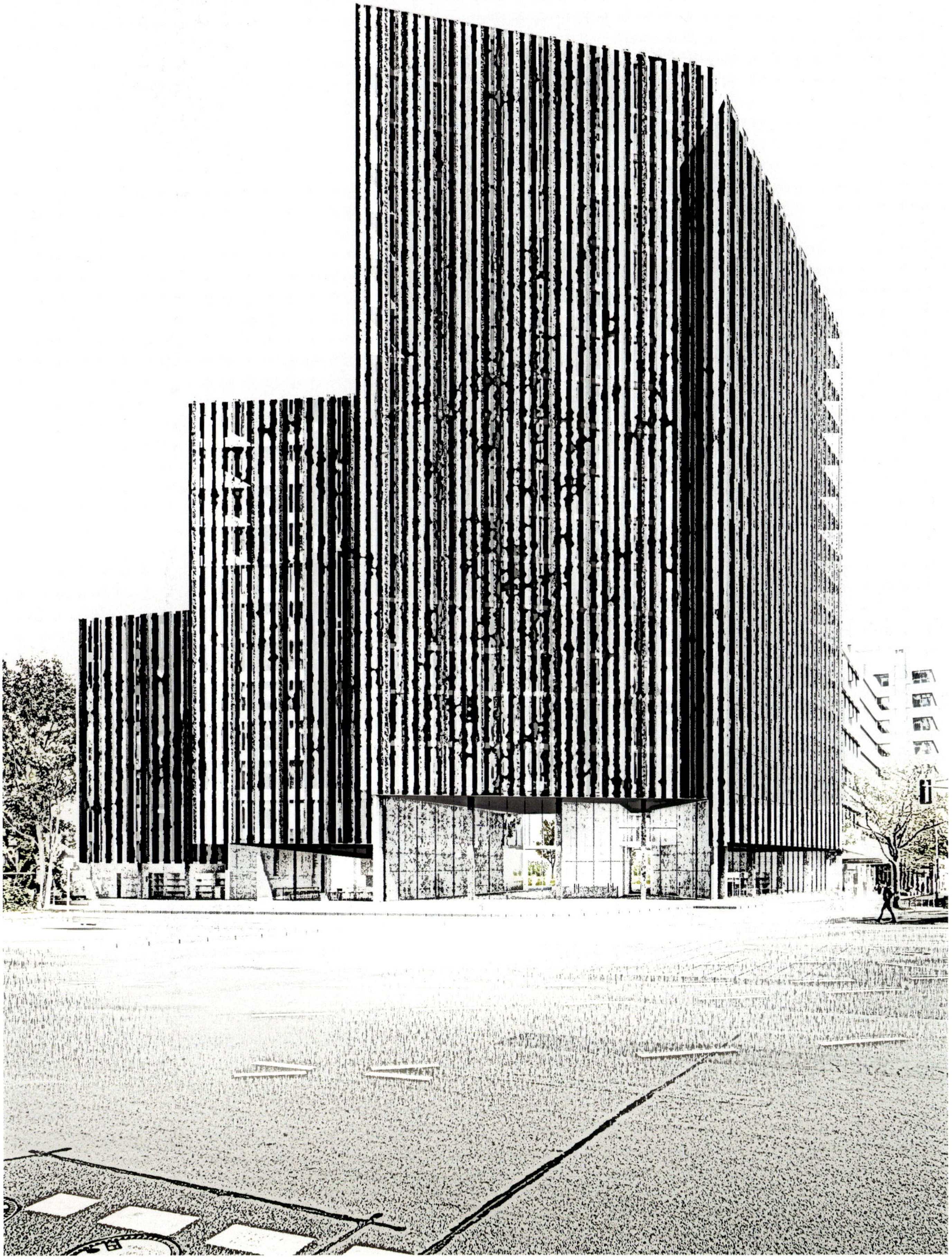

Liminal Facade

166

Over the years, we've discovered that facade-making is an aspect of architecture that largely determines a building's identity, image, and performance – it's also an area in which architects often have some operational freedom.

167

Liminal facade locates the skin of a building – the conditions that demarcate the exterior from the interior, as an architectural site of relative autonomy. The facade is emphasized as a means of determining the visual identity of a building and its performative range, defining its spatial threshold.

Liminality speaks of an architectural boundary that is physical and spatial – site that, though it can be occupied, exists foremost in-between.

How does one disguise infrastructure? Our parking structure for the TRUMPF Campus in Ditzingen provided a much-needed facility for both conventional and electric vehicles.

The building was constructed of prefabricated steel and concrete elements as a kit-of-parts and was customized with a screen facade – produced by TRUMPF – which imbues the volumes with an ephemeral lightness, as a function of the application of 3,600 vertical metal tubes axially laser cut.

The visual effect? That of a moiré: the structure changes as passersby walk or drive, transforming the building into an abstract object. It is thanks to this very dynamic facade that the building takes on an identity that is much more provocative than that of a standard parking structure.

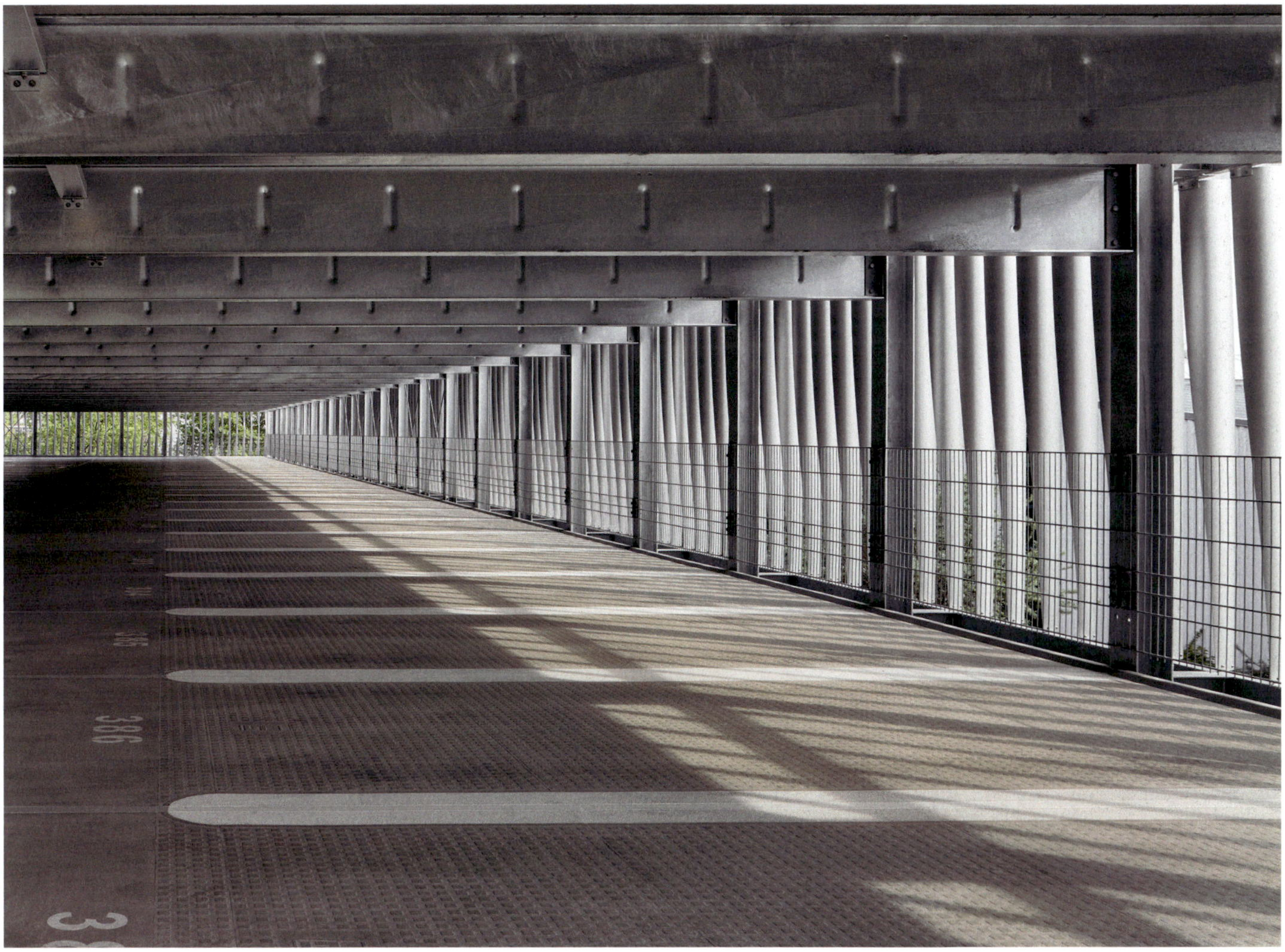

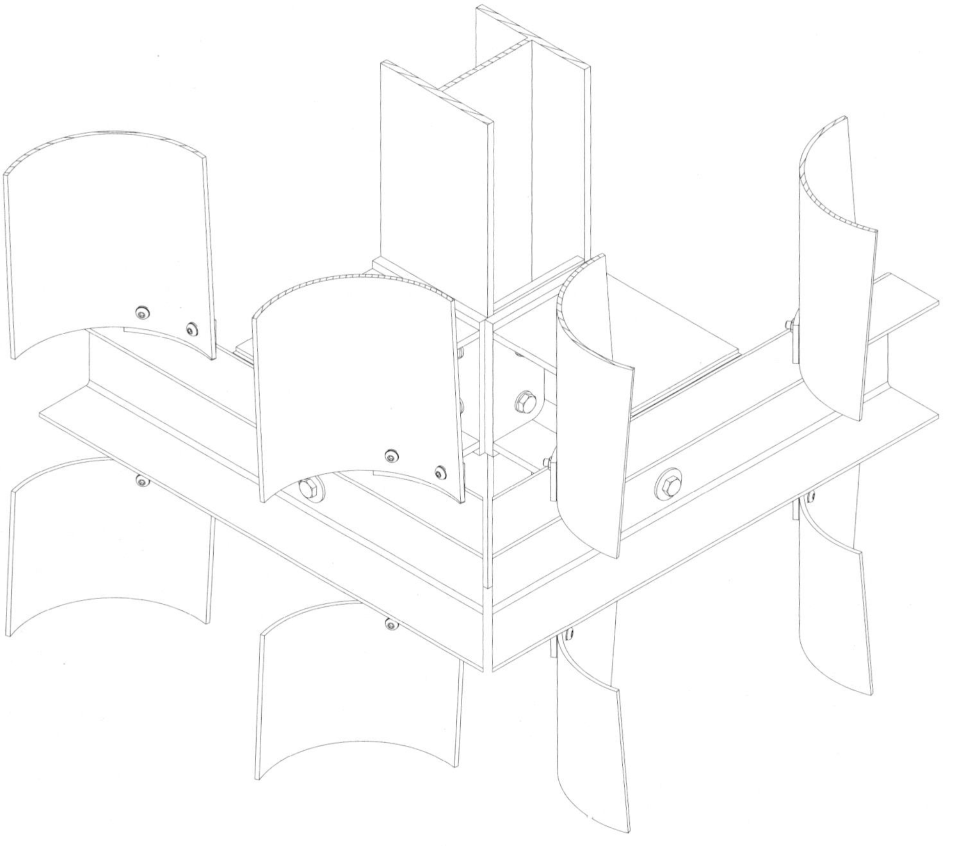

174

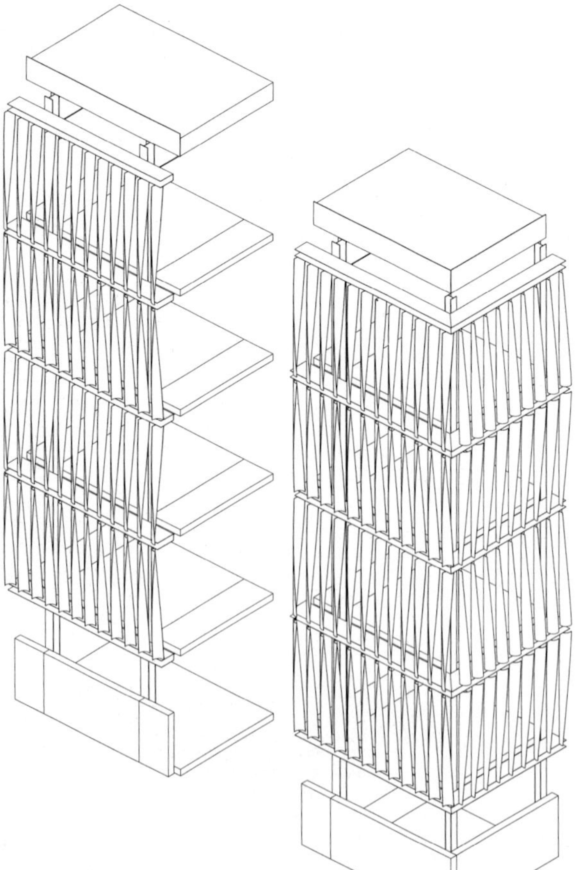

175

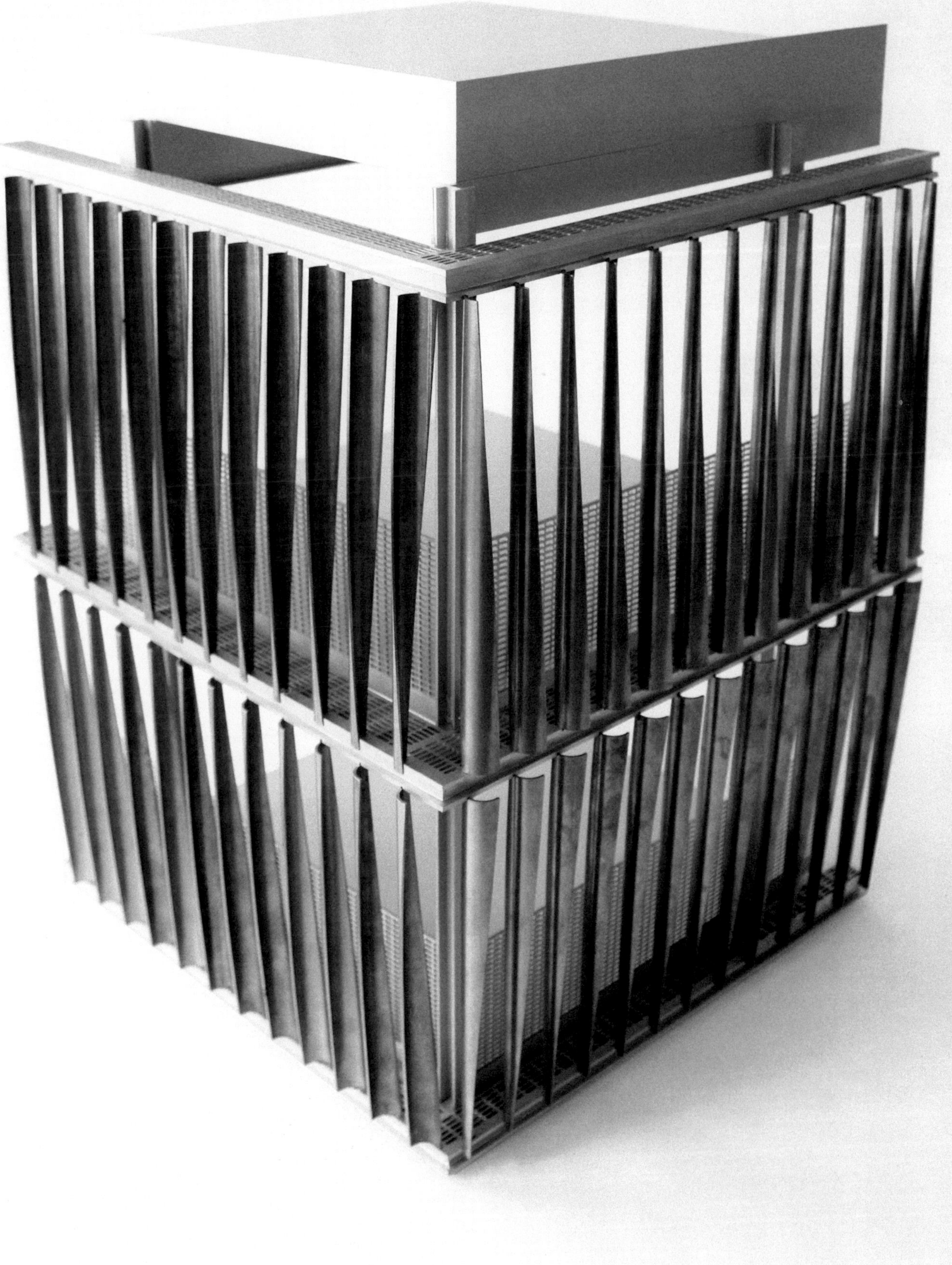

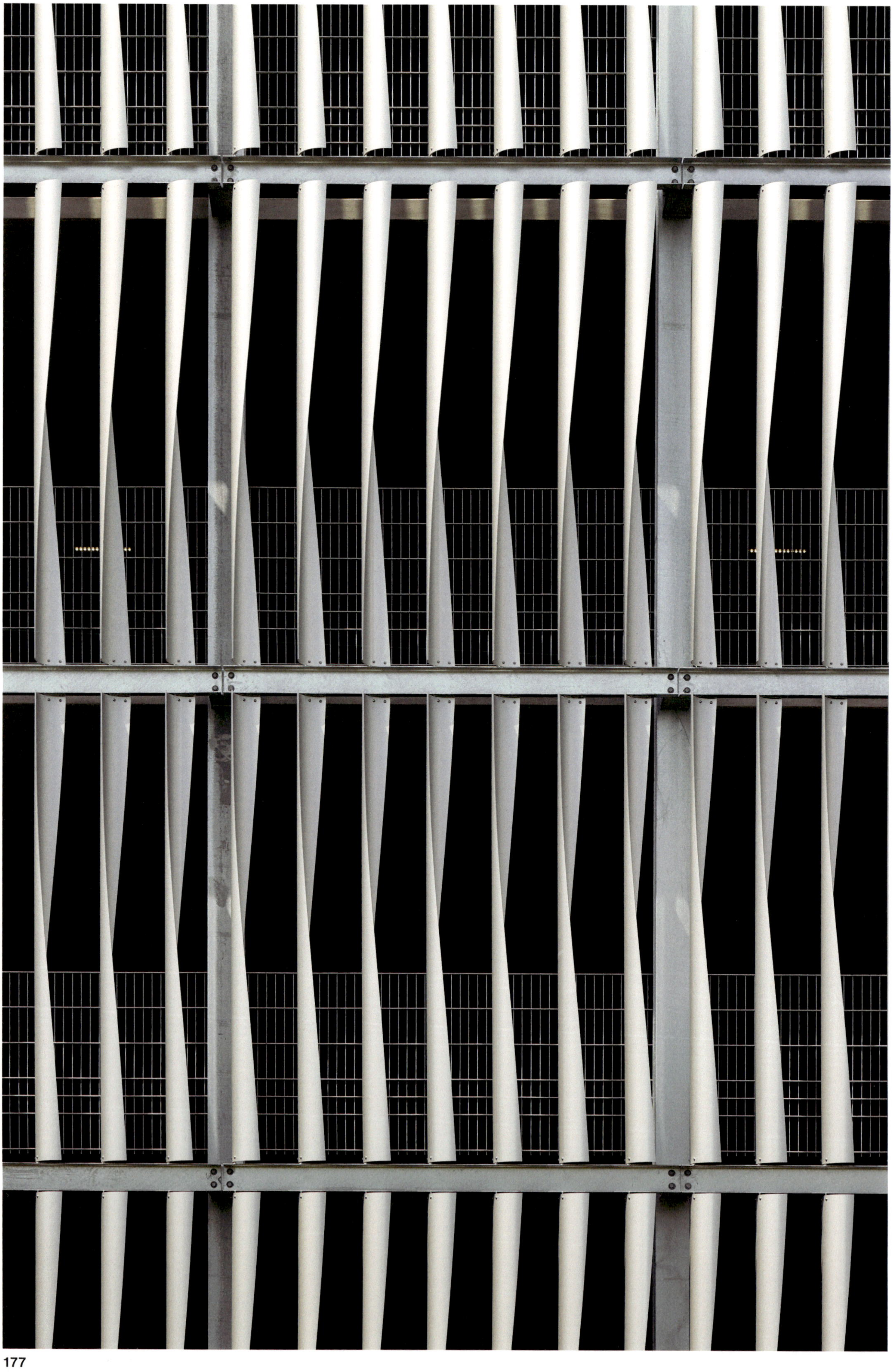

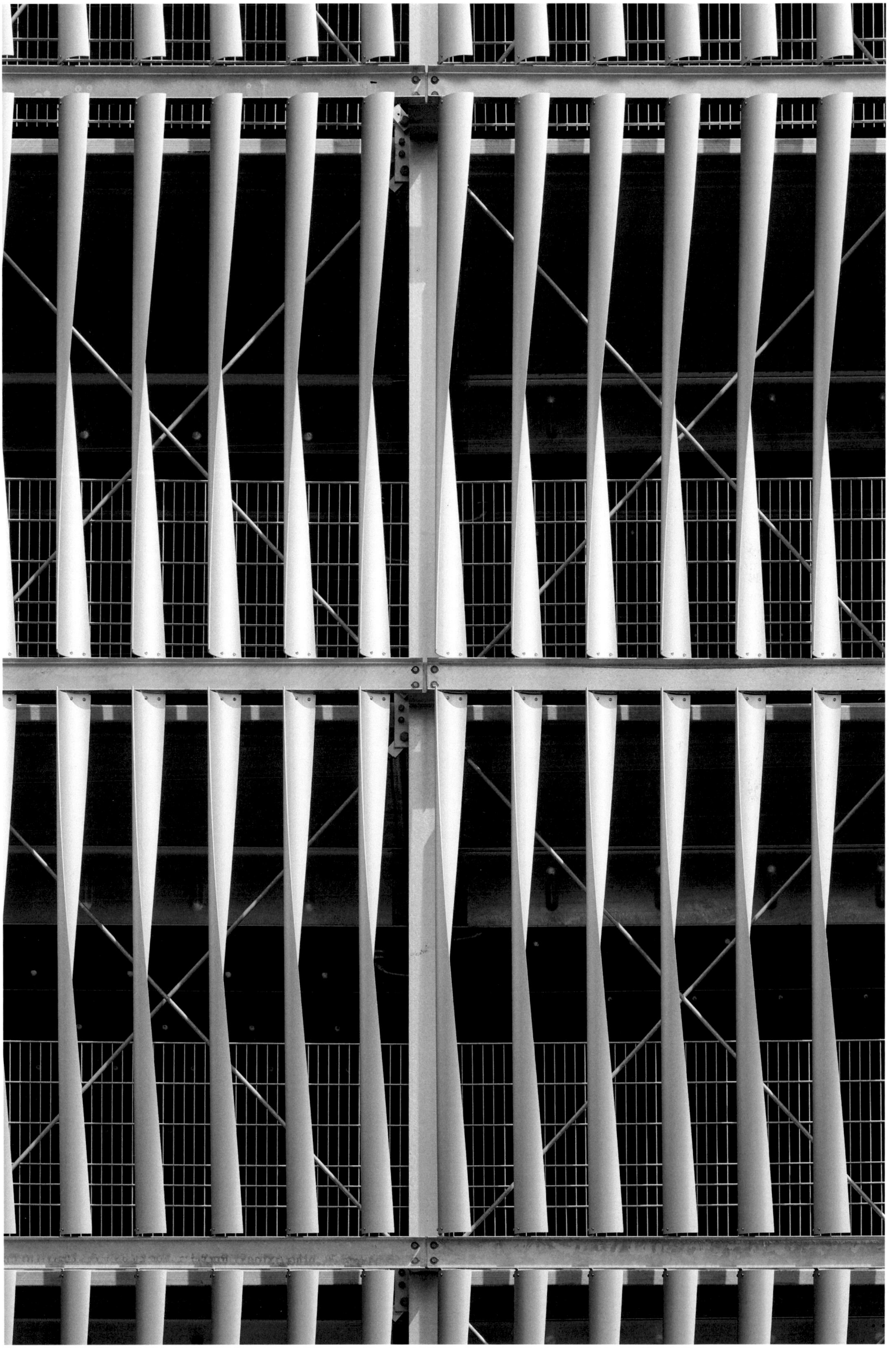

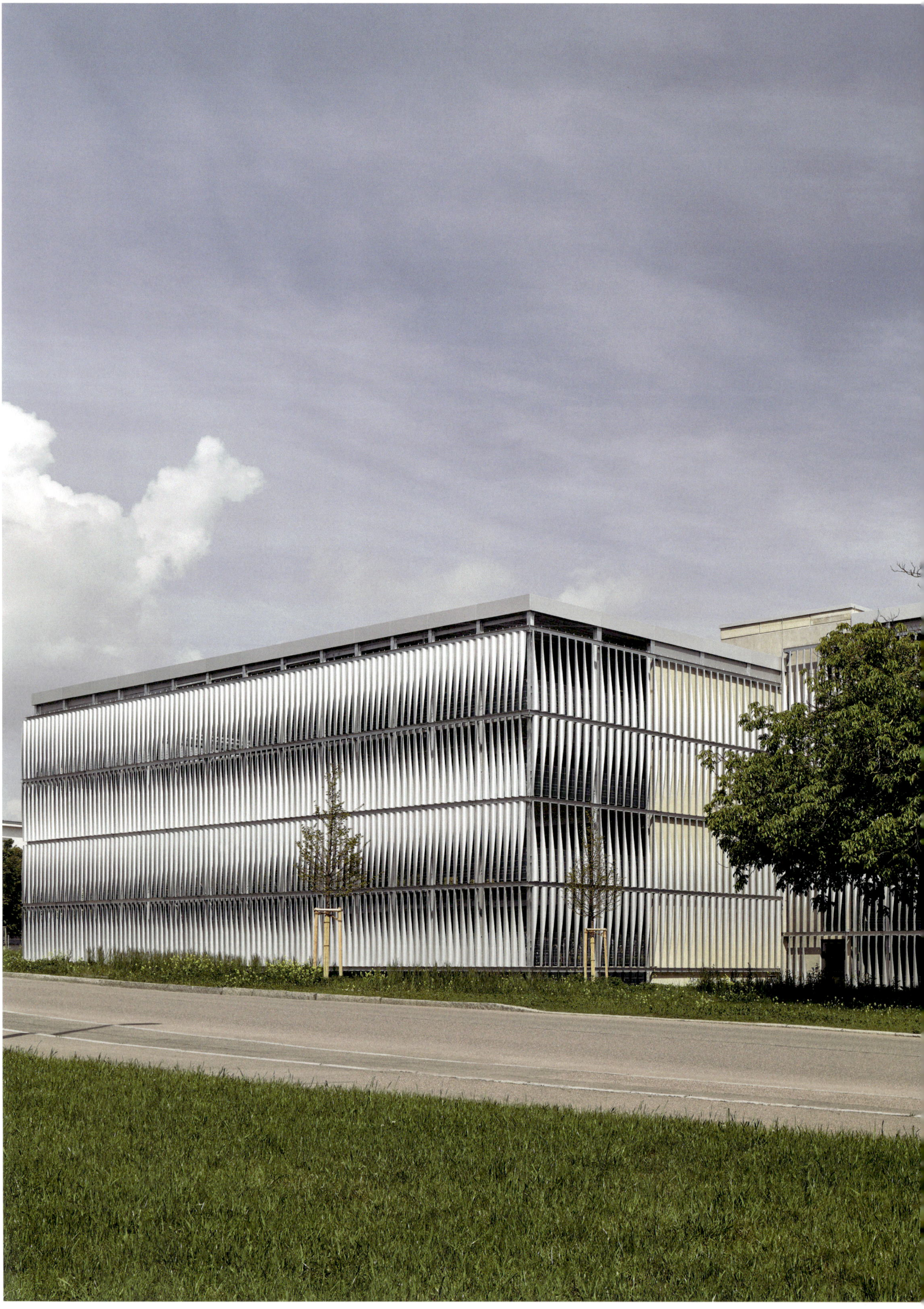

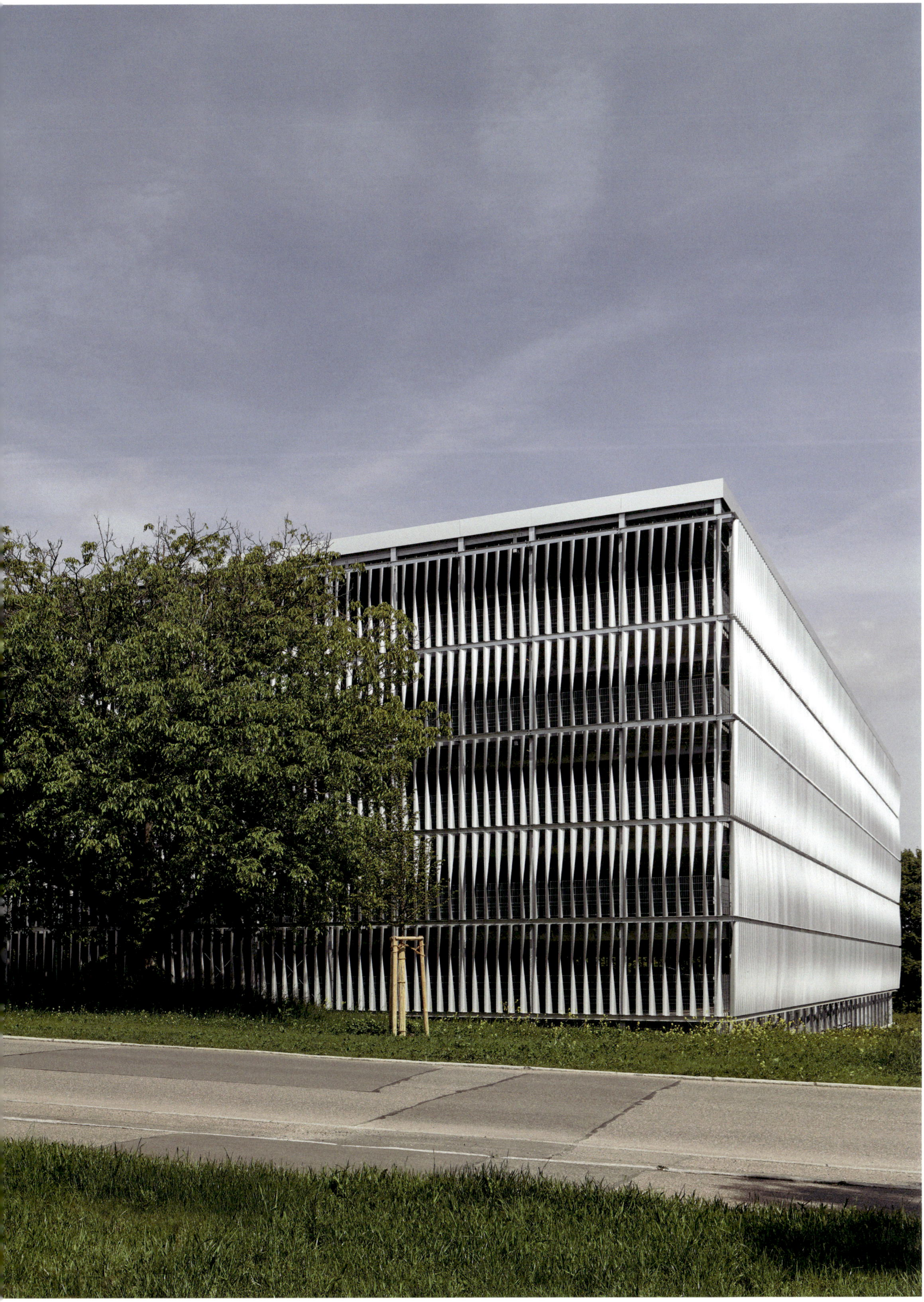

TRUMPF Poland's new headquarters in Warsaw, which houses offices and exhibition areas, is a prototype for an economically constructed industrial building that combines a simple form with complex surfaces and differentiated interior spaces.

The facade fins were laser-cut and folded to widen and narrow over their 10-m length, reflecting the company's technology to the outside world. Here, the facade manages to project outward the potential of the technologies housed within the building.

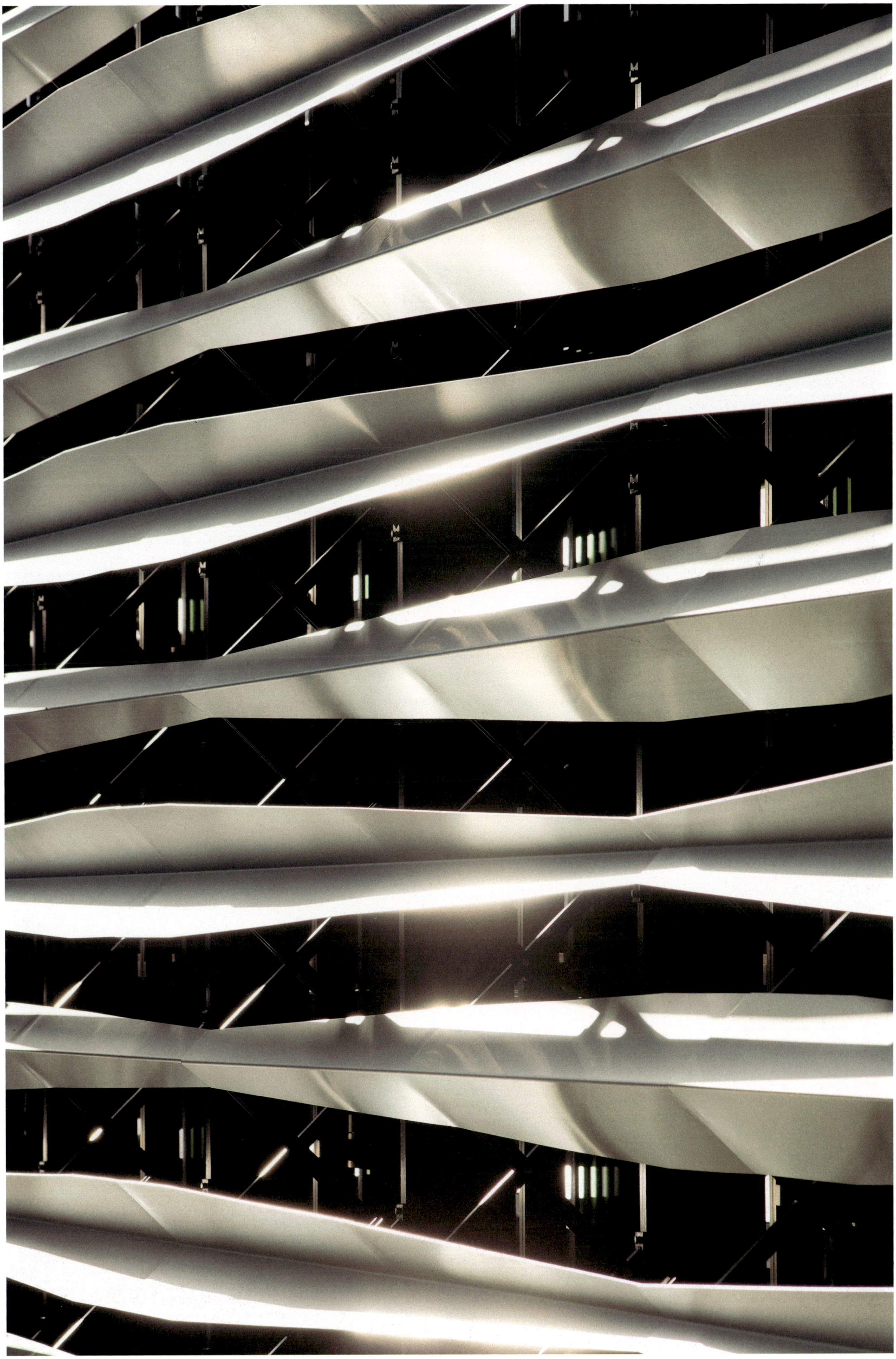

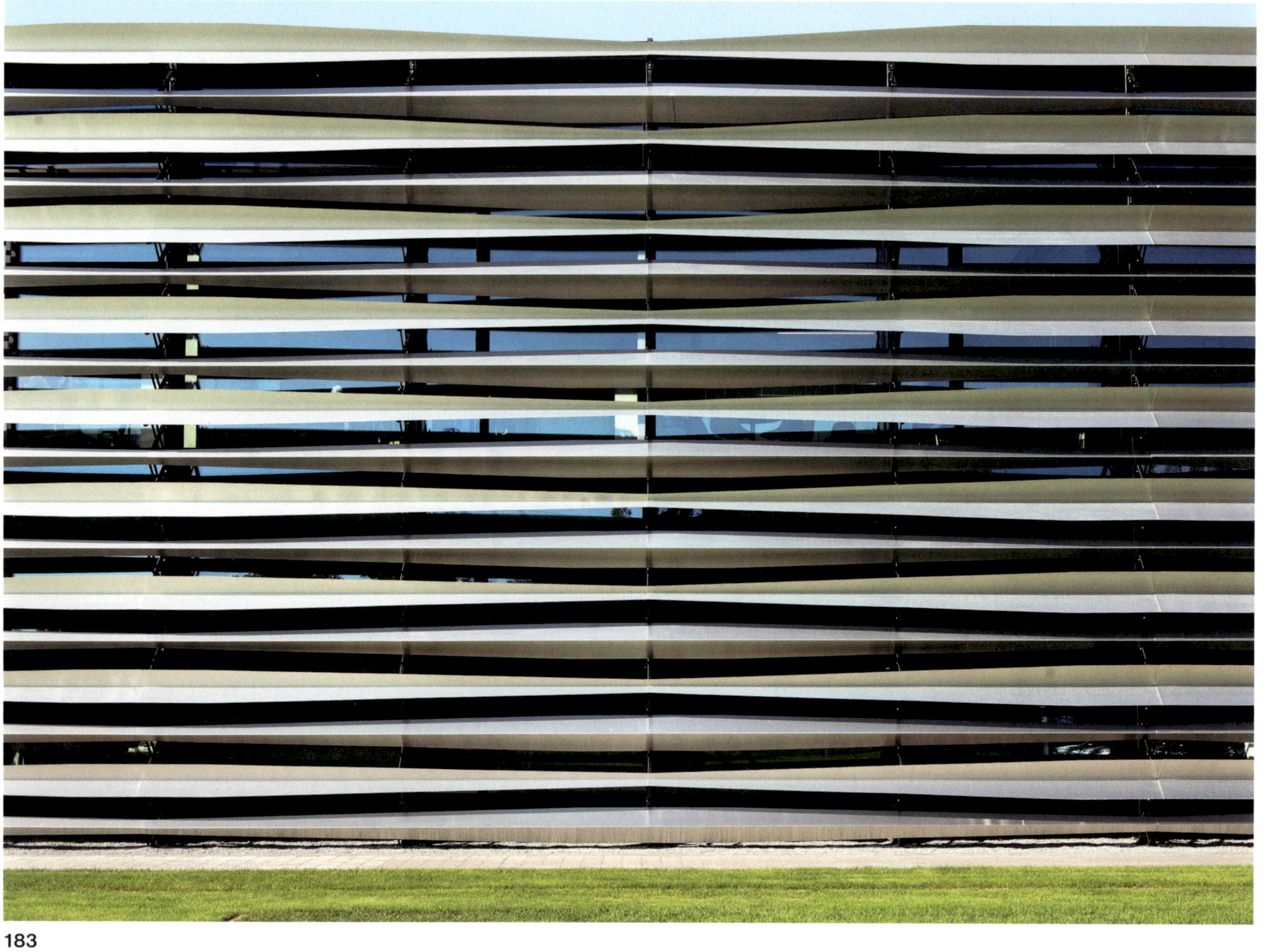

183

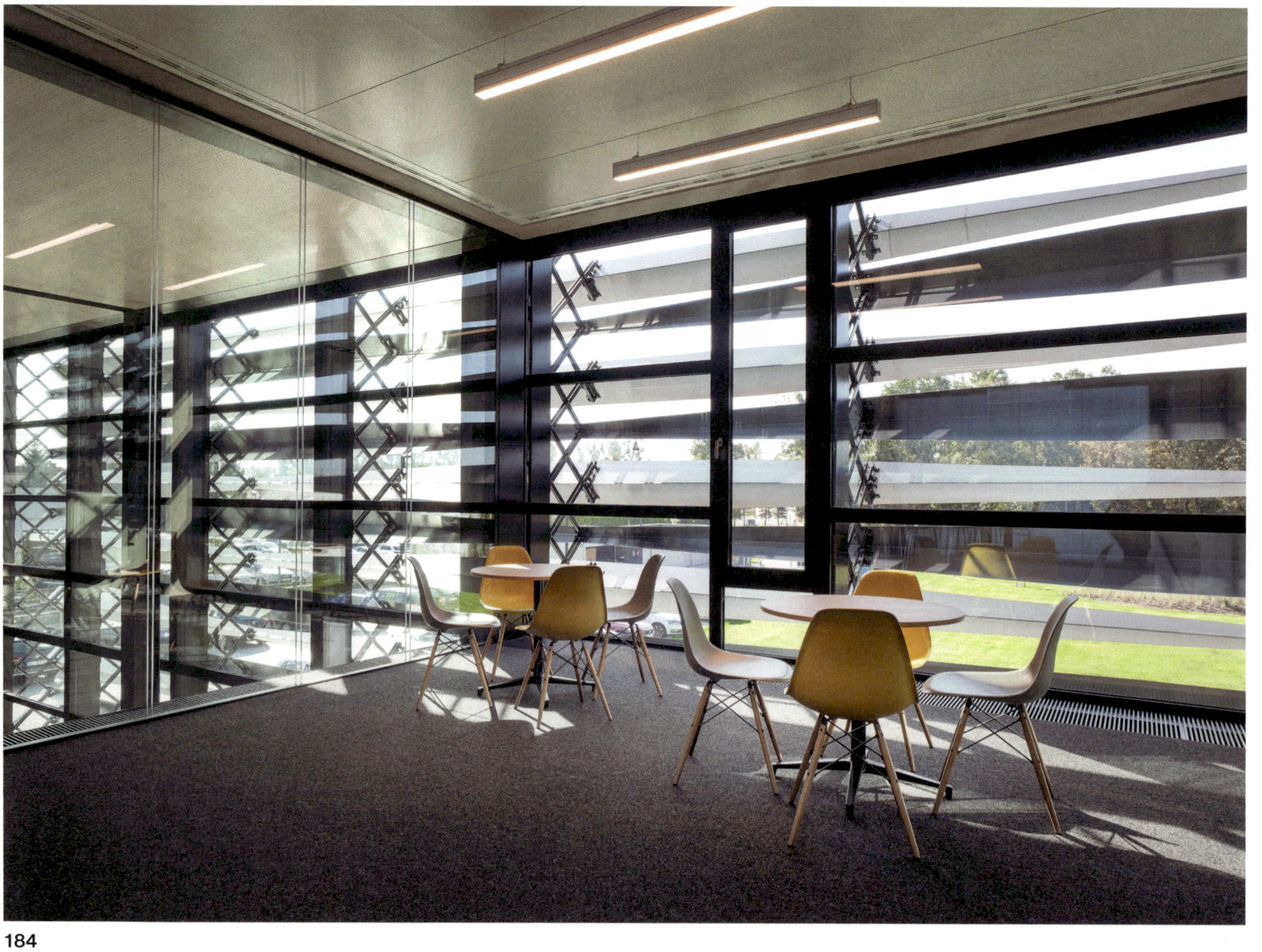

184

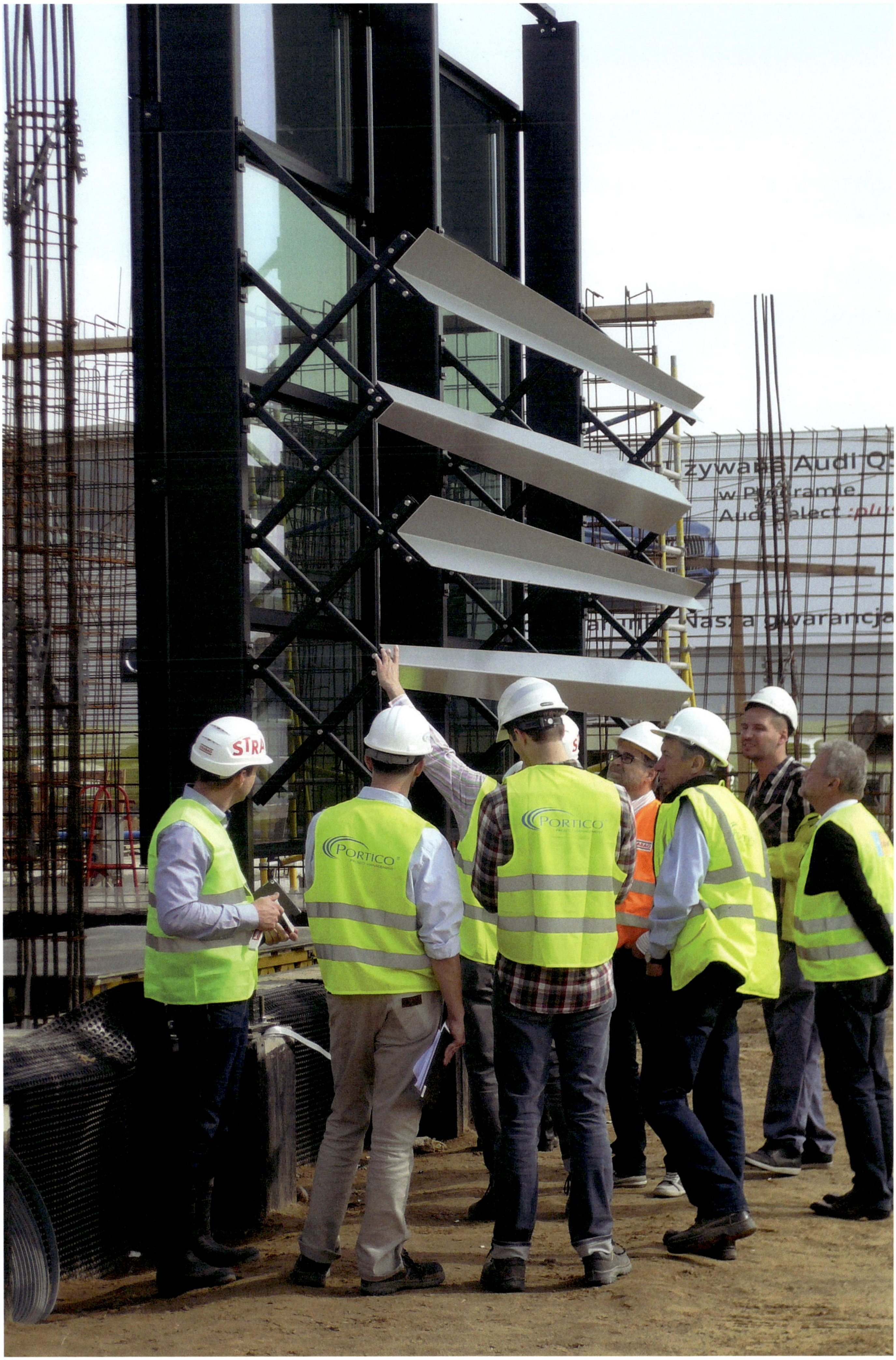
PORTICO
PORTICO
PORTICO
Nasza gwarancja

The glass curtain wall, pioneered in the 1918 Hallidie Building in San Francisco by Willis Polk, was recently deemed energy inefficient by New York's then-mayor Bill de Blasio. We took this as a challenge: Jannowitzbrücke defies this assumption.

At a location not far from Berlin's Alexanderplatz, we designed a triangulated office high-rise that reacts to its context in all directions, exploiting views and orientations. This strong form is clad in a glass skin of multilayered opaque, transparent, and photo-voltaic glass.

The highly performative structure proved the enduring aesthetic viability of the curtain wall as a construction type. Light and ephemeral, layered and nuanced, the building glows like a lantern at night as a function of its multilayered surfaces.

This atmospheric, dematerialized high-rise confirms the essential force of the liminal facade on the character of the building and its surrounding context.

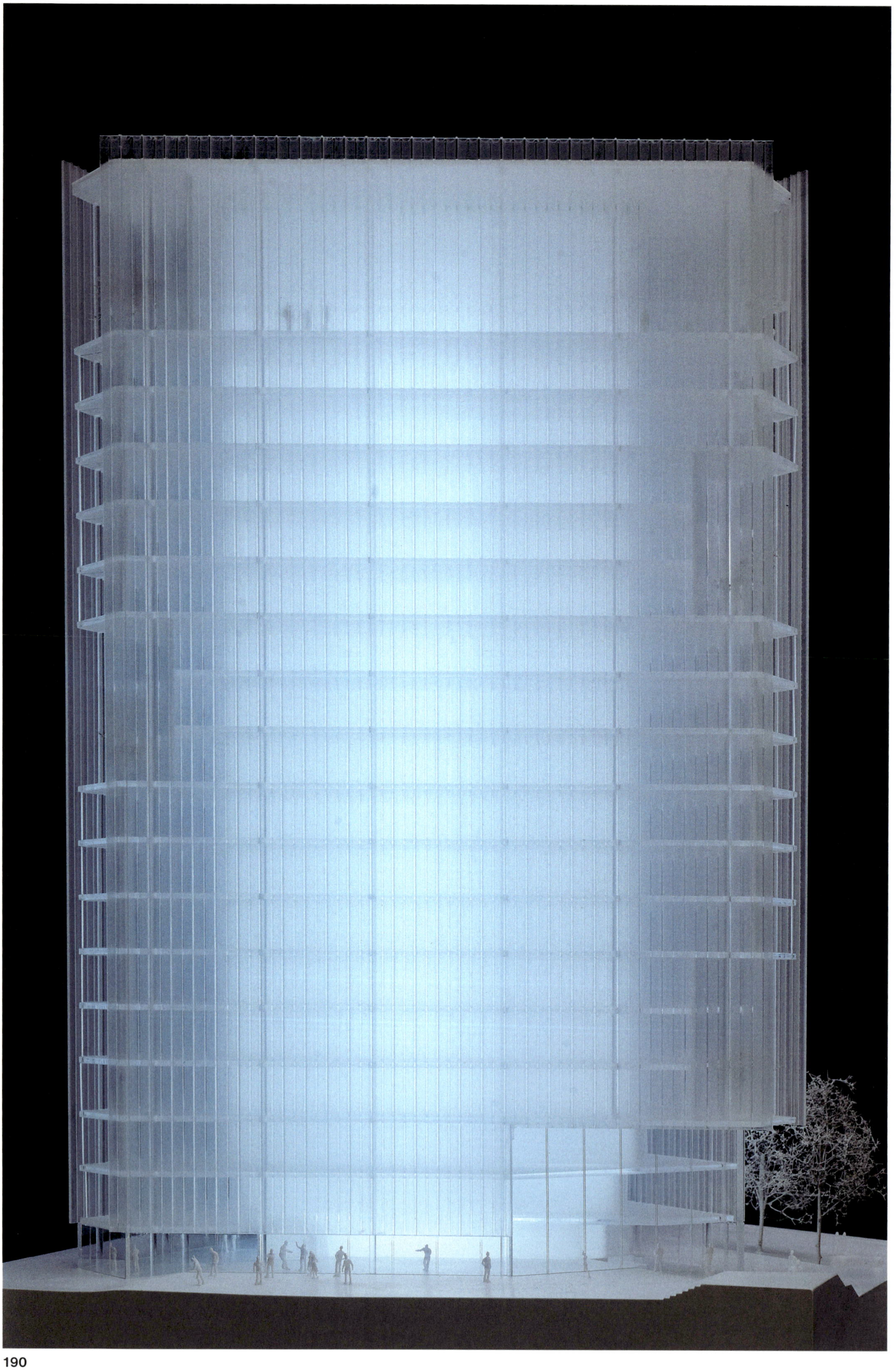

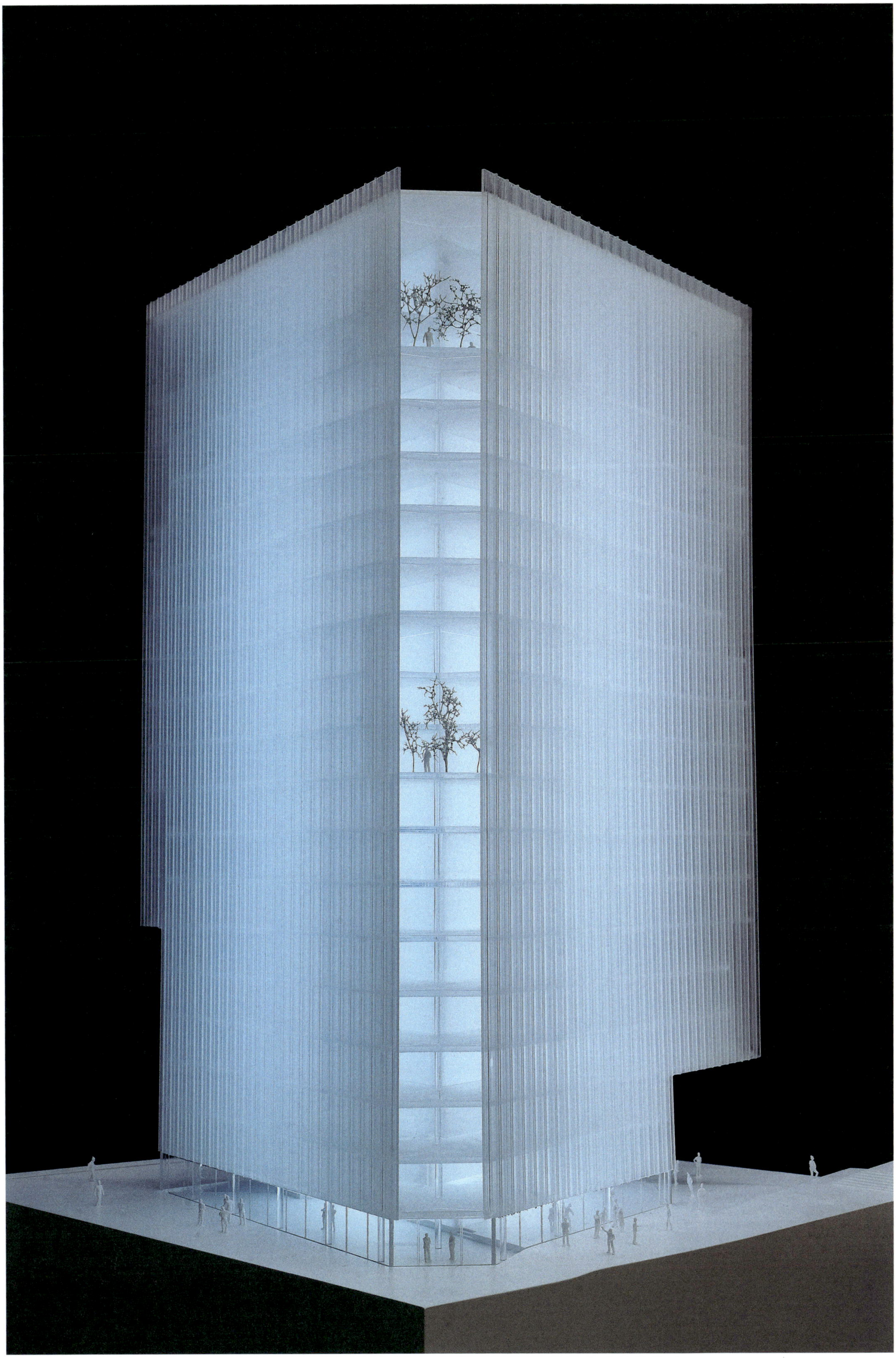

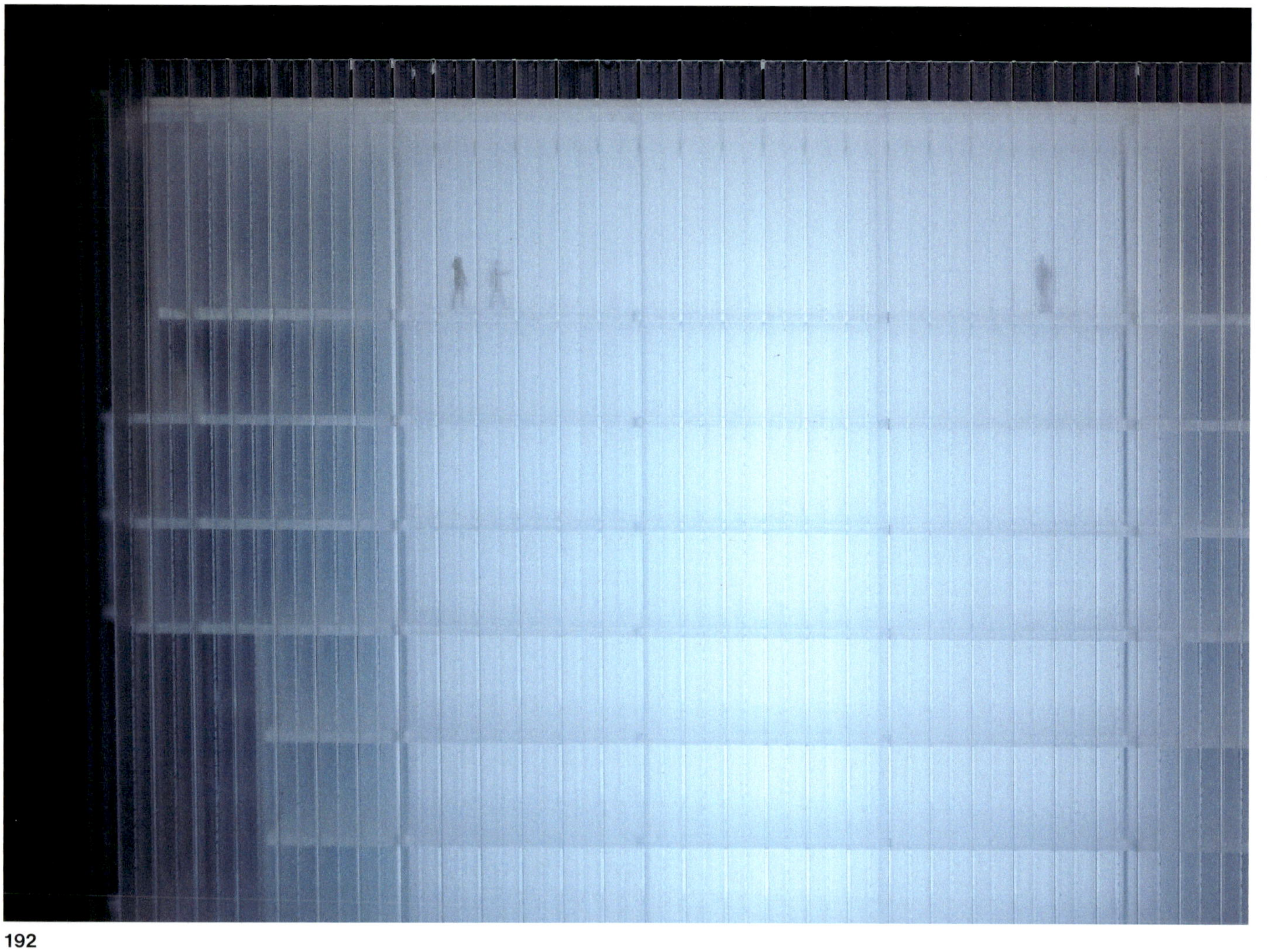
192
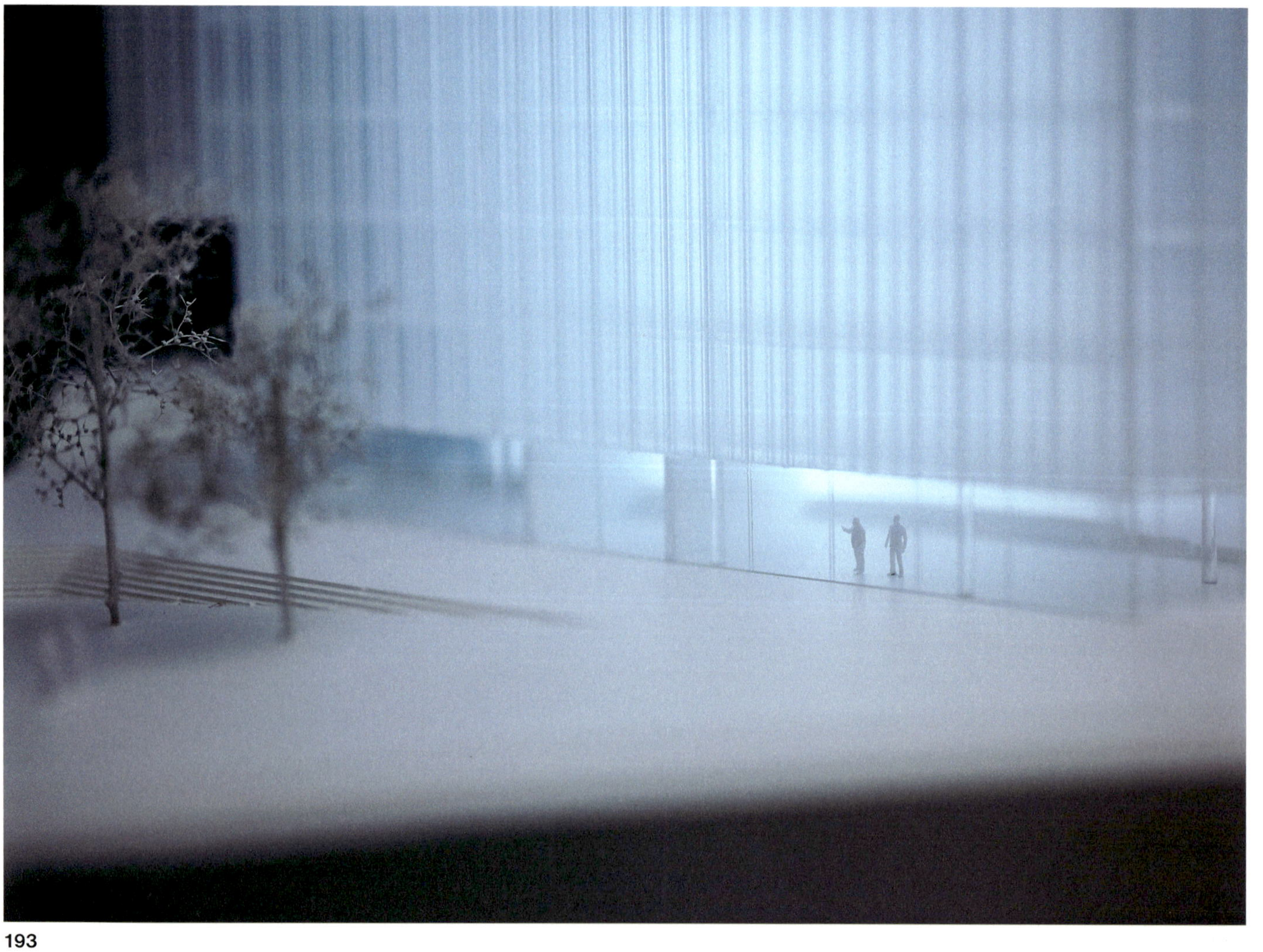
193

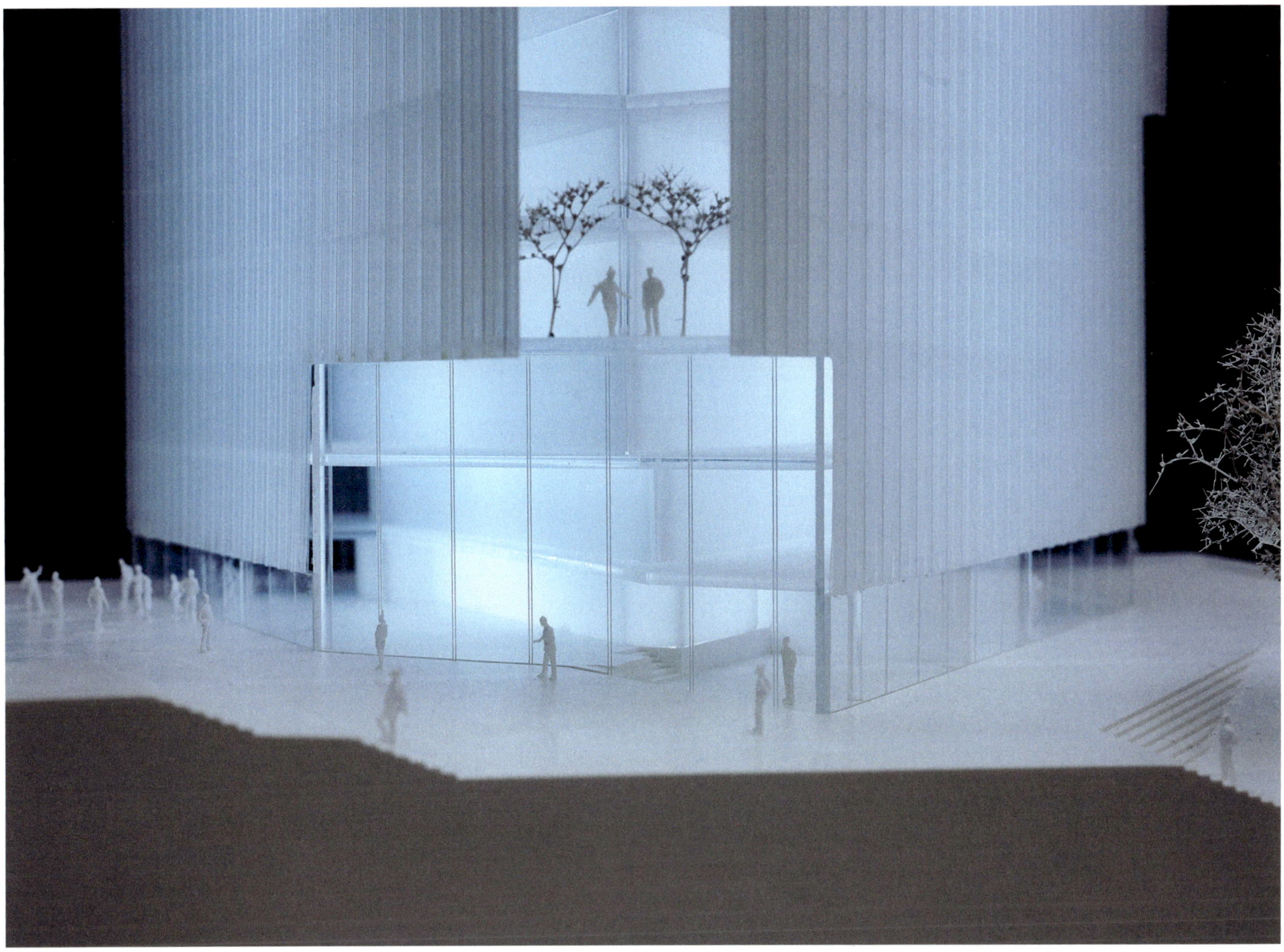

194

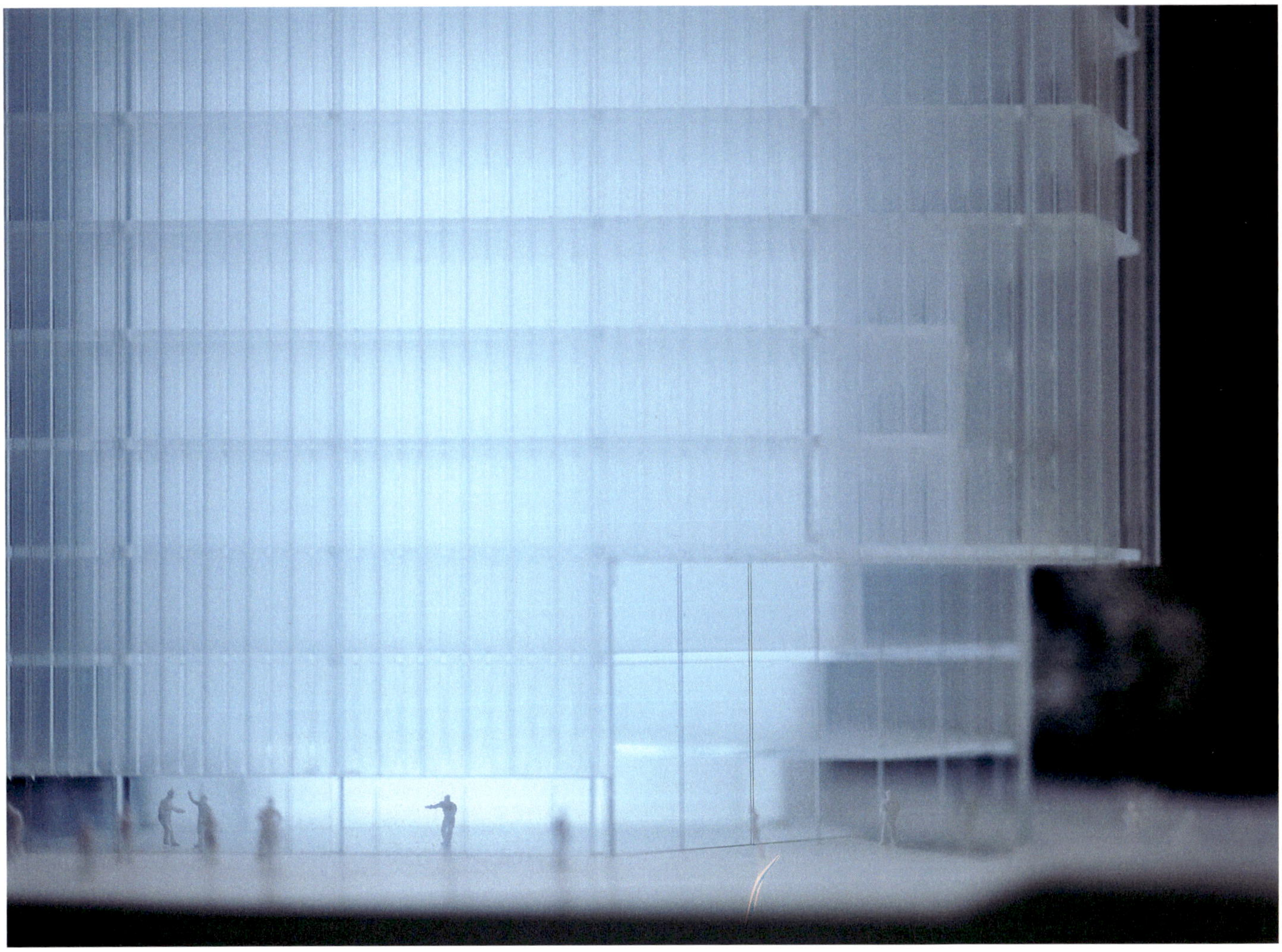

195

Can stone appear to be light? Bertha, a large block of office space located near Berlin's central train station, became an opportunity to experiment with an answer. The space is situated within a masterplan initiated by O.M. Ungers, which later refined by Auer Weber.

Allowances for architectural freedom were confined to the facade of the building and its split-level site. This small allowance nevertheless proved to be a rich site for experimentation. Travertine, a stone that has historically clad many Berlin facades with punched widows, is closed and opaque.

This stone, which is normally thought of as heavy, generates a lightweight veiled facade. The depth of the stone fins varies, producing a visual effect that transforms the building into an abstract urban object. Scale and purpose are obscured; architectural surface becomes an optical field.

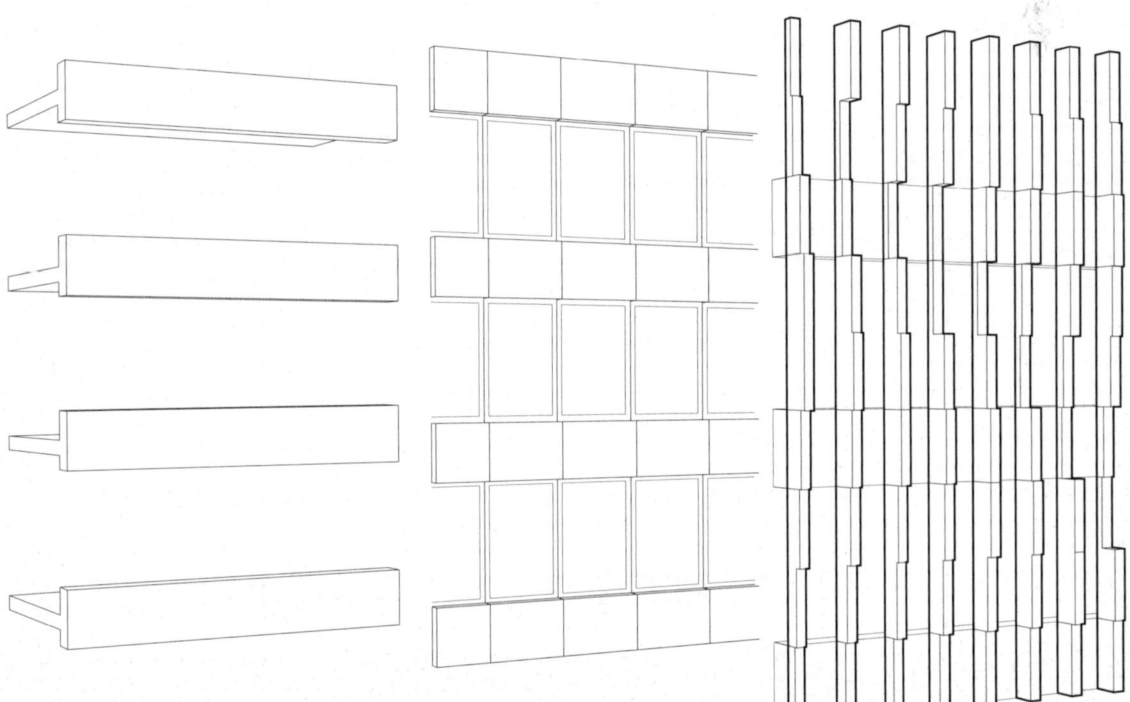

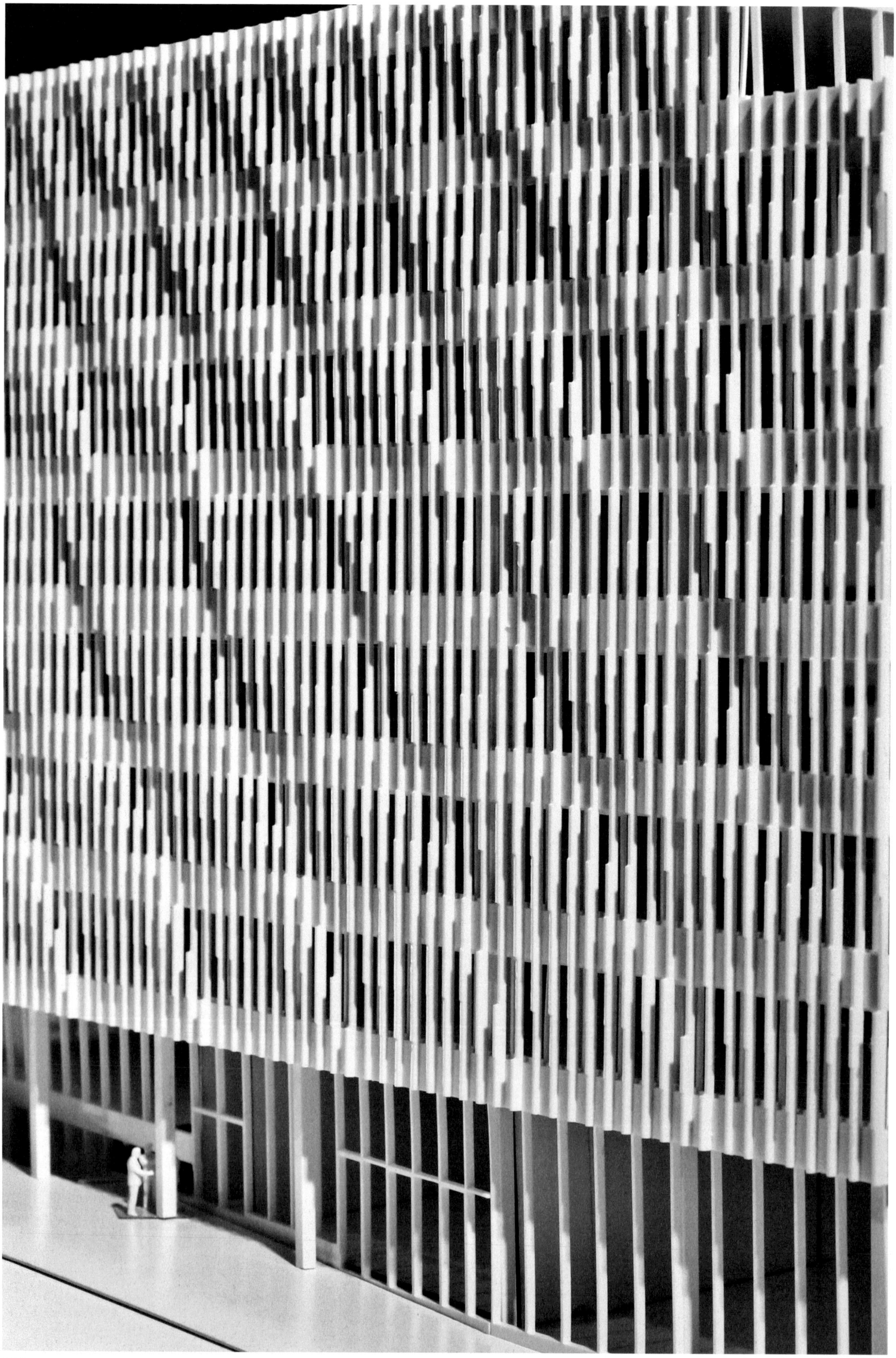

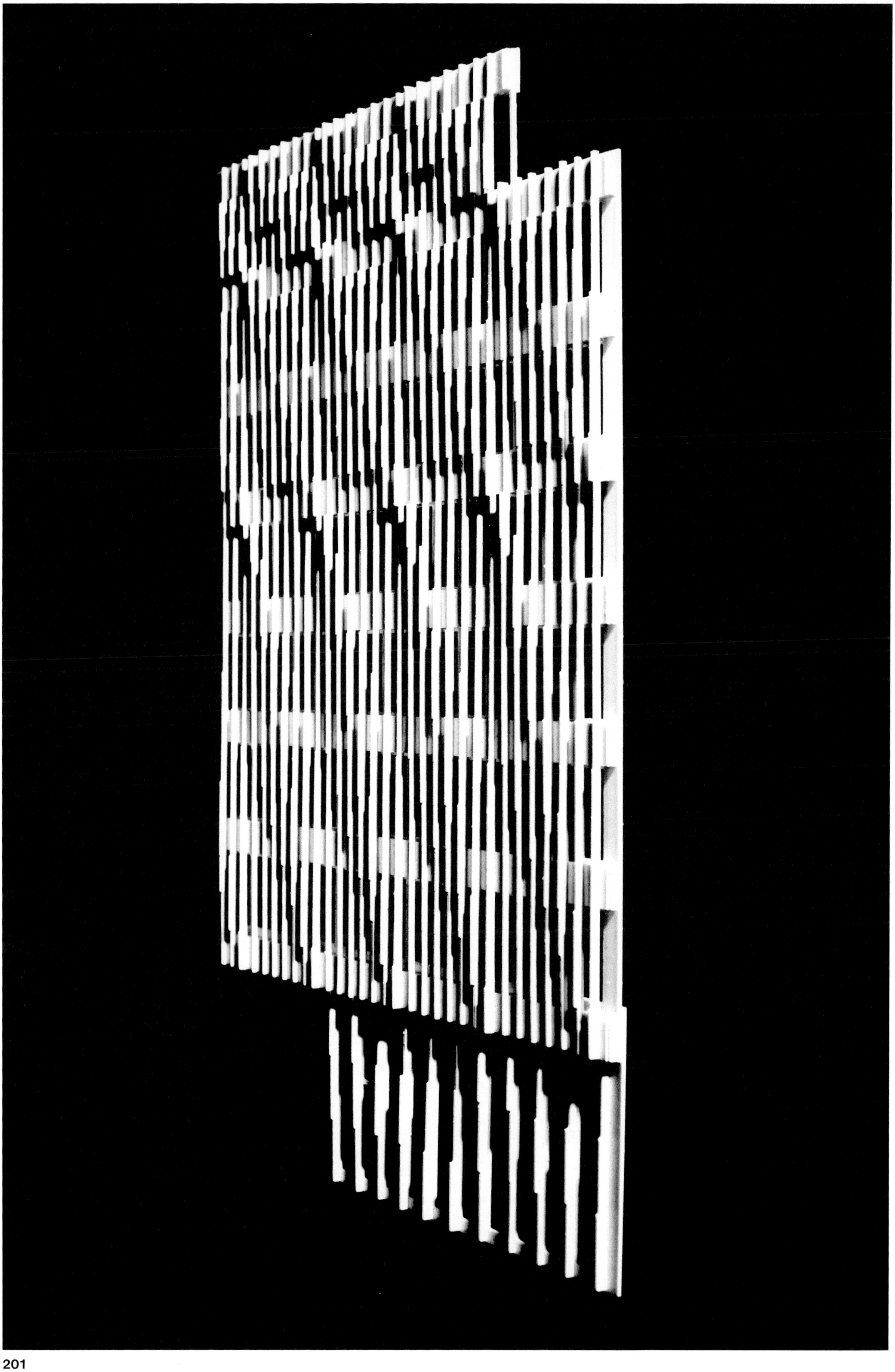

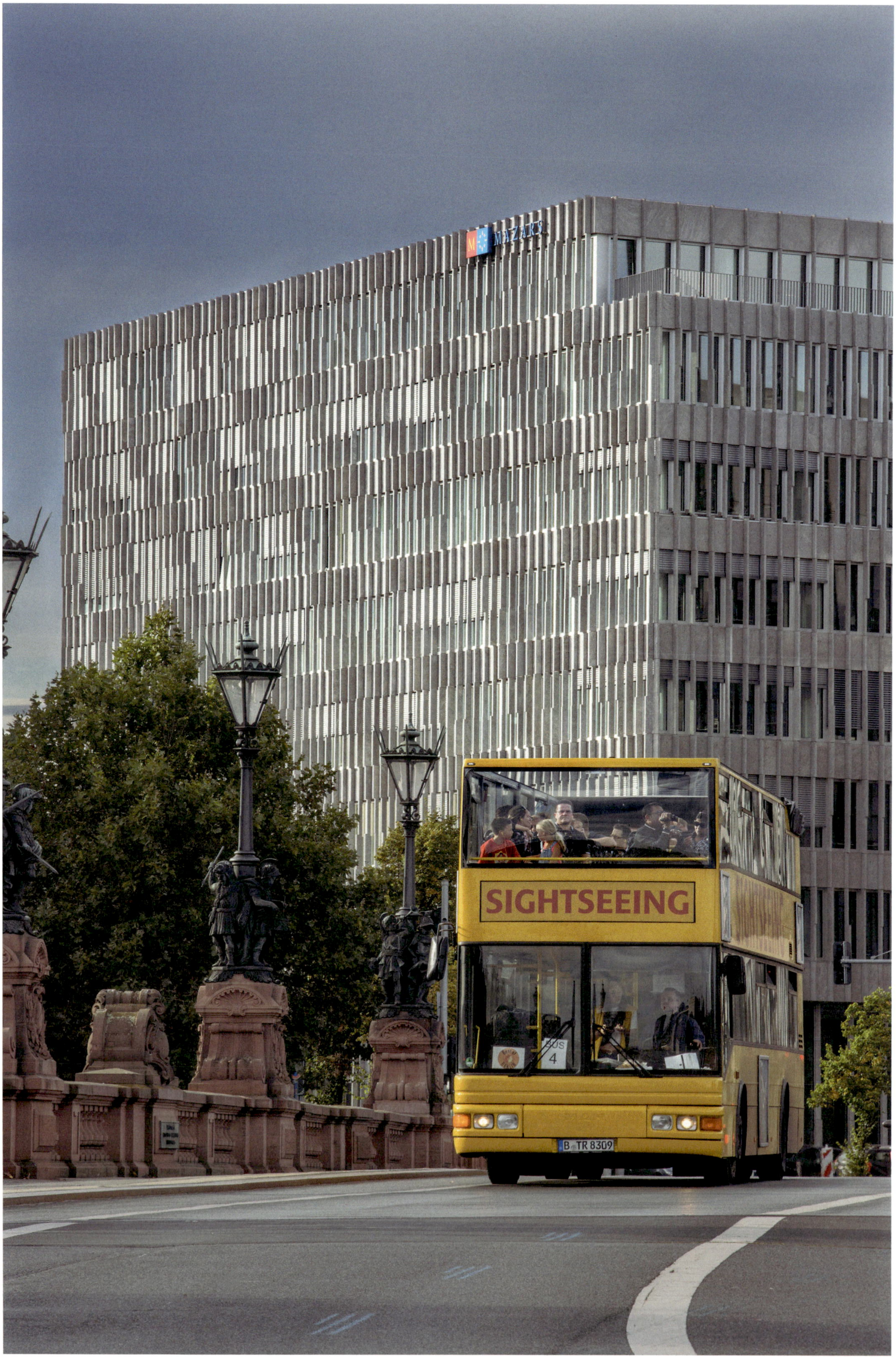
MAZARS
SIGHTSEEING
BUS
4
B TR 8309

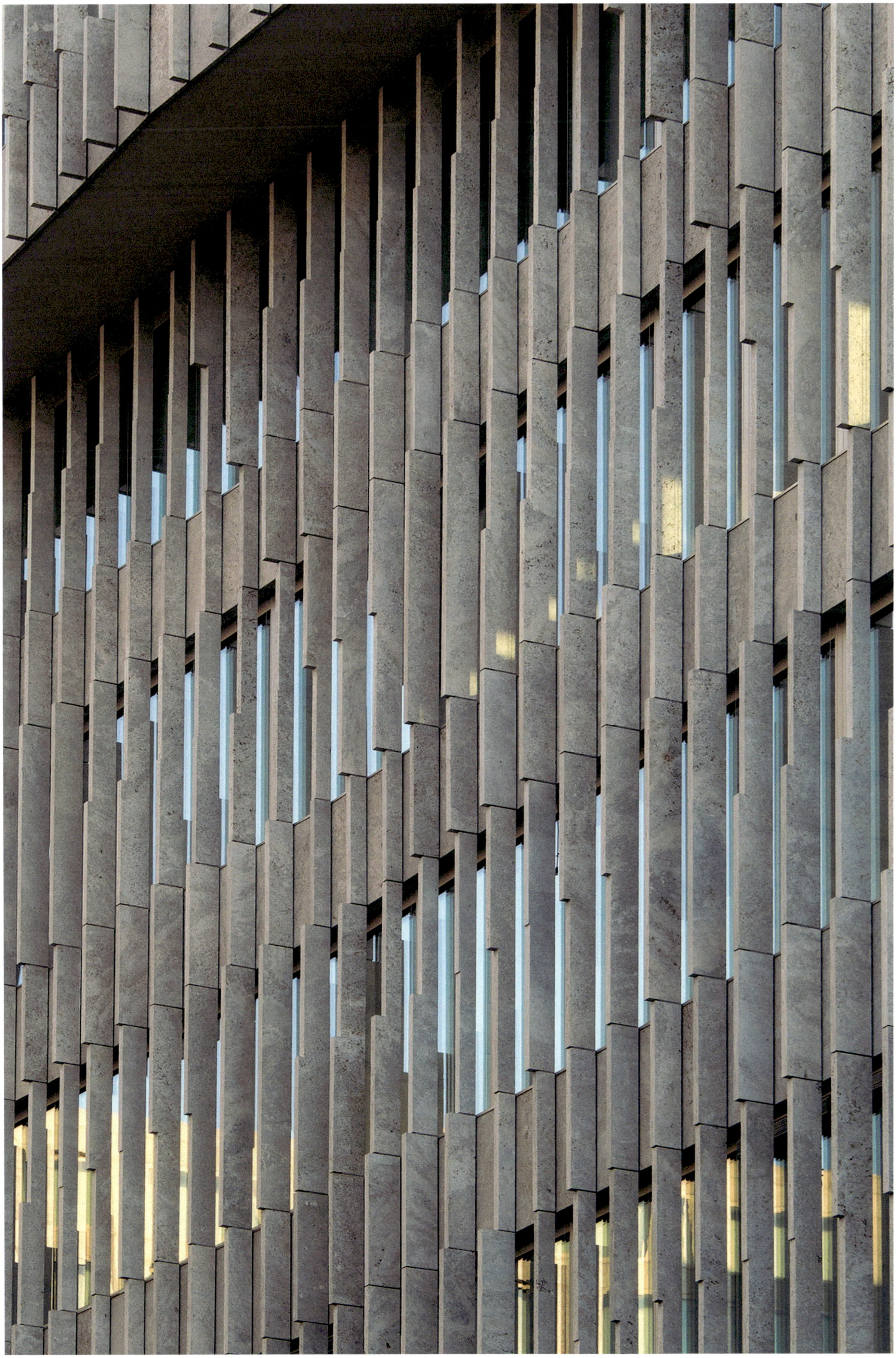

Ultrastructural

204

Ultrastructural is an architectural approach by which structural engineering, as an extreme or ultra-condition, informs and dominates an architectural outcome. *Ultra* refers to structure as design and as a tool that defines an architectural idea rather than simply or only supporting it.

205

Our structures are spatialized and occupiable. Structure is expressed rather than repressed. To this end, historical models are modified and updated, and emerging technologies are exploited and expanded.

In response to the prompt from the 2017 Chicago Architecture Biennial – namely, to consider what it means to 'make new history' – we reimagined the unrealized 1945 project for a chapel for Fritz Bultman, conceived by artist/ architect Tony Smith.

We imagined the project as a timber construction (three columns and floor plates), completed by steel tension cables and nylon cord woven throughout to create a new skin for the structure. The result? A new modular prototype for housing, as an aggregating composite structure.

This is a concept that lacks a particular site or place; the building's structural expressiveness is foregrounded above all else. Does such an emphasis on the structural nature of a building come at the expense of site-specificity?

Just the opposite: the elemental character of the structure derives from the use of an economy of means (elements that can be added or subtracted). Employing this economy of means make the structure, in turn, adaptable to a multitude of sites.

210

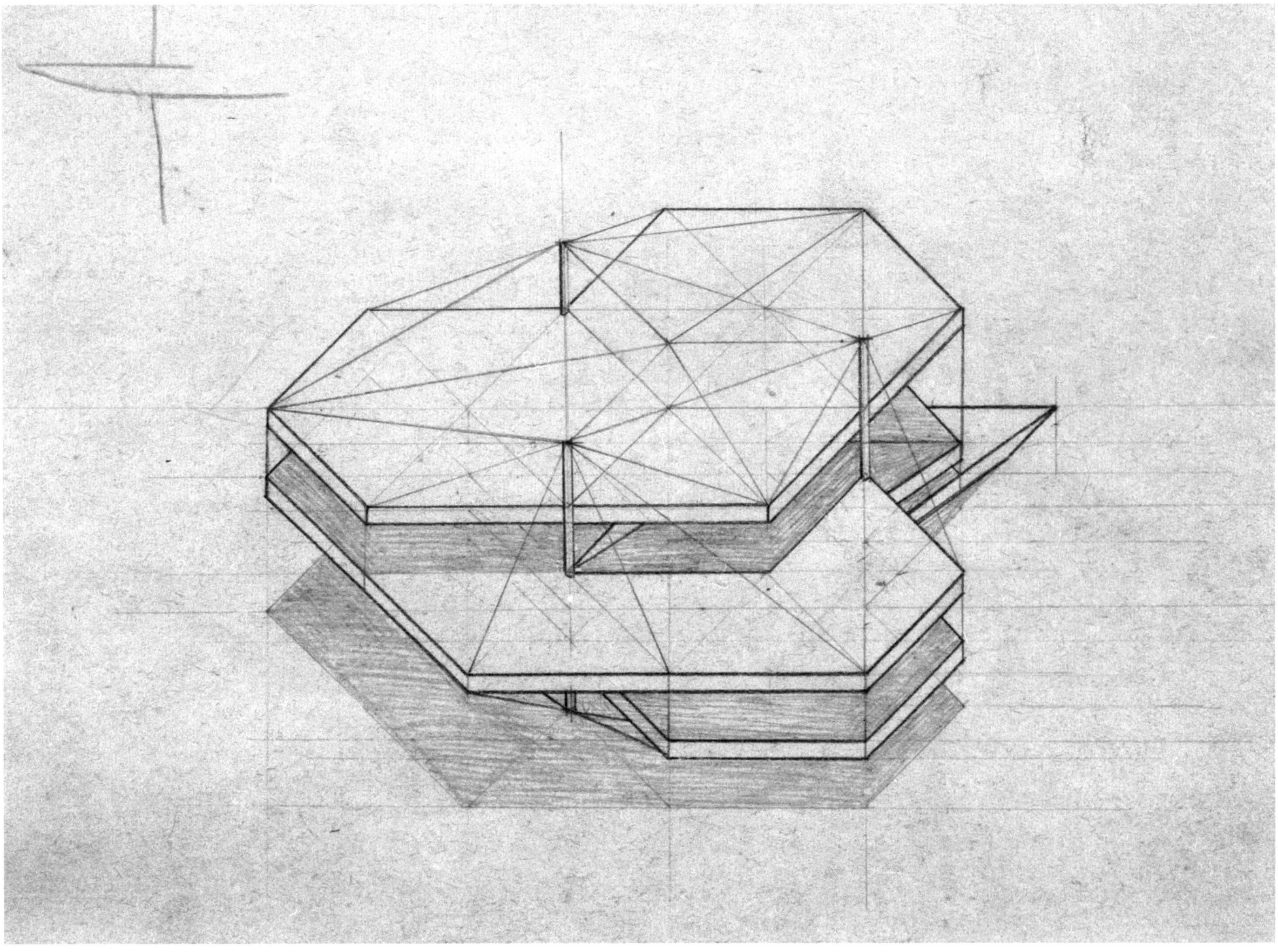

211

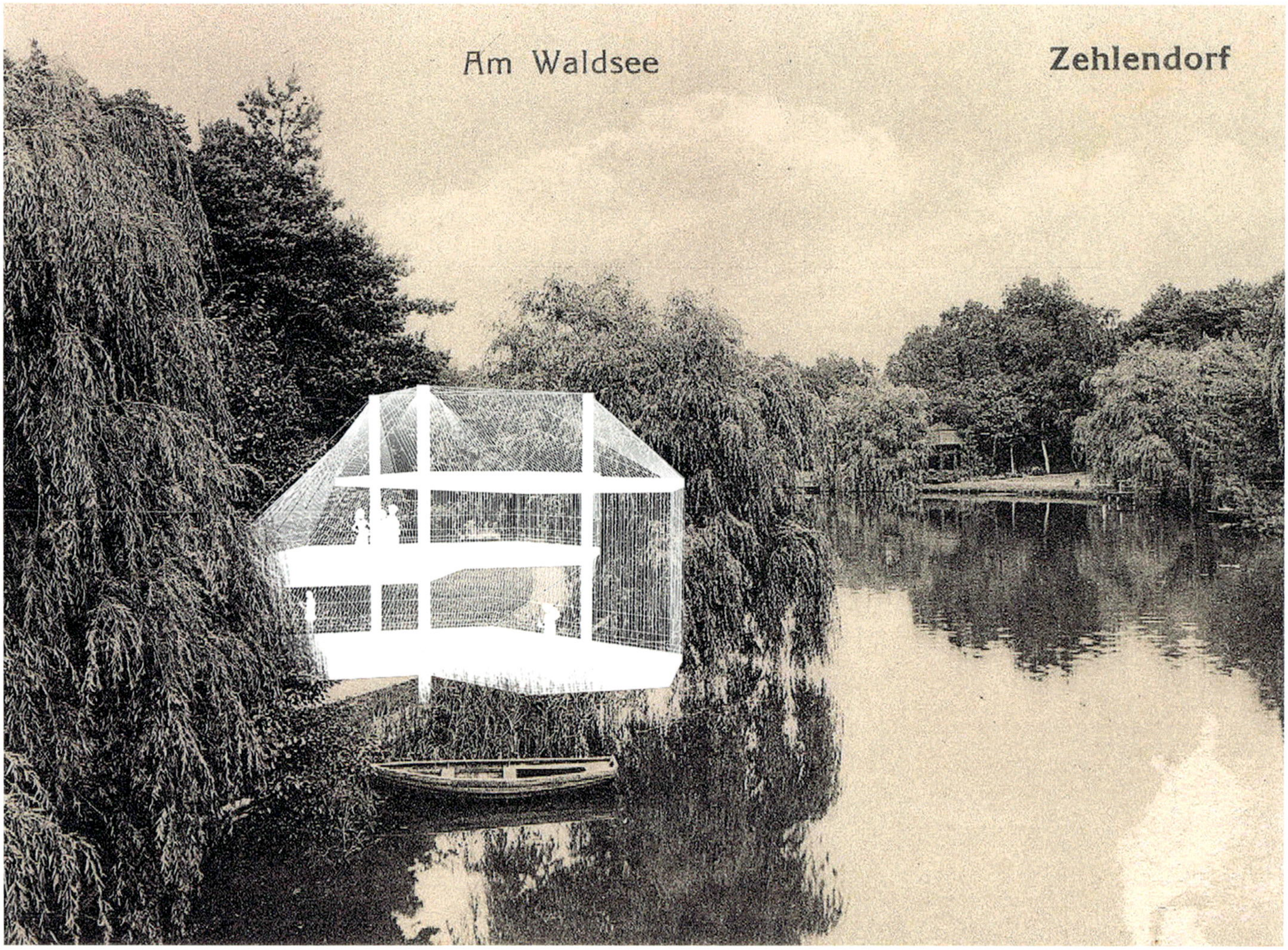

212

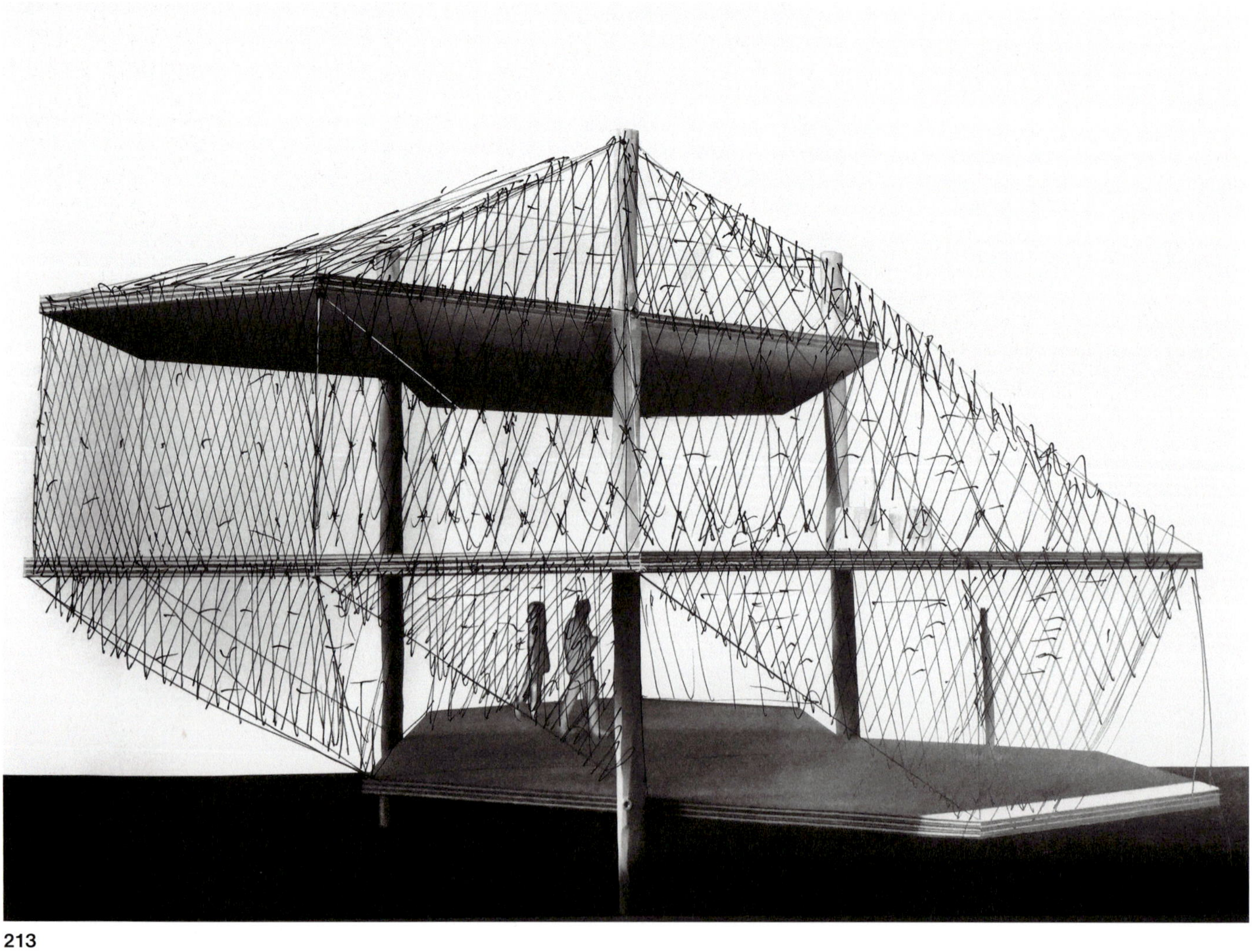

213

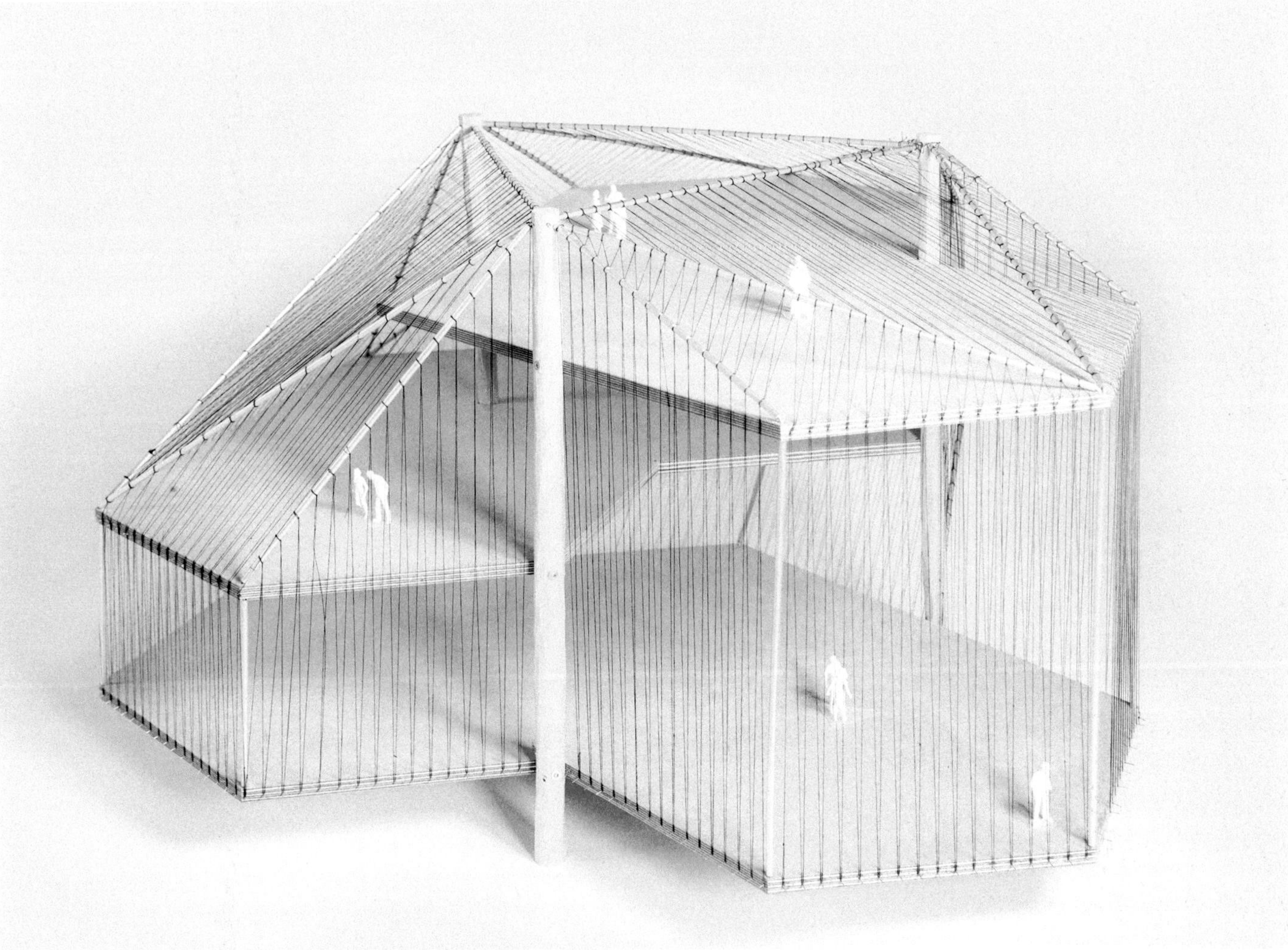

214

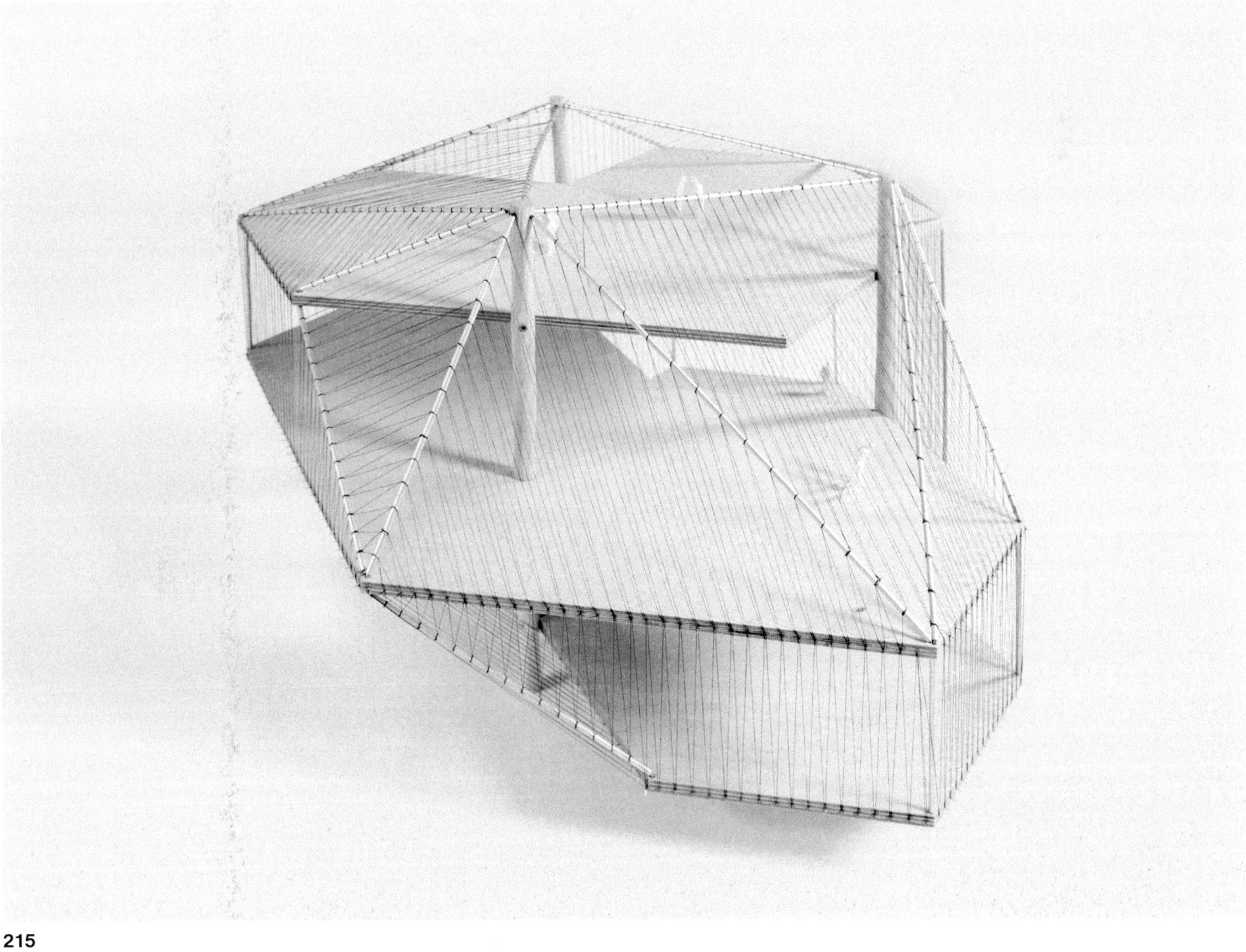

215

Loom Hyperbolic, our installation for the 2012 Marrakech Biennale, is constructed of yarn and timber. For this structure, we exploited 'ruled geometry' to form hyperbolic surfaces that created enclosures near the Kutubiyya Mosque, encouraging public encounters.

As a prelude or study, Loom Hyperbolic would later inform the garden pavilion we designed for the American Academy in Berlin, which offers additional workspaces for the fellows there. The concept of the loom was adopted and updated (from string to steel) for the roof of the American Academy pavilion.

In this version, we conceived a lightweight prefabricated glass and steel structure. Central to the enclosed space of seven study carrels is a roof of steel beams, which forms a hyperbolic surface that covers the spaces below.

The structural engineering of the roof is expressive, and it encloses and defines the cellular spaces below. The Fellows Pavilion is ultra as both a concept of complex geometry and as a referent to the hipped roofs of the adjacent and historical villa of the main American Academy building.

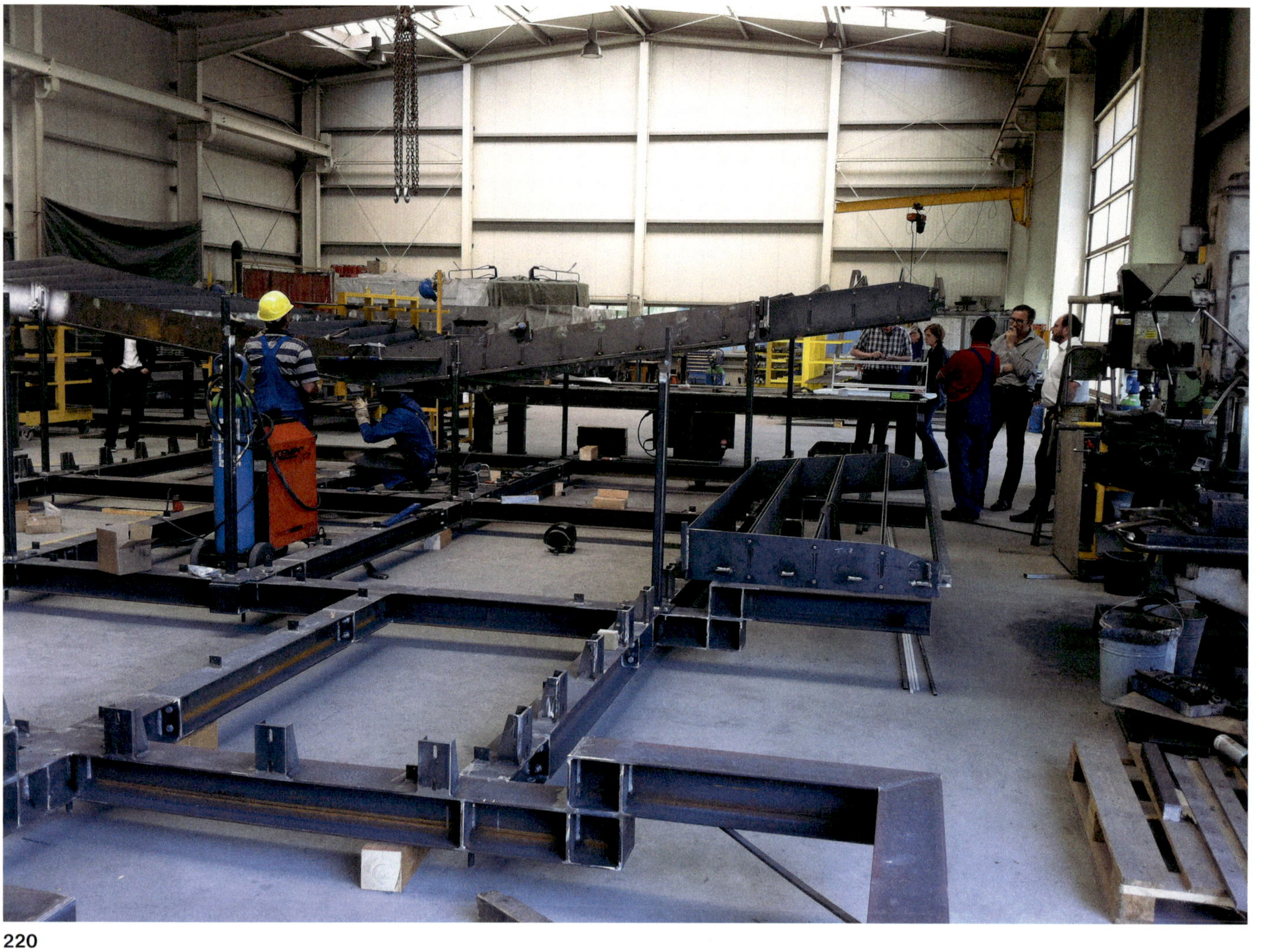

220

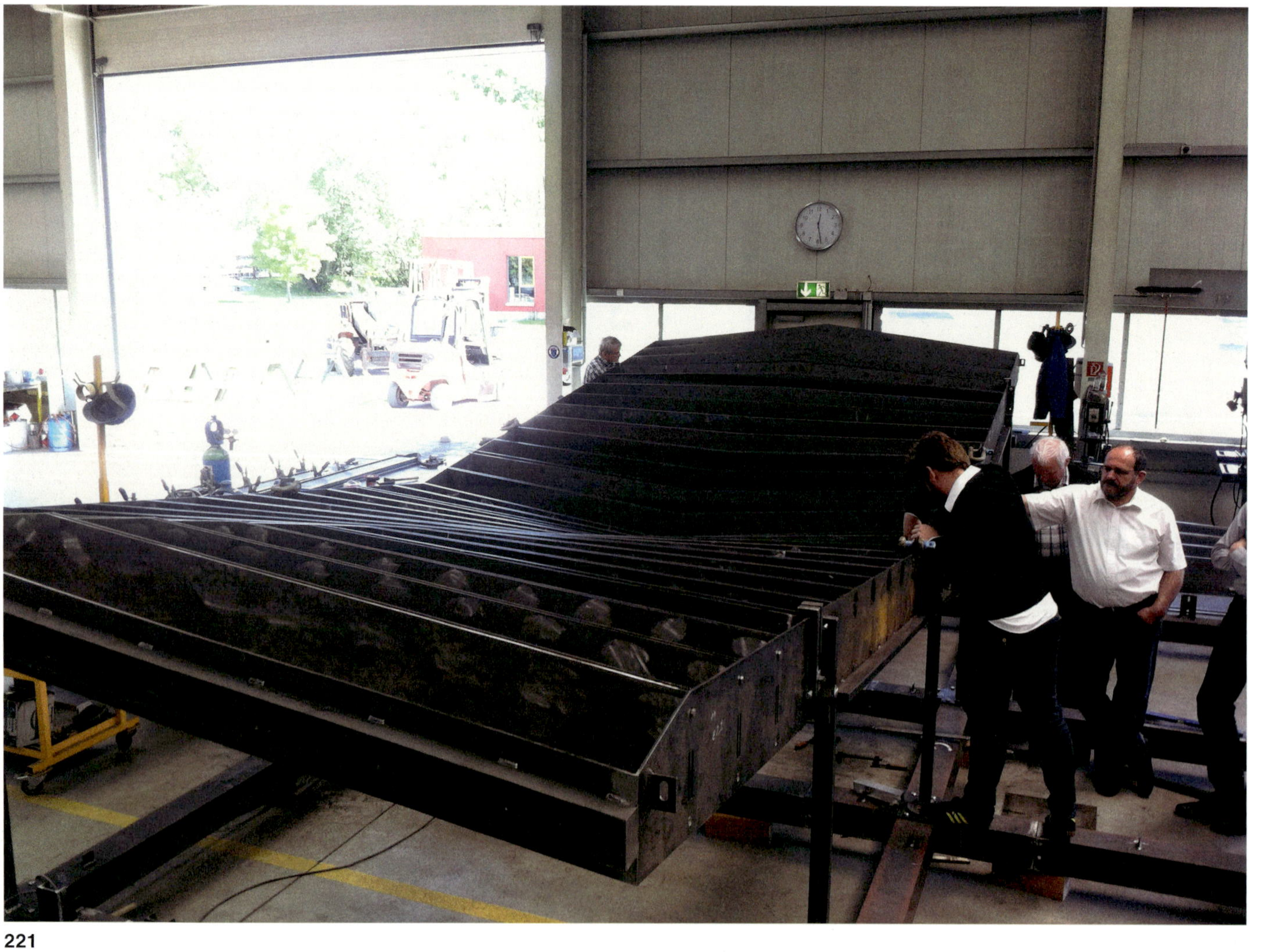

221

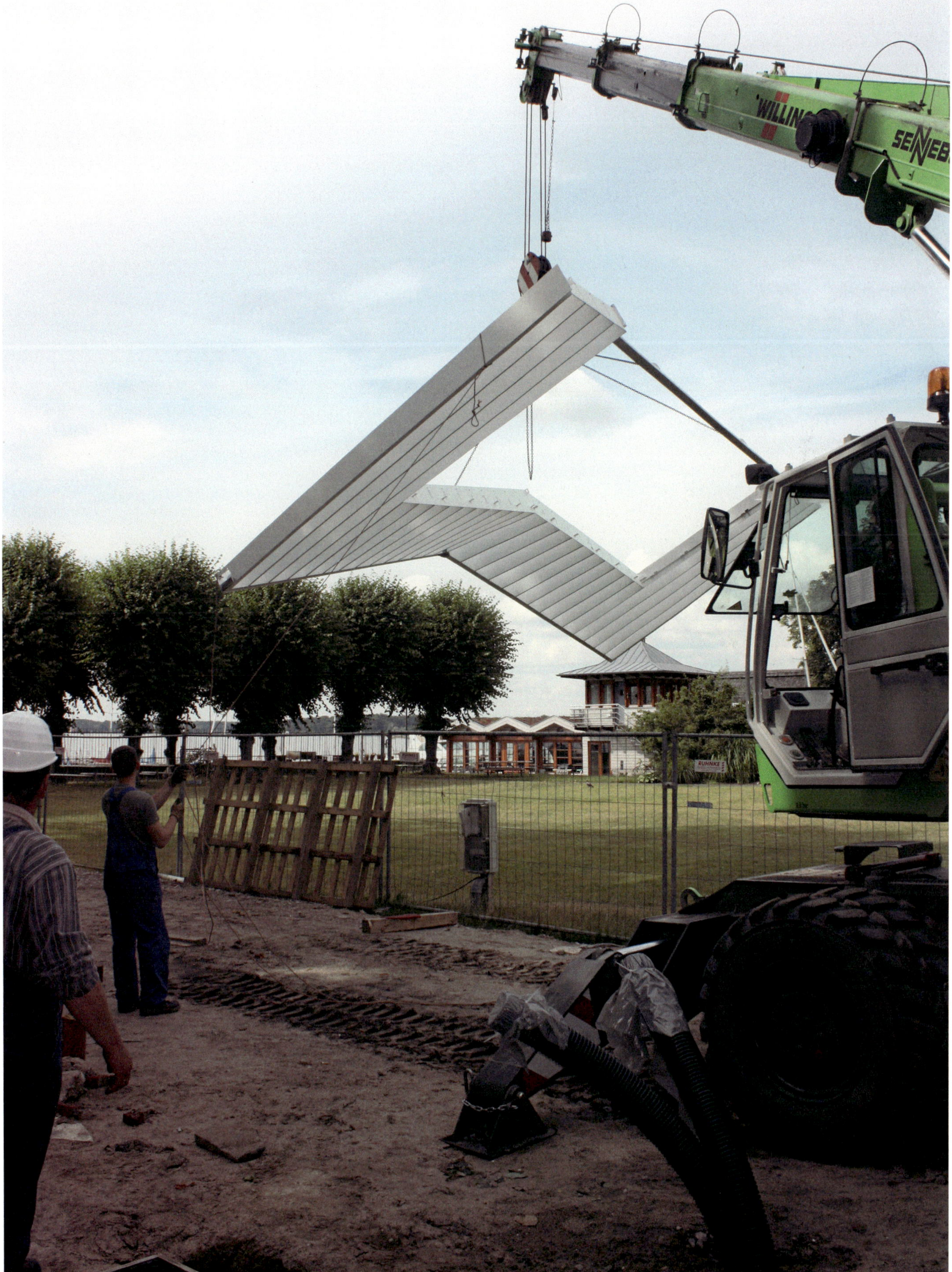

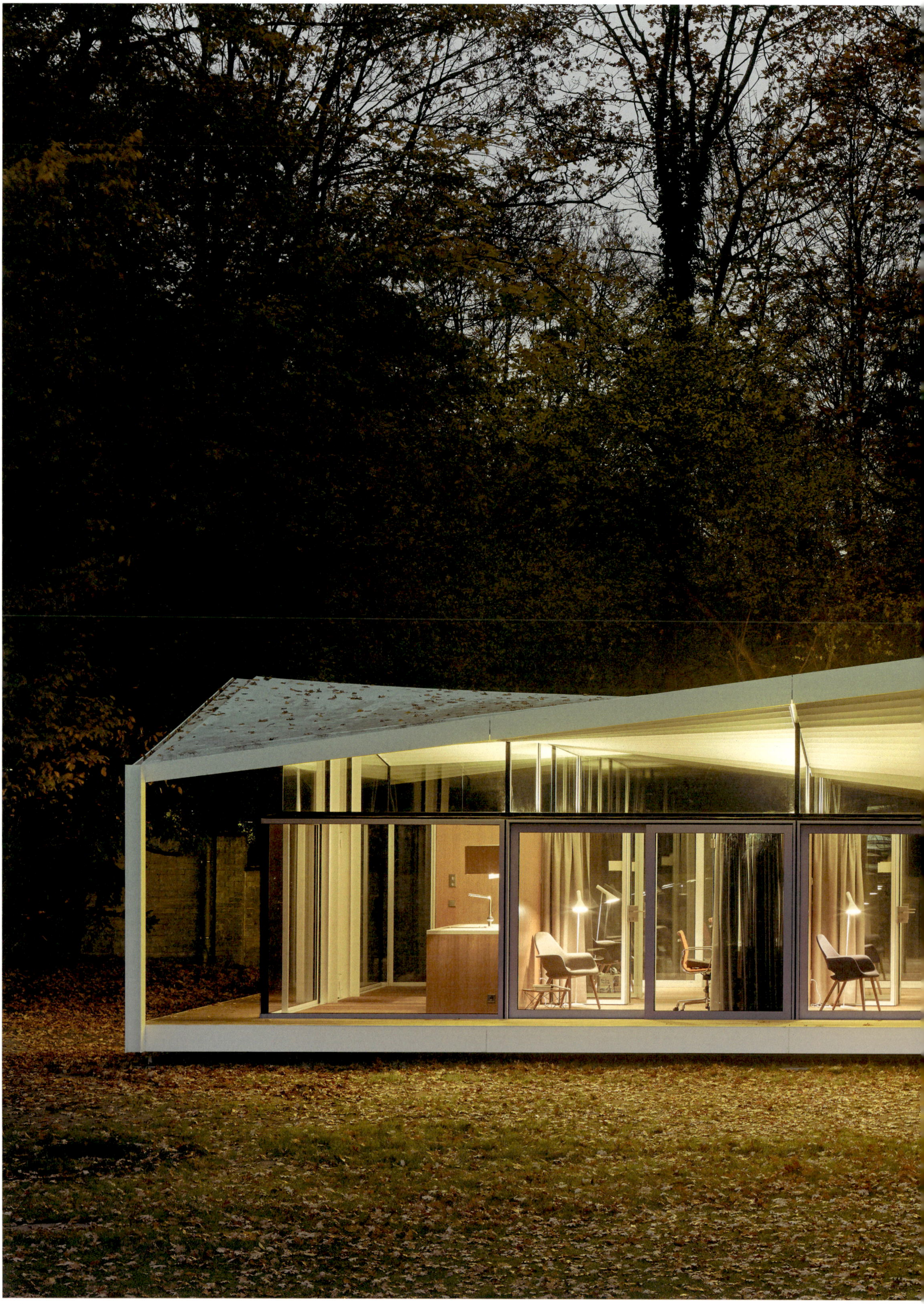

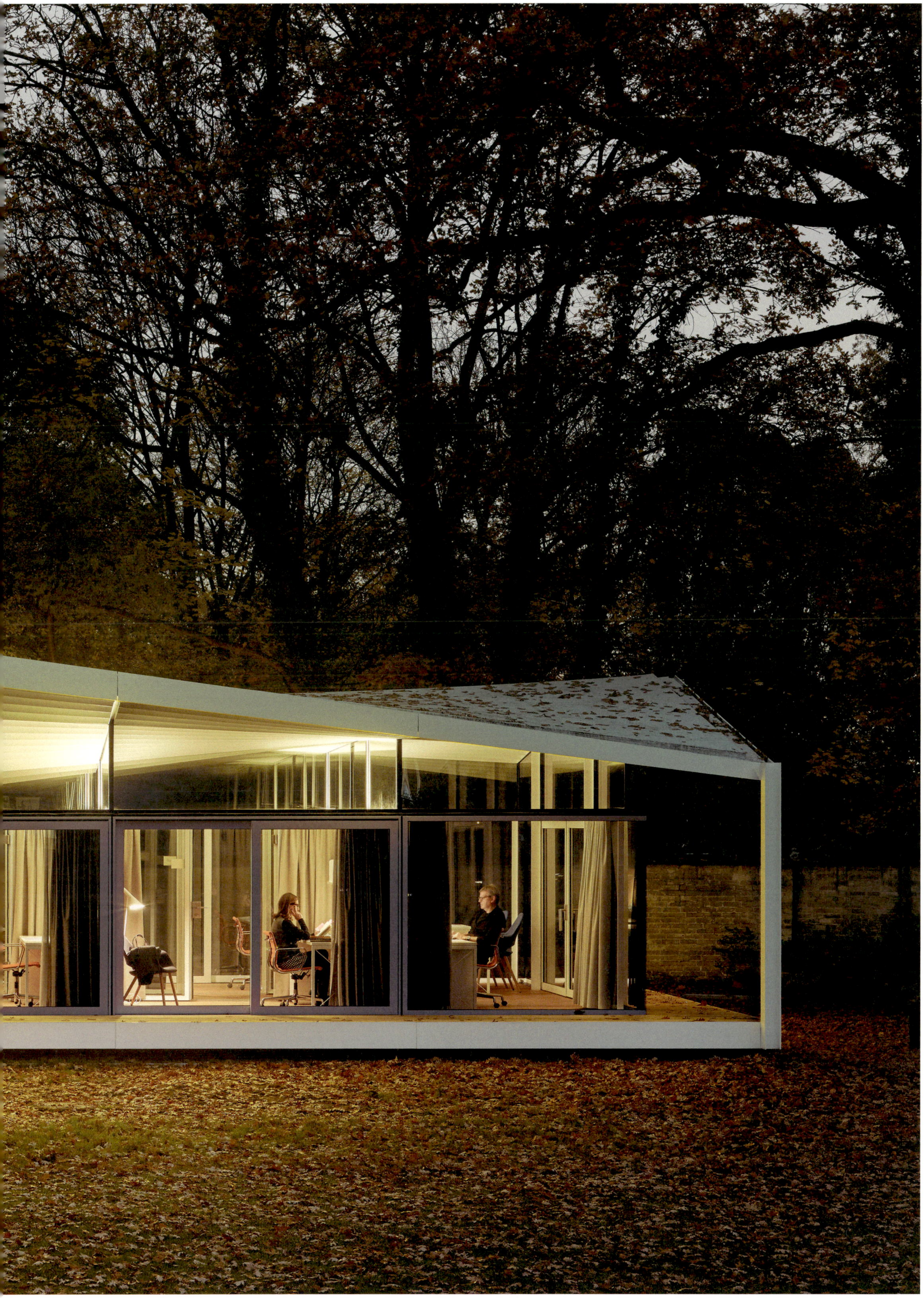

The building industry is responsible for 50 to 60 percent of global resource usage, around 35 percent of carbon dioxide emissions, 35 percent of energy consumption, and 50 percent of the production of mass waste.

In light of these facts, we used building components made of structurally gradient concrete – concrete made porous in response to structural loading to reduce weight and material. Installed at Akademie der Künste, Urbainable explored gradient concrete through suspended structures.

The gradiation of the material property is achieved either by varying the porosity of a material or by altering the mixing ratio in response to structural loading.

Such a principle can be found in nature: bone structures for example whose sponge-like variation responds to the structural economy of bone trabeculae. Translating this logic to concrete saves material, reduces mass, and improves recyclability.

The prototype designs developed for the exhibition demonstrated the spatial and architectural potential of this construction logic. These prototypes addressed transferring the principle of a construction section to three-dimensional space – from sectional to spatial thinking.

What are the architectural consequences of gradiated concrete for columns, beams, or decking, and what could this new lightness bring strategically? models tested configurations in which lightweight, cable-suspended floor plates offered new spatial possibilities for living or working.

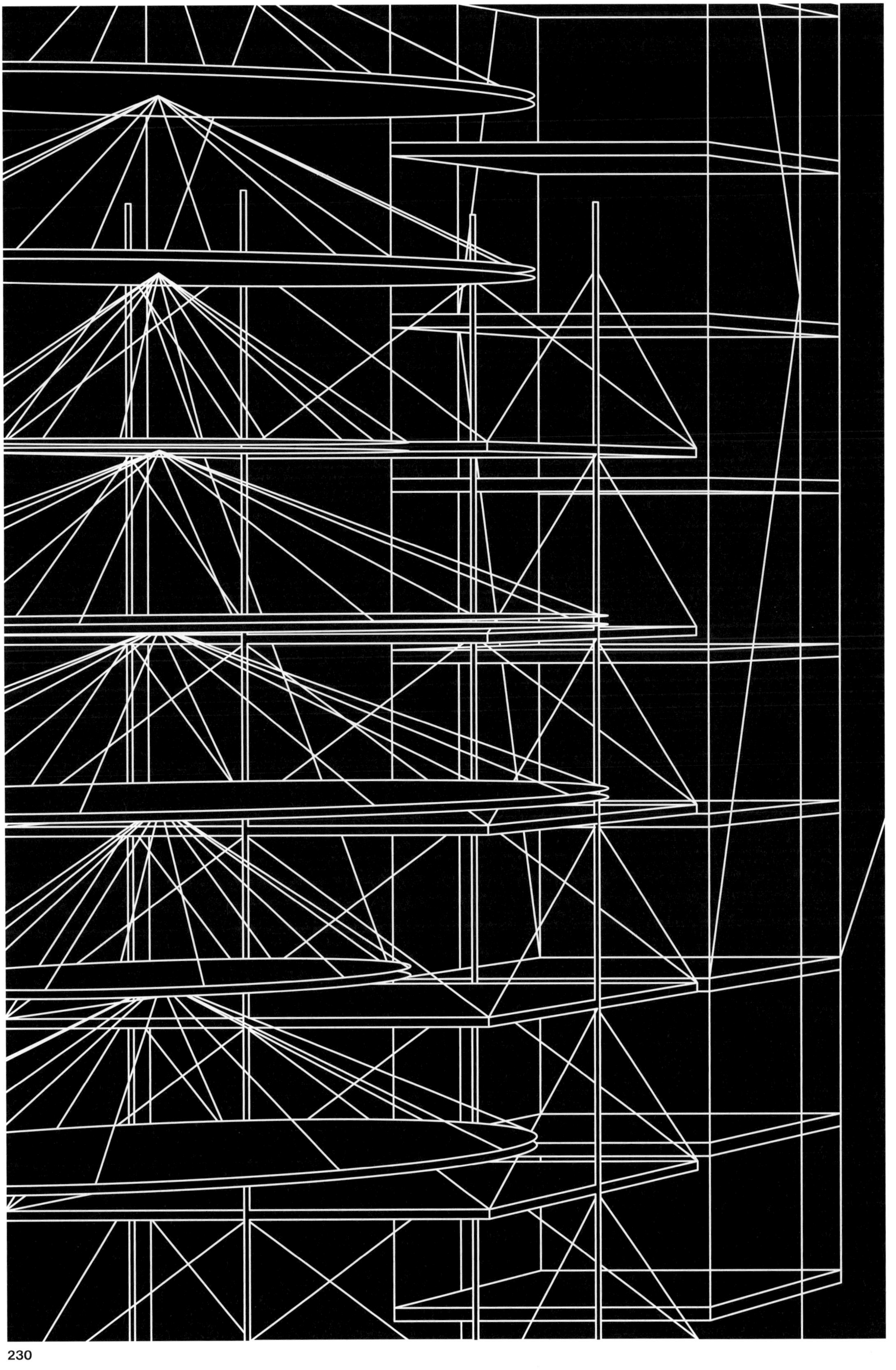

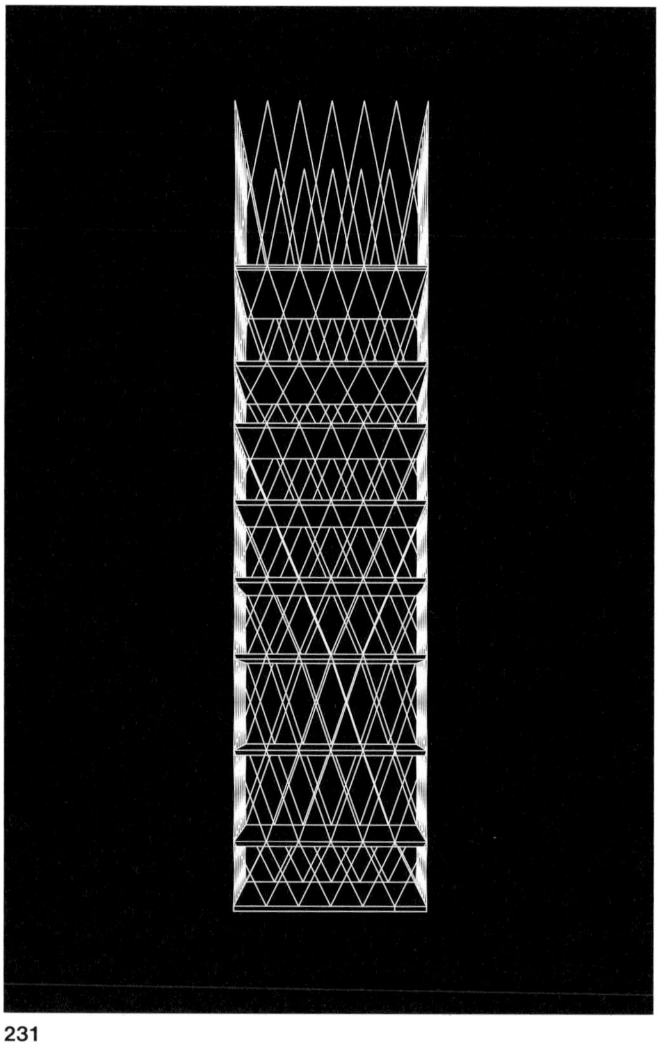
231

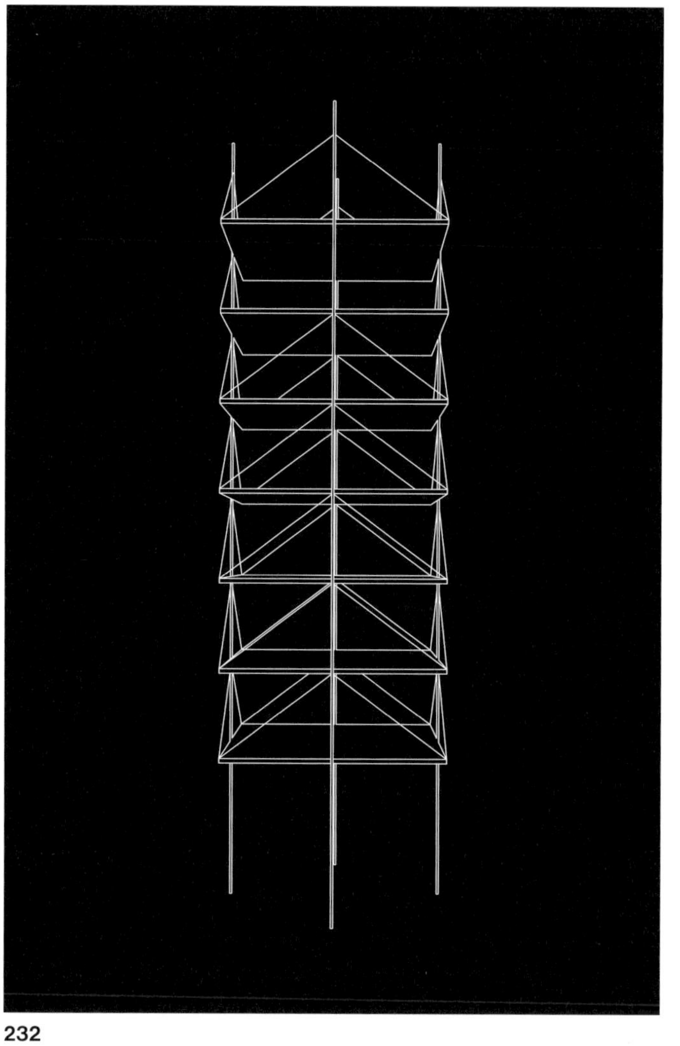
232

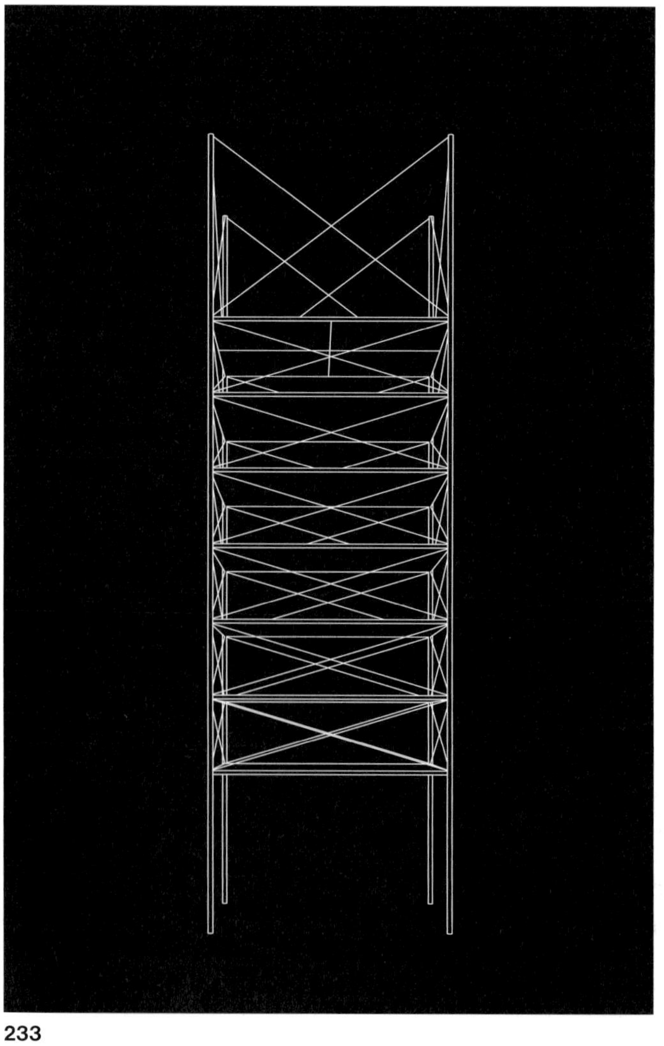
233

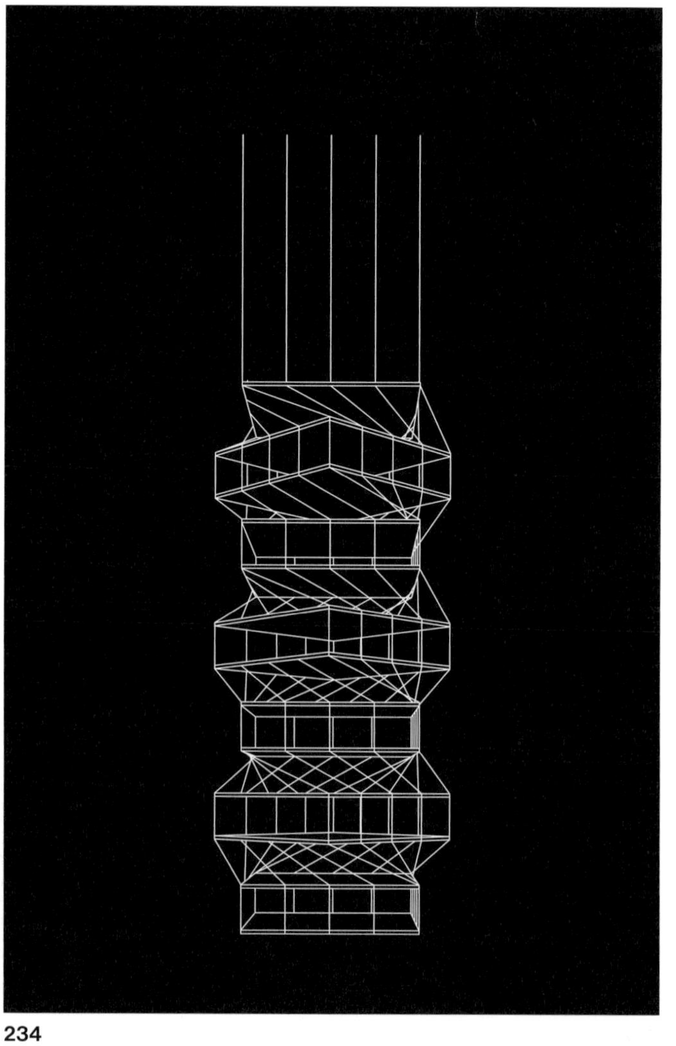
234

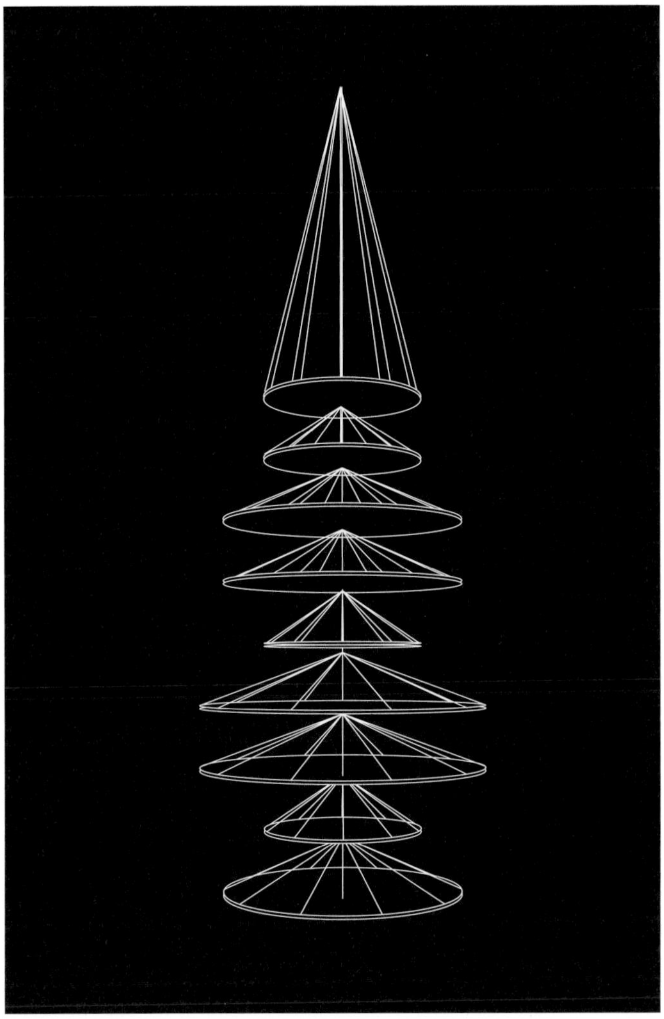
235

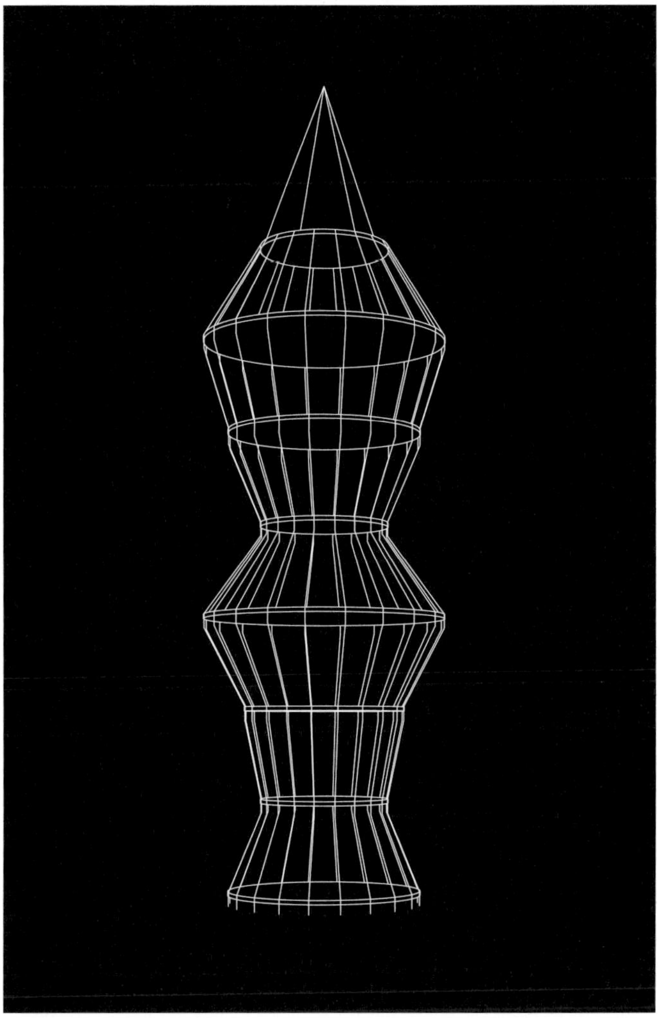
236

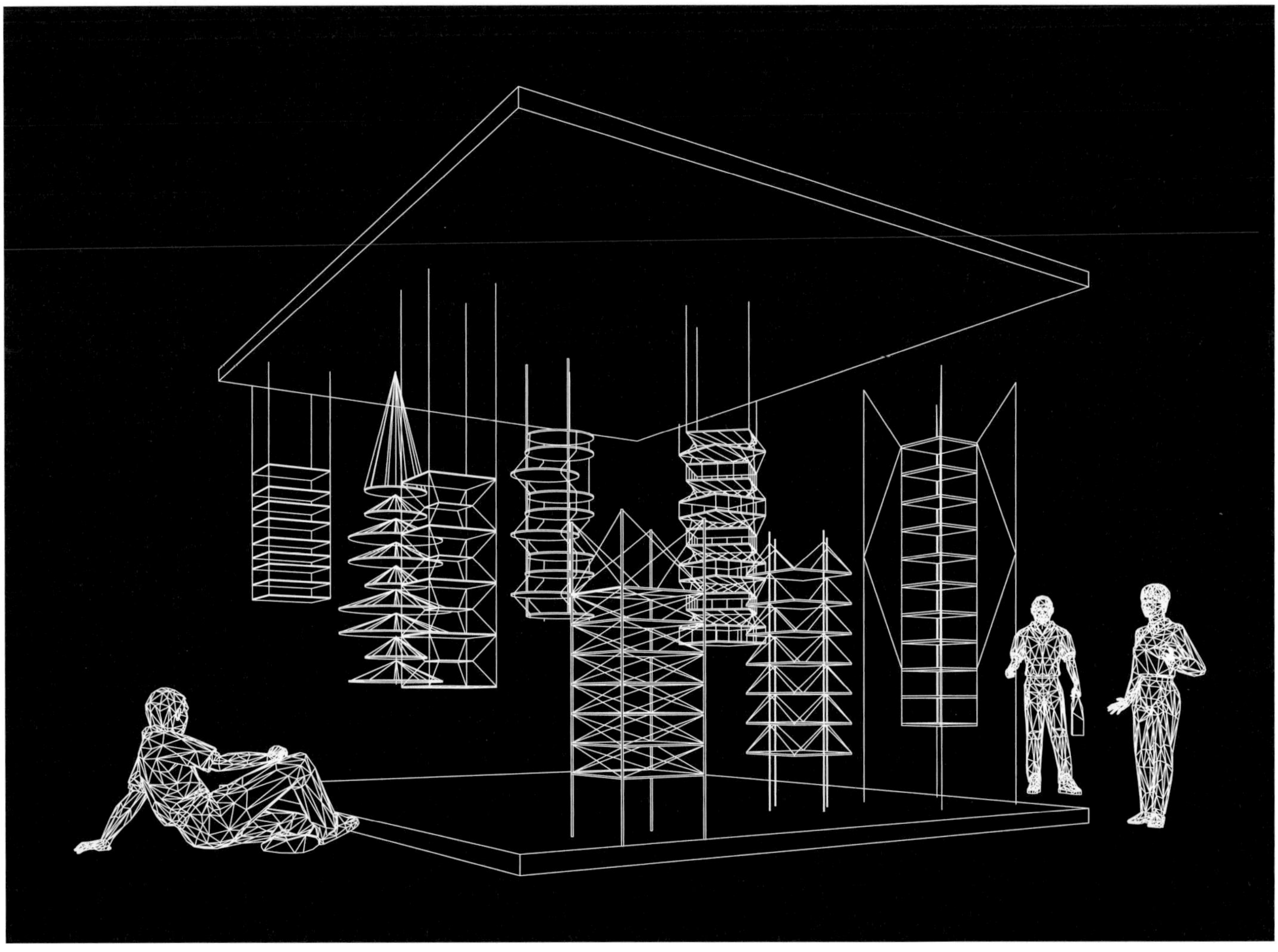
237

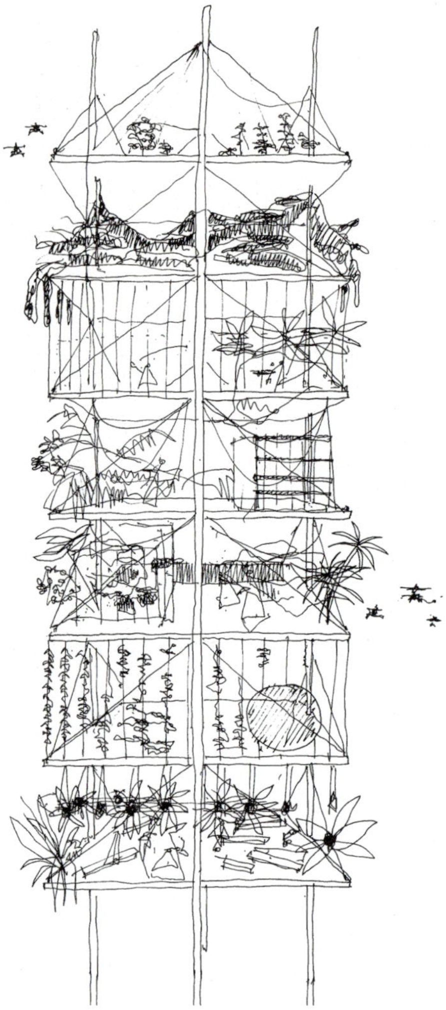

238

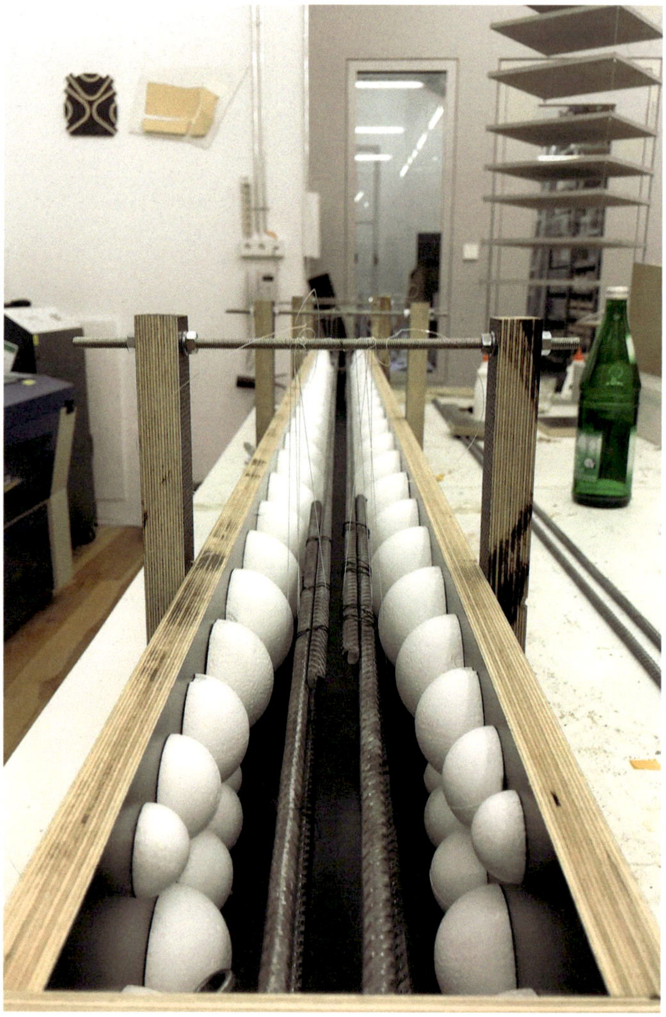

239

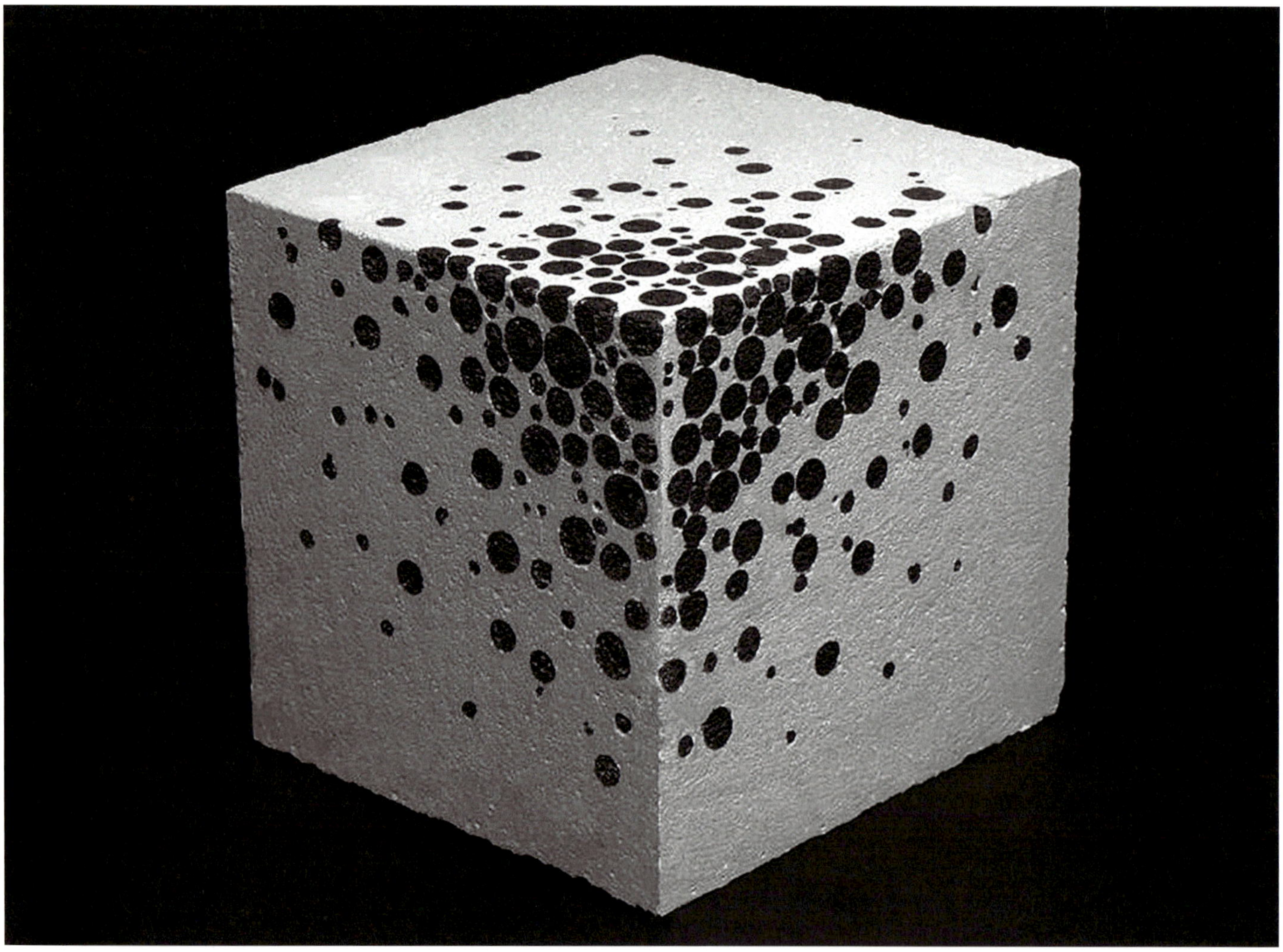

240

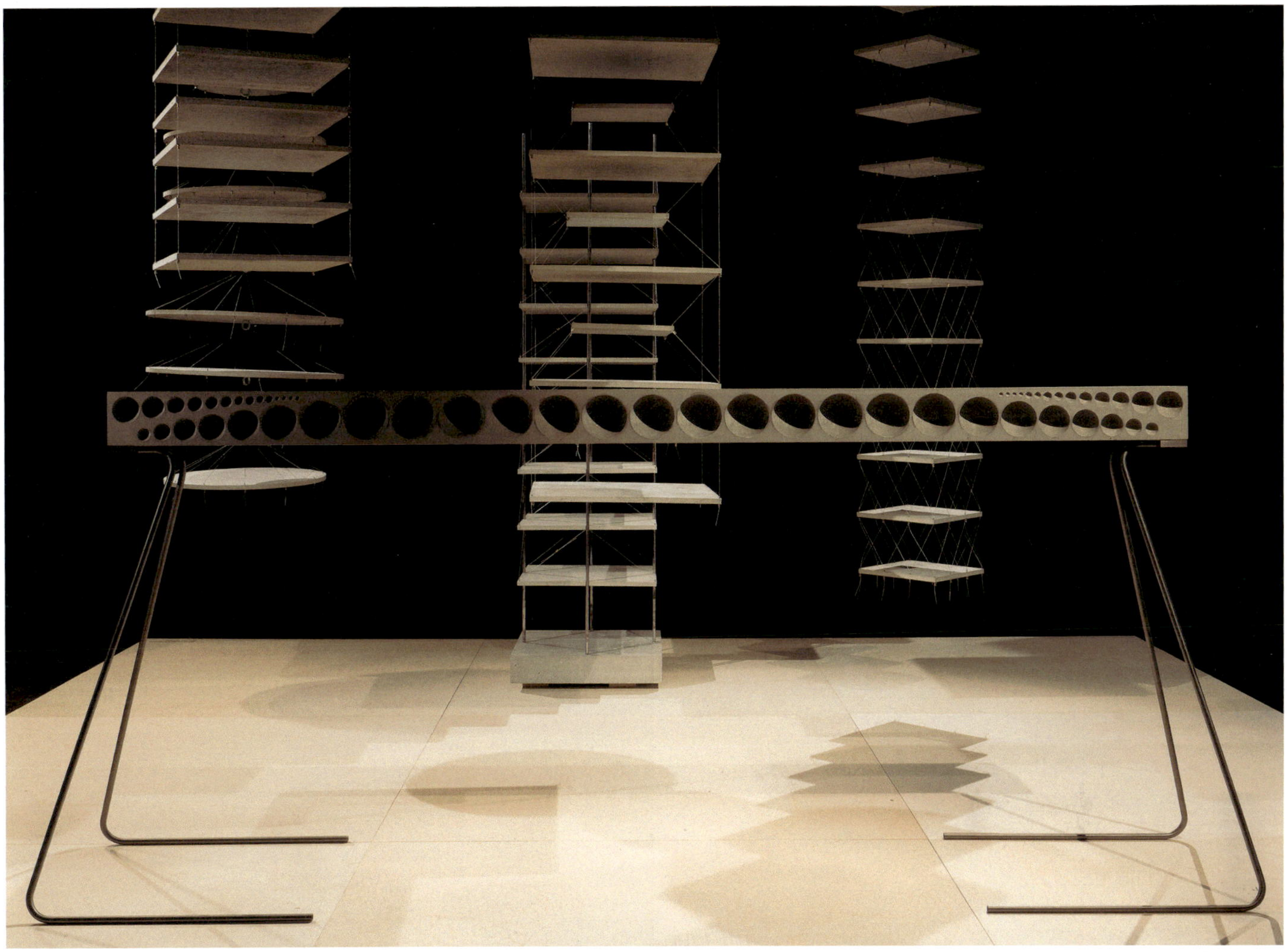

241

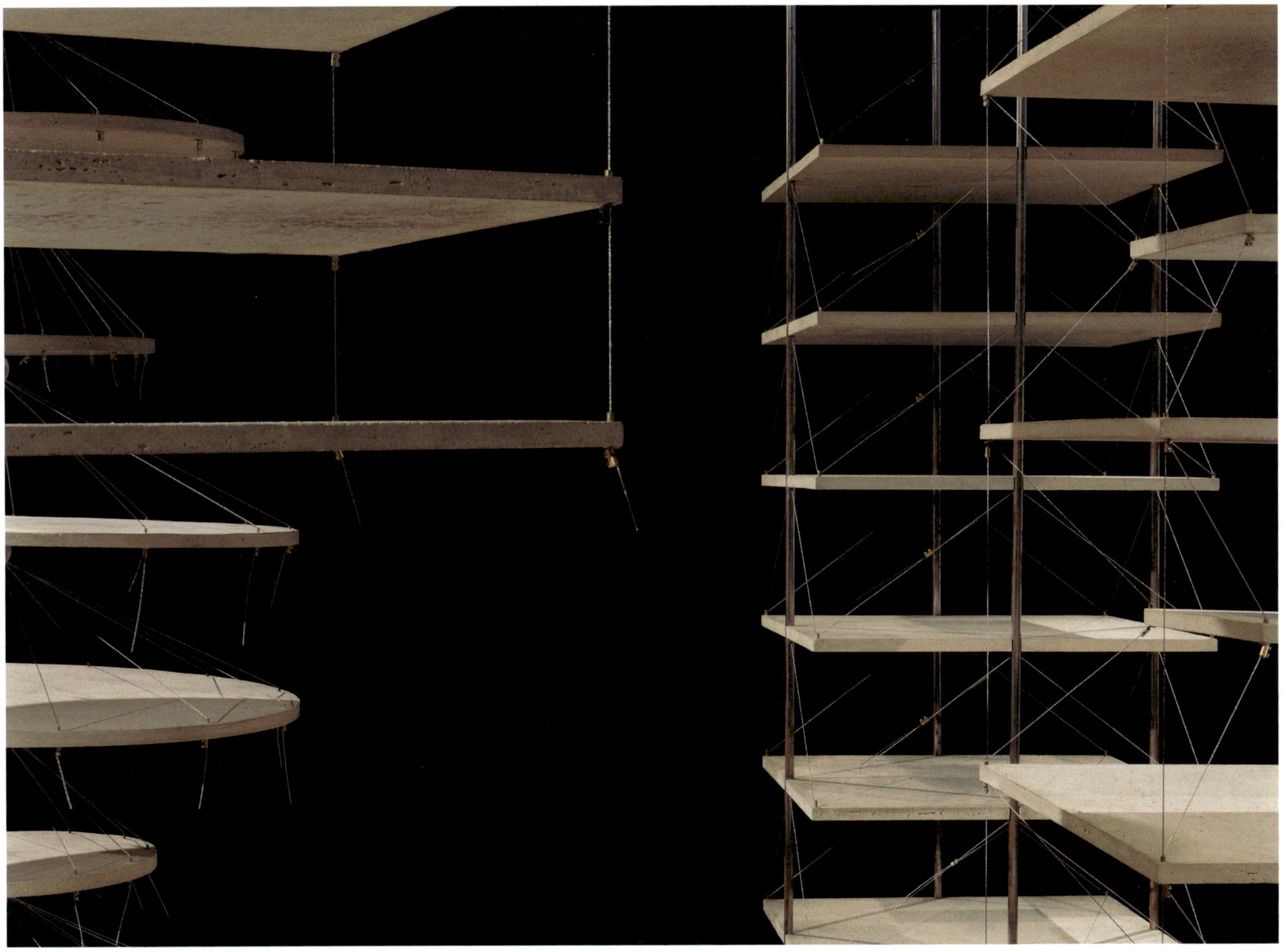

242

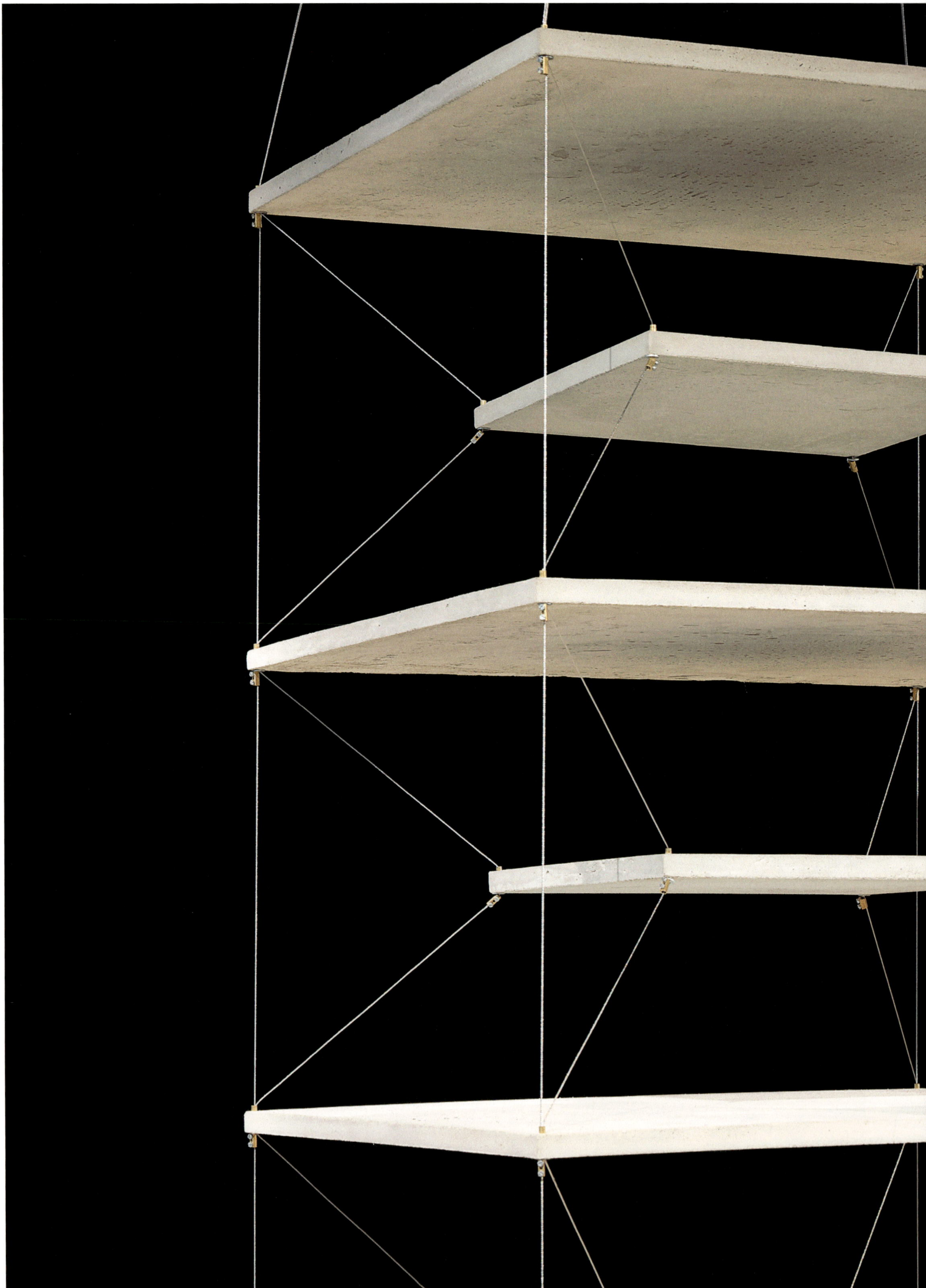

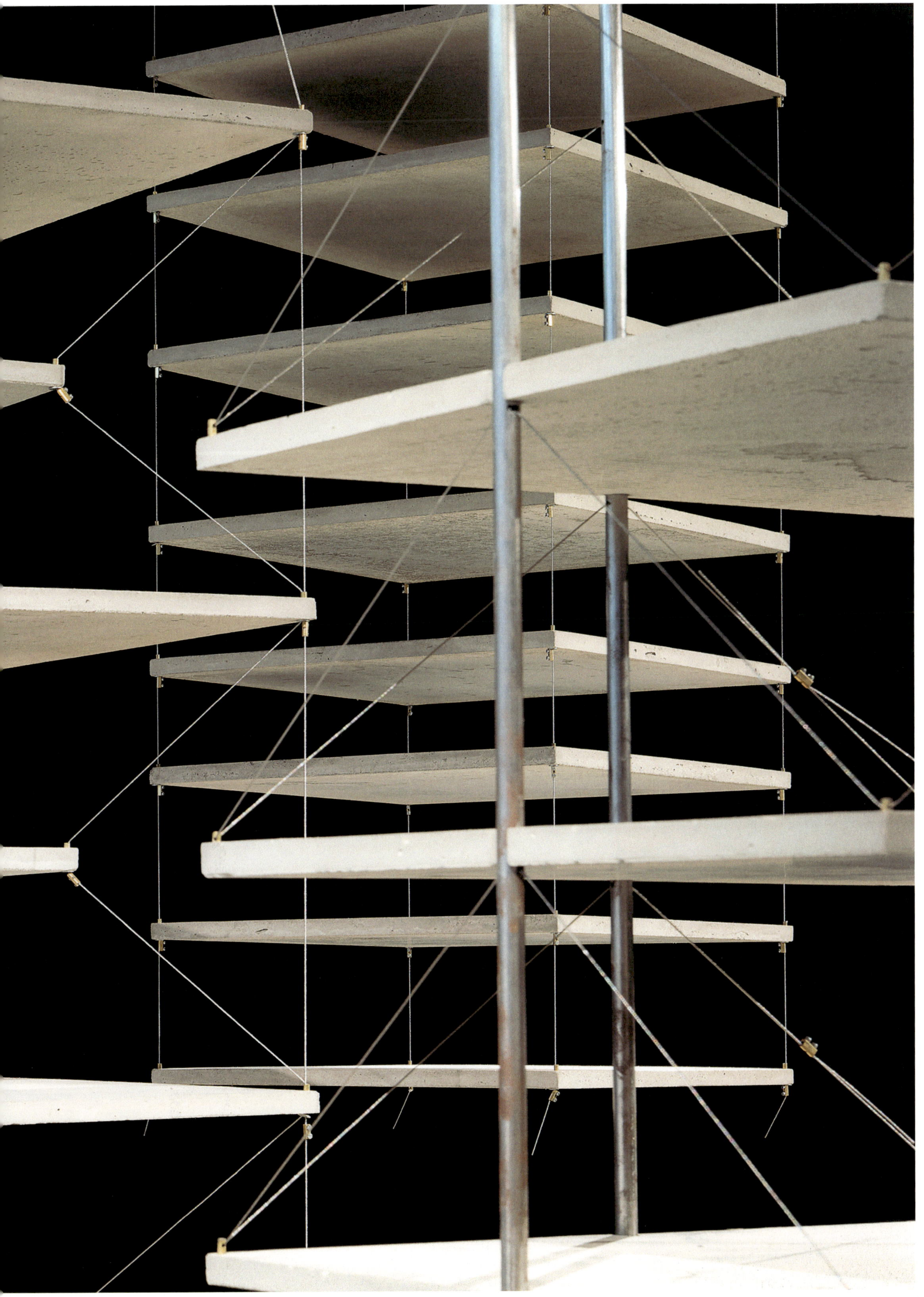

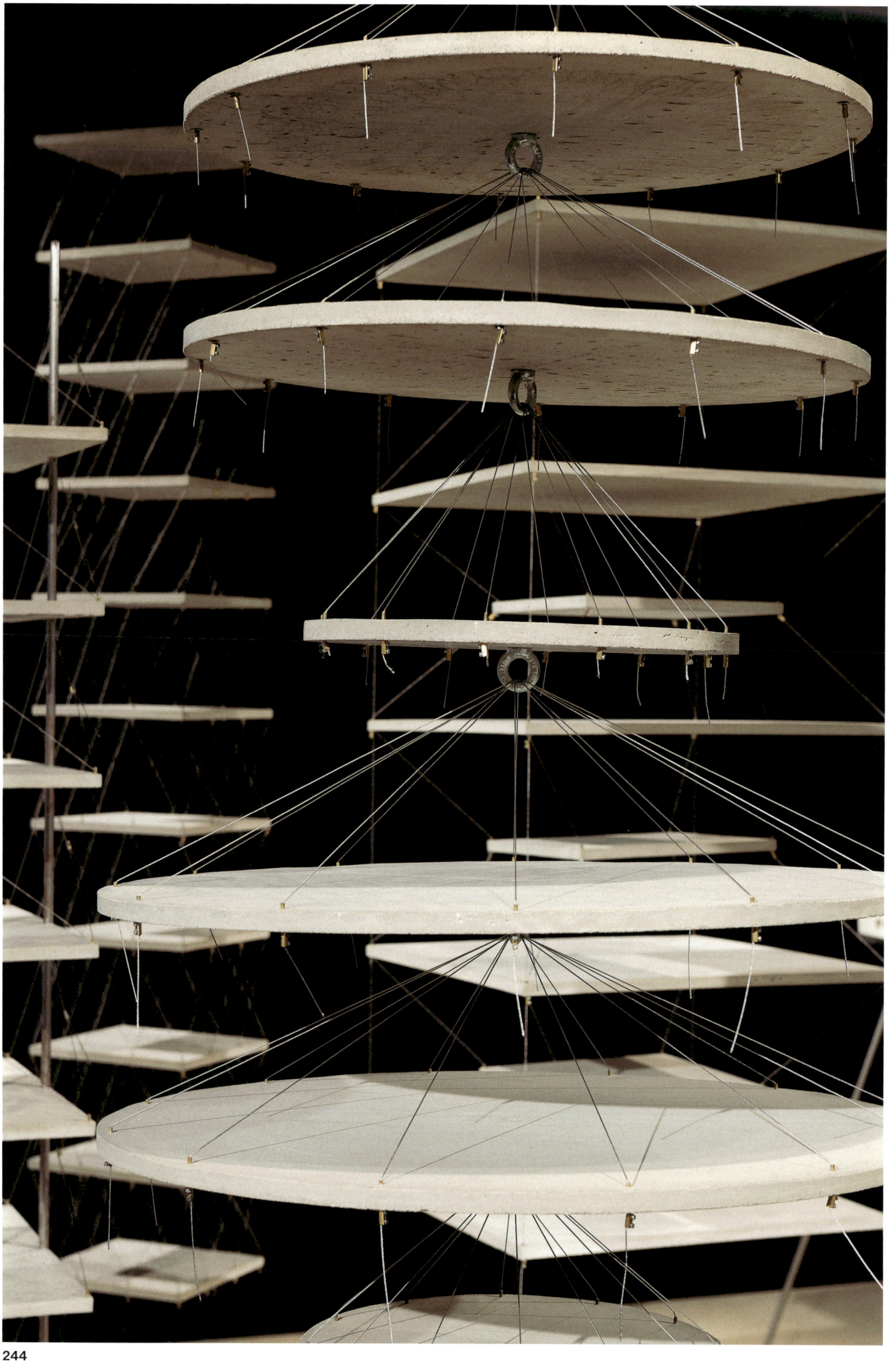

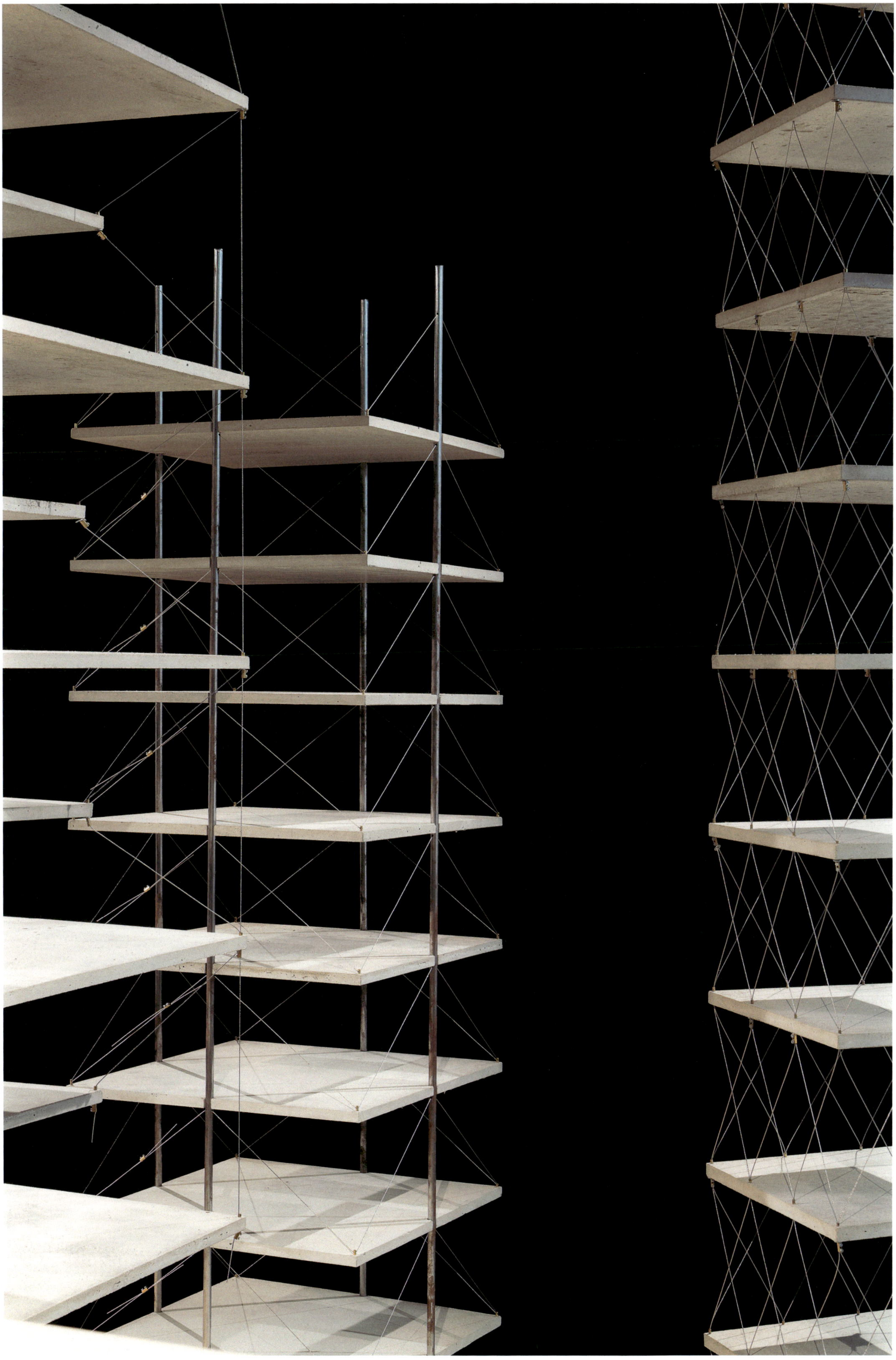

The new in architecture is best driven by its response to evolving technologies or cultural prompts. That is the question of our research here.

246

An education center for the TRUMPF machine-tool factory for their young apprentices was both temporary and made using lightweight timber construction. This is one of the most sustainable methods for building, especially mass timber elements prefabricated off-site.

Complementing the 2008 Kantine, the building geometry is a hexagon roofed over by a timber-beam radial and pin-wheeled construction with an oculus/skylight at its center. An undulating facade of glazed or opaque panels define the exterior of the building.

A central auditorium of wood seating connects the ground level to the upper mezzanine. A series of meeting spaces and machine rooms are situated around the auditorium on one or two levels.

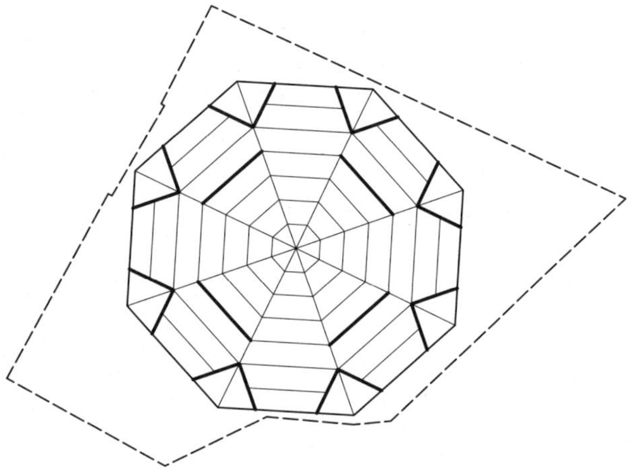

250

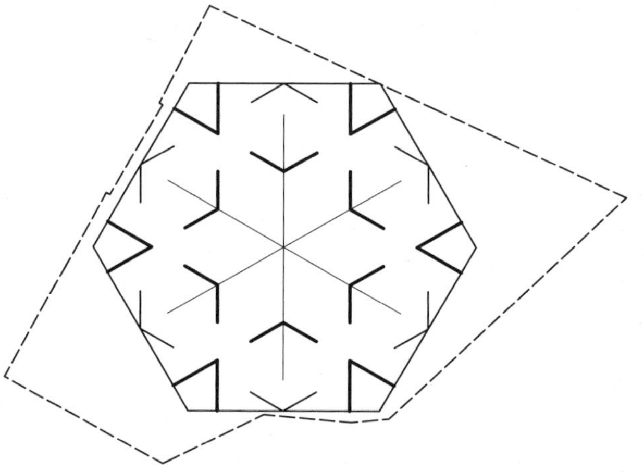

251

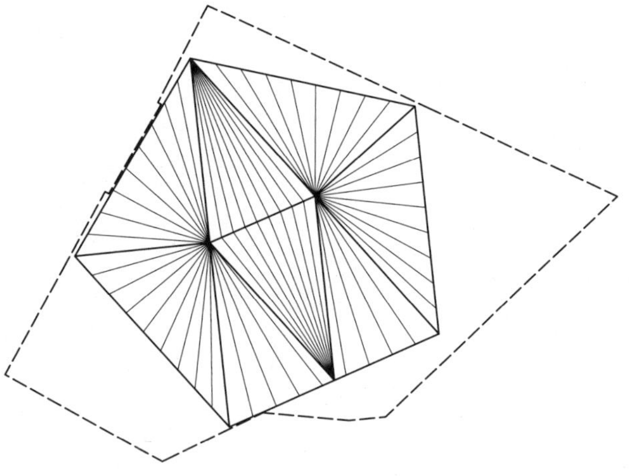

252

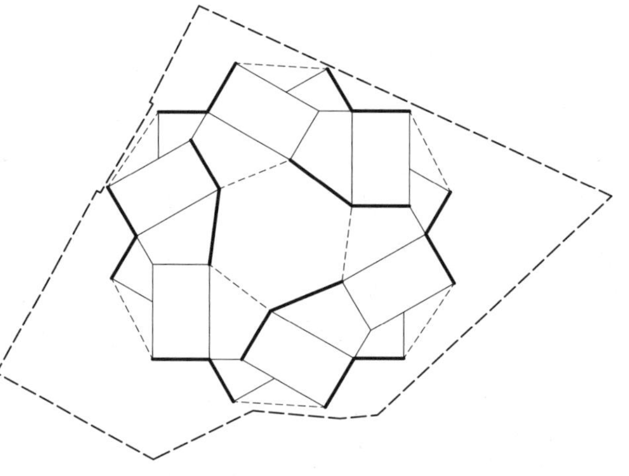

253

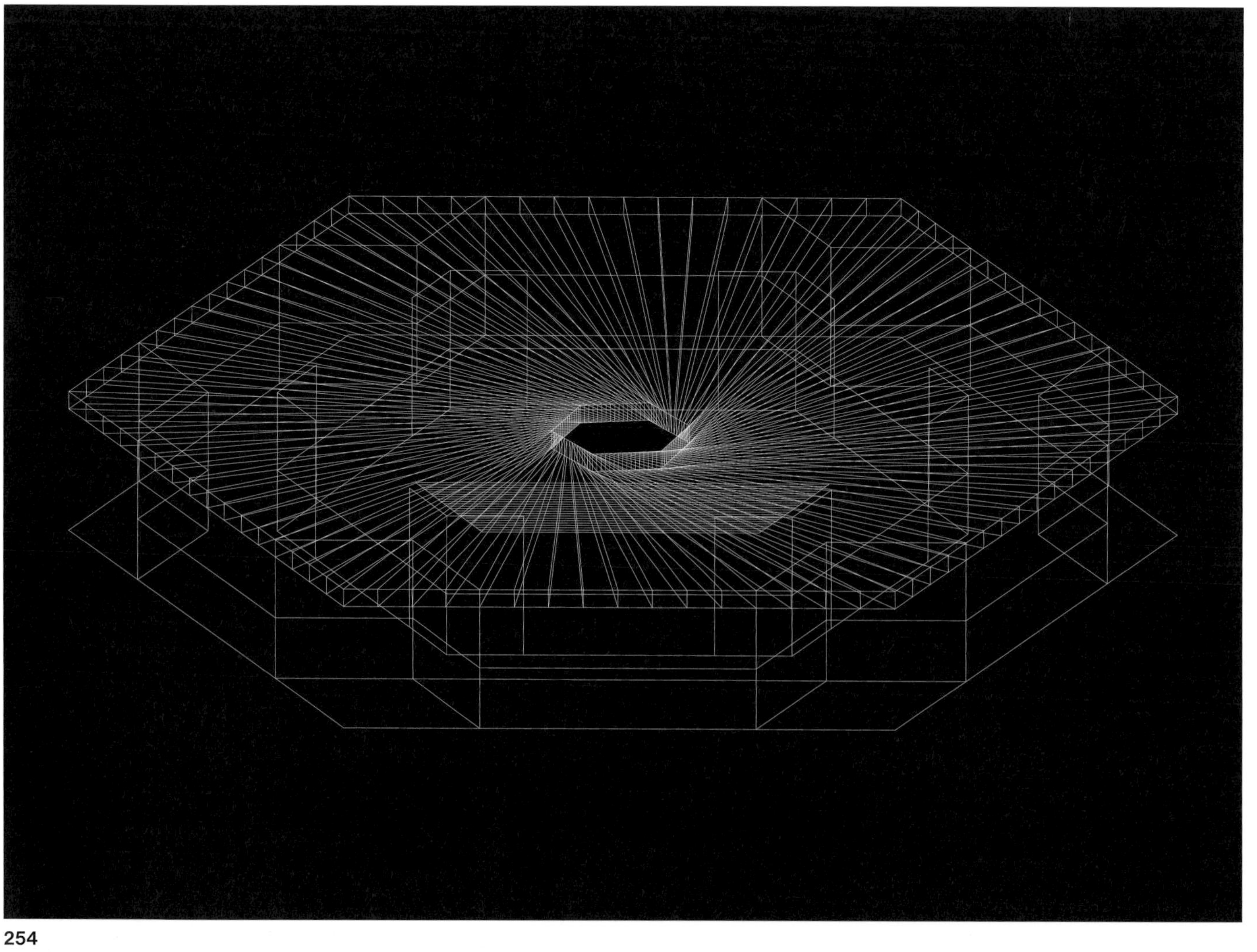

254

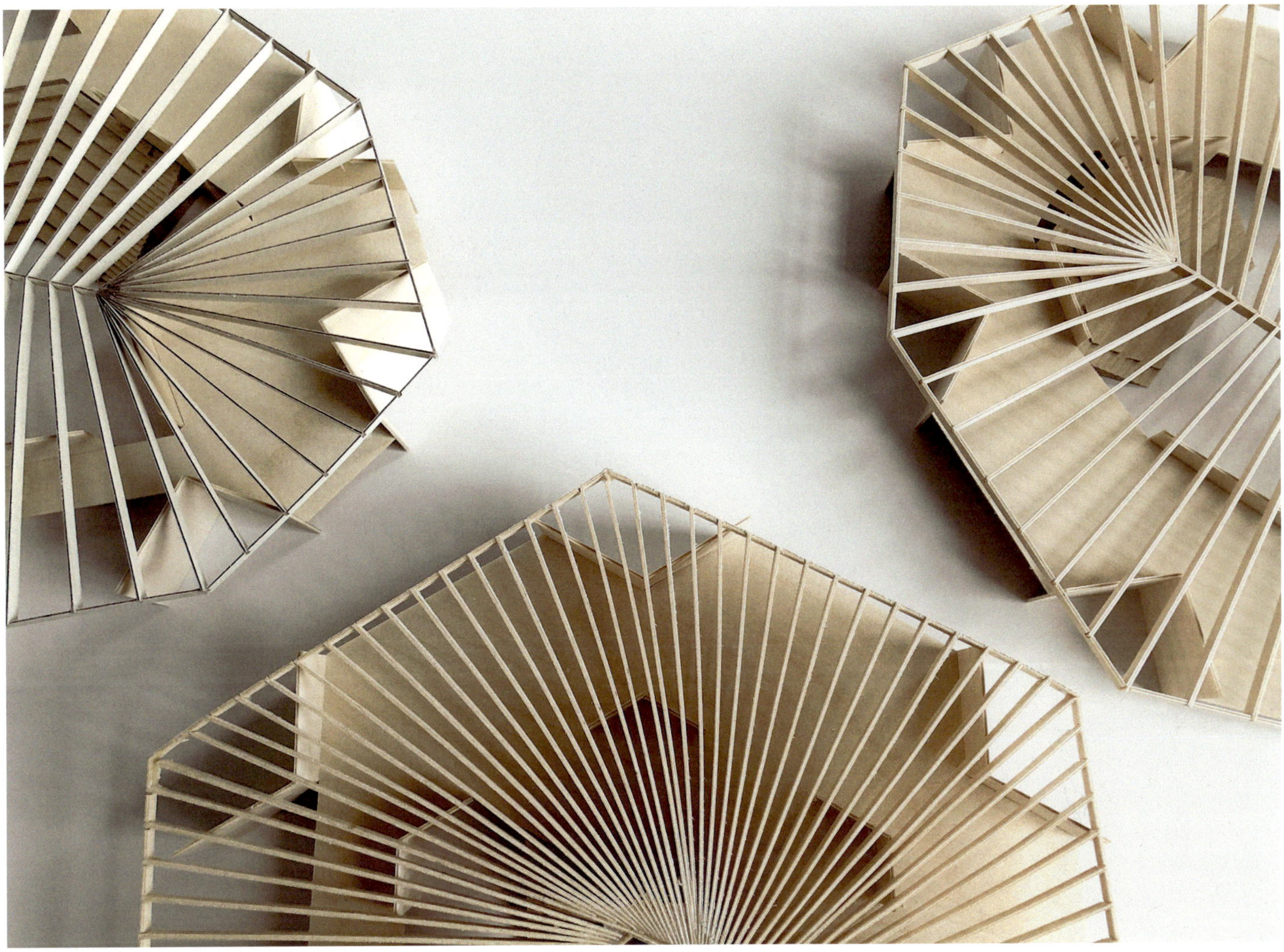

255

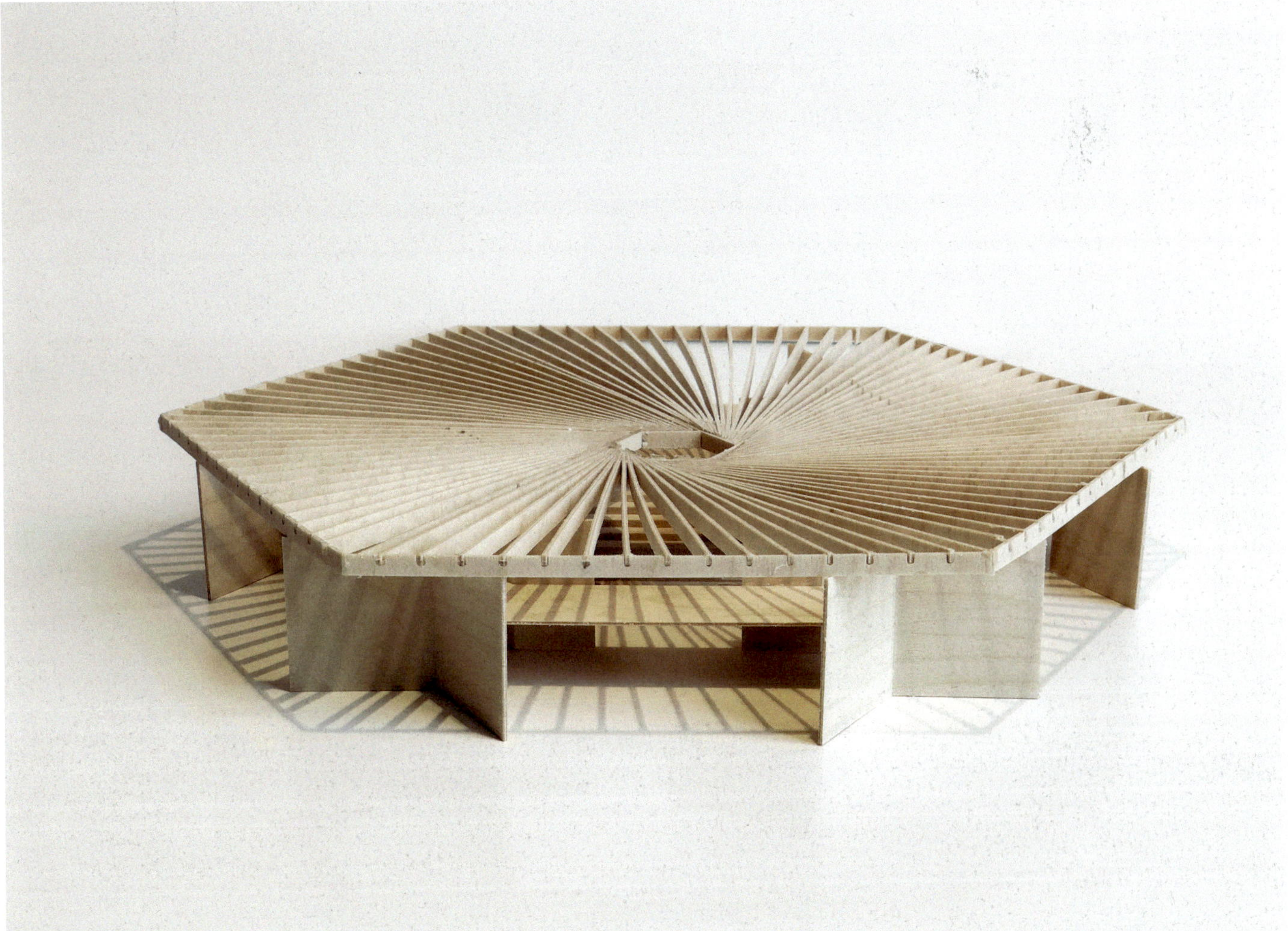

256

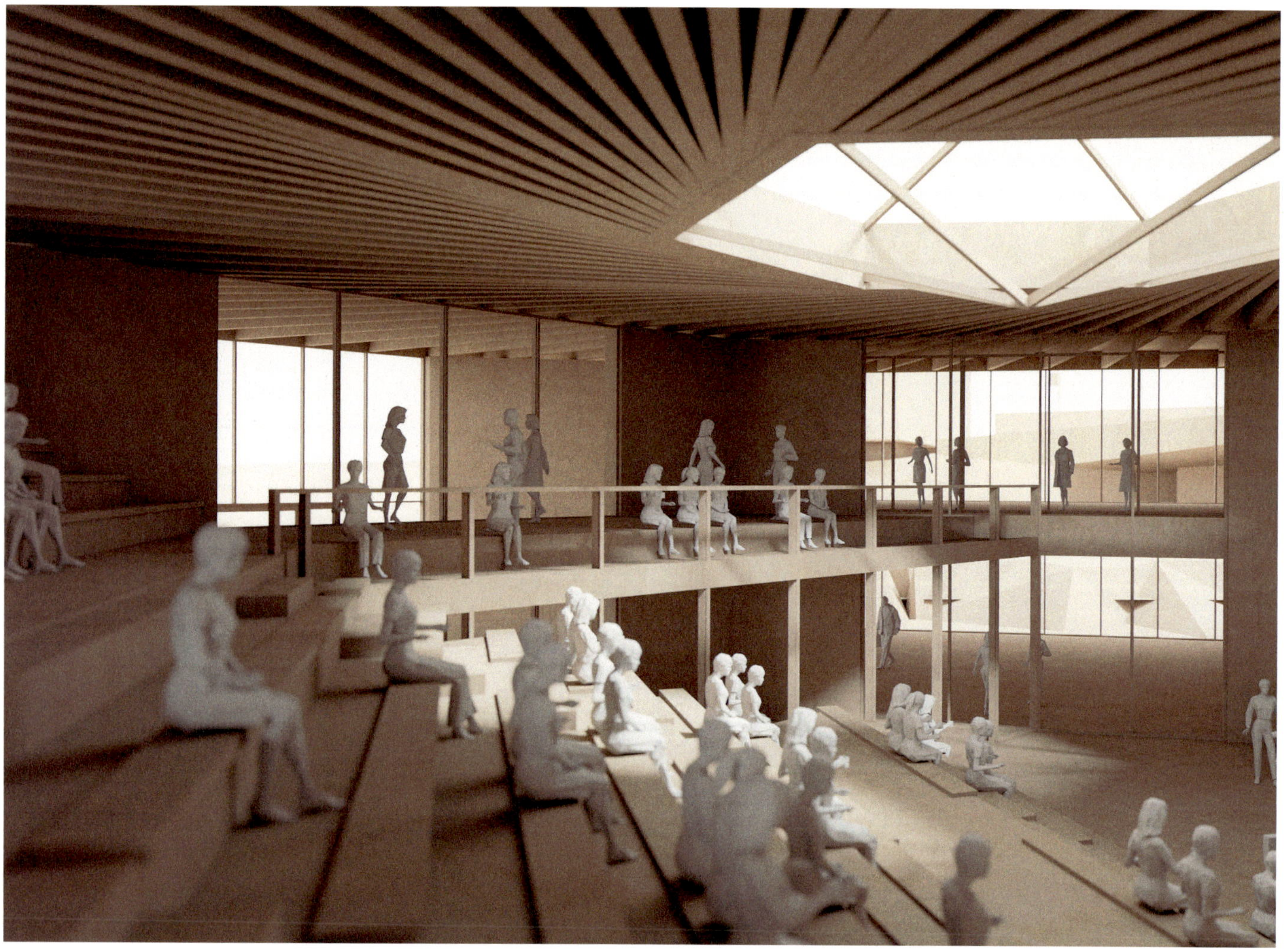
257

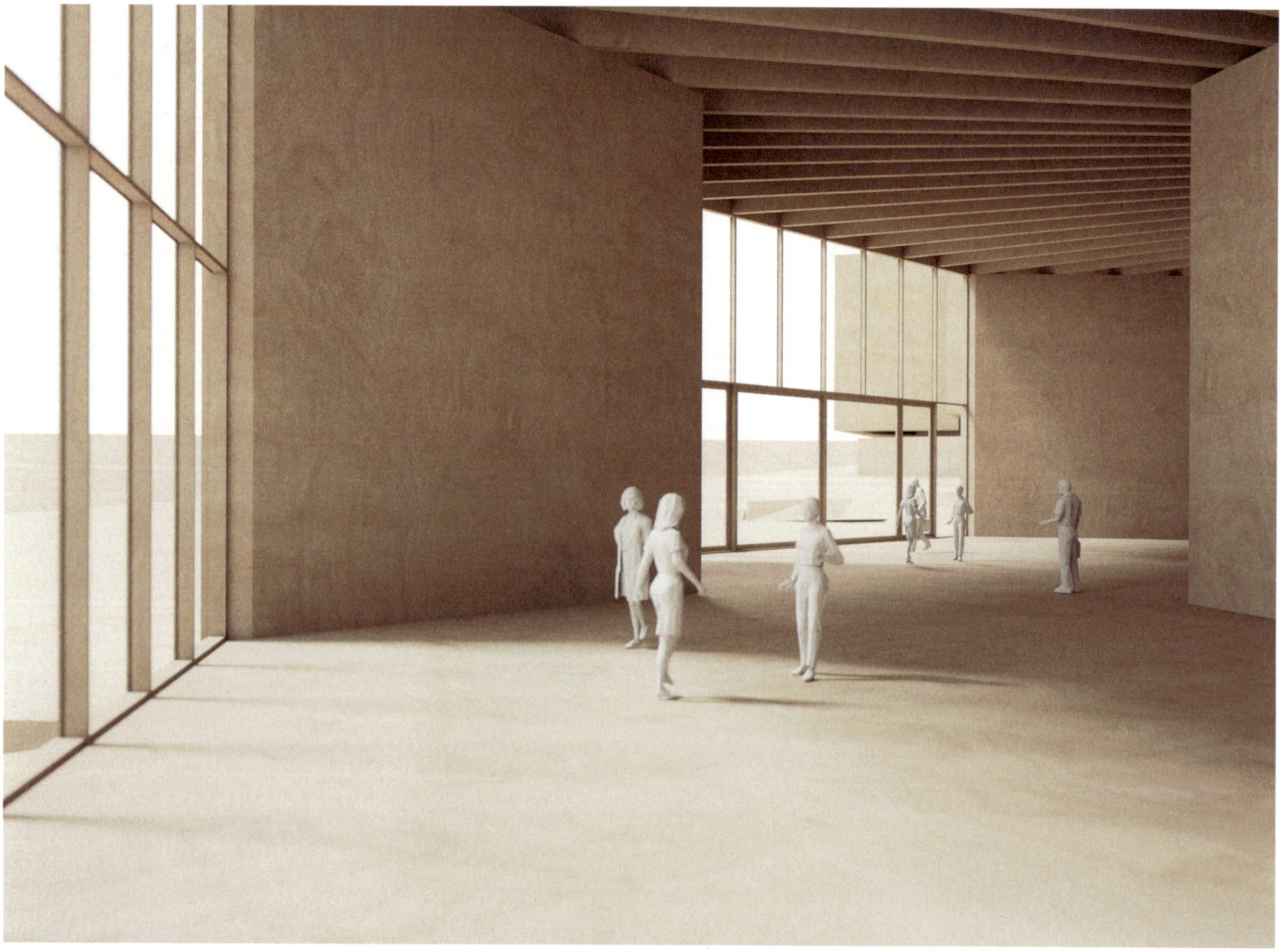
258

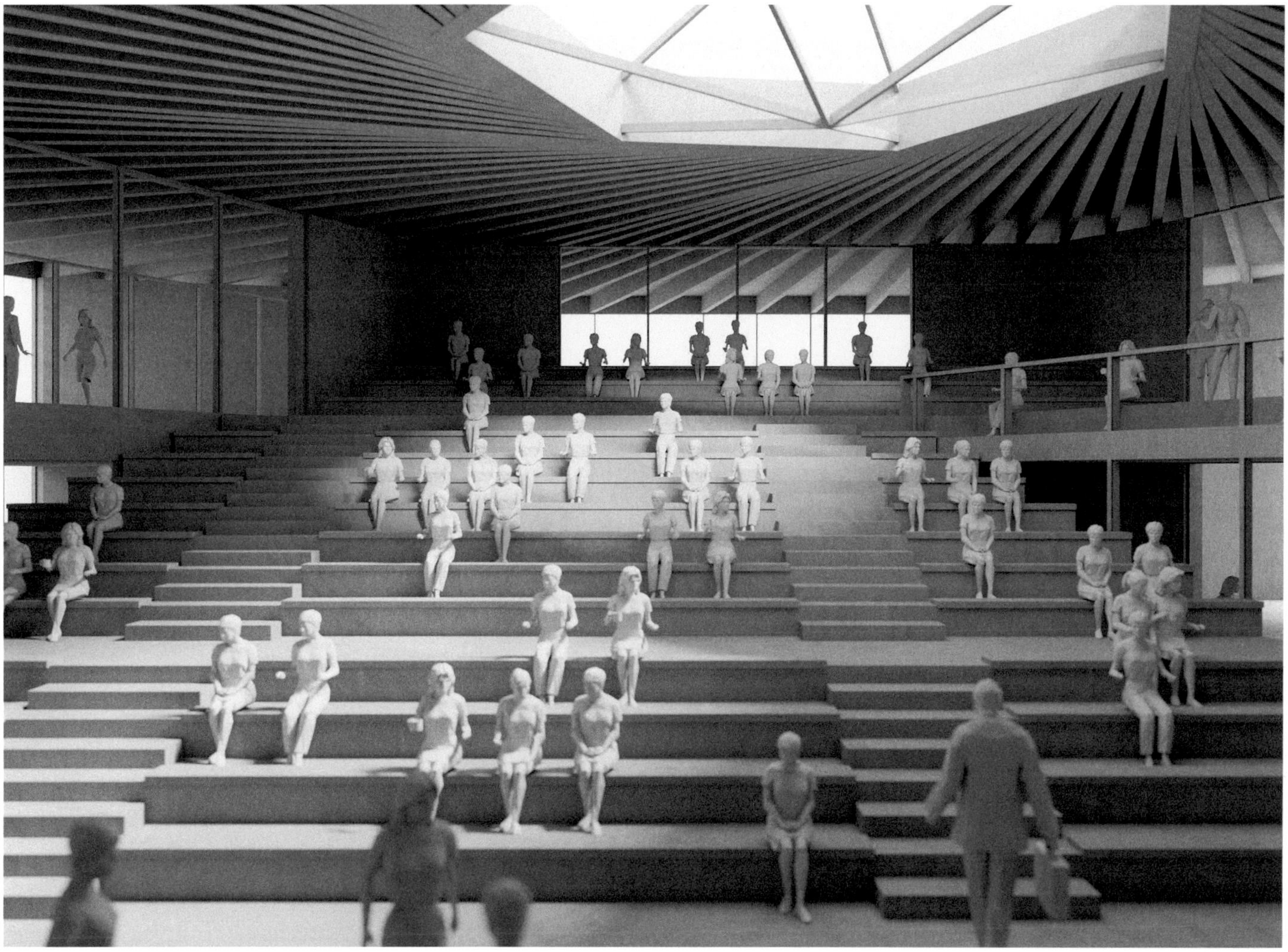

259

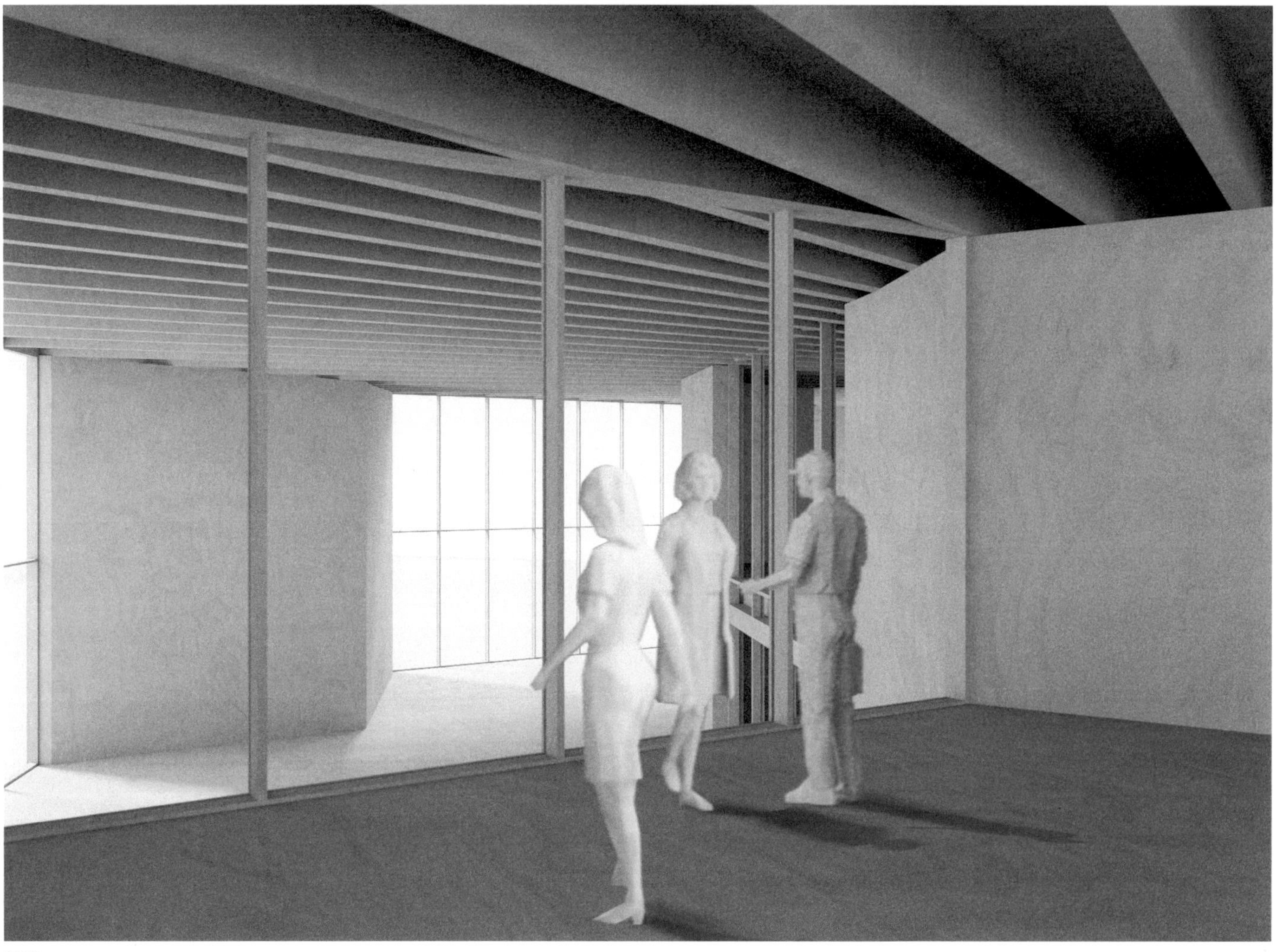

260

Frames

261

Frames are structural elements that refer to space frames, trusses, arches, and beams: they are structural elements that generate spatial, visual, and structural fields through their serial repetition. They may or may not be physically occupiable. Importantly, frames often refer to historical precedents.

262

Yet these legacies can also be technologically and materially updated. Frames can, moreover, be constructed from almost any material; they are versatile. Significantly, frames also have the ability to delimit views and orientations – to frame a view or to frame a landscape, for example.

Recalling the earlier work of the visionary architect and engineer Konrad Wachsmann and his Grapevine Structure from the 1950s, our pavilion for the Remstal Gartenschau was an interpretation of this unrealized project.

The roof construction, designed in cooperation with Werner Sobek, rested on slim pillars and was made of steel rods, intertwined to form a space frame. The frame functioned as a shelter for hikers and bikers along a path that connects a series of architect-designed pavilions.

Through the construction of the Harvard University ArtLab, we created urgently needed space on campus for the arts. The lightweight building structure consists of steel columns and prefabricated steel trusses, clad with transclucent polycarbonate partially glazed panels.

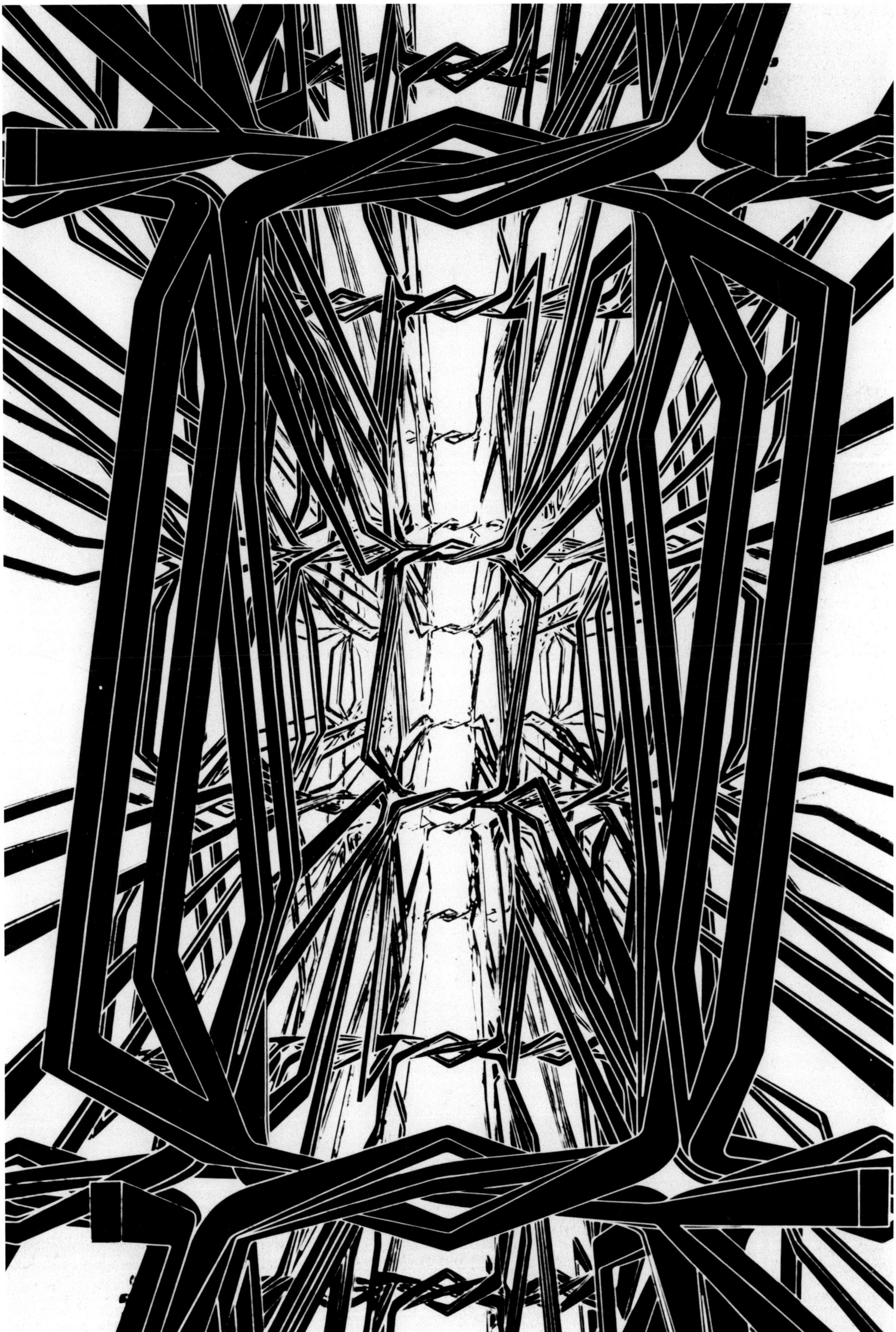

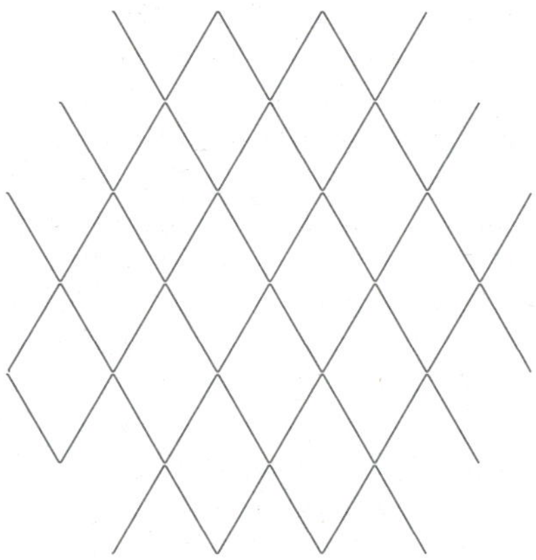

268

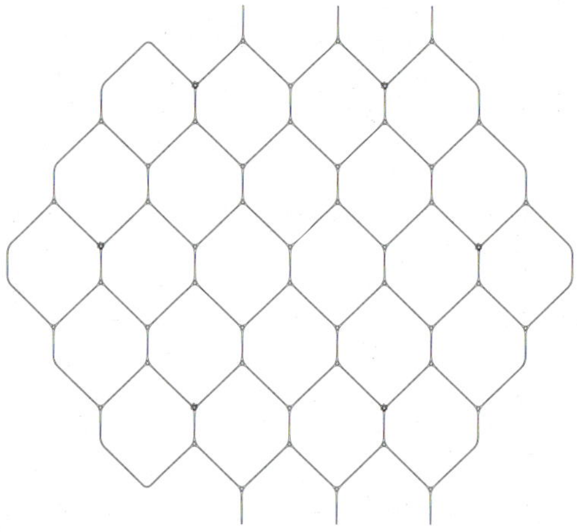

269

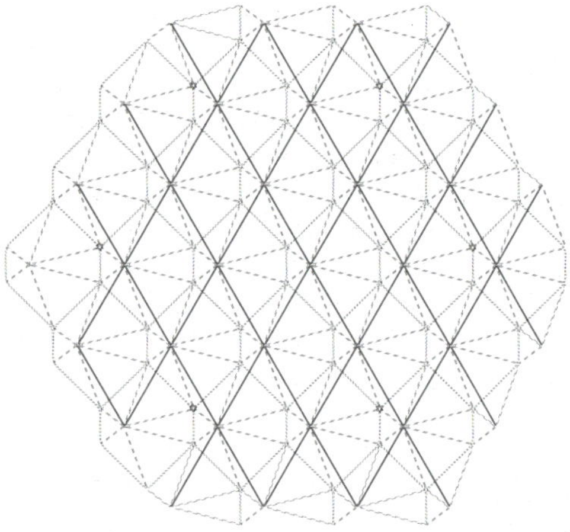

270

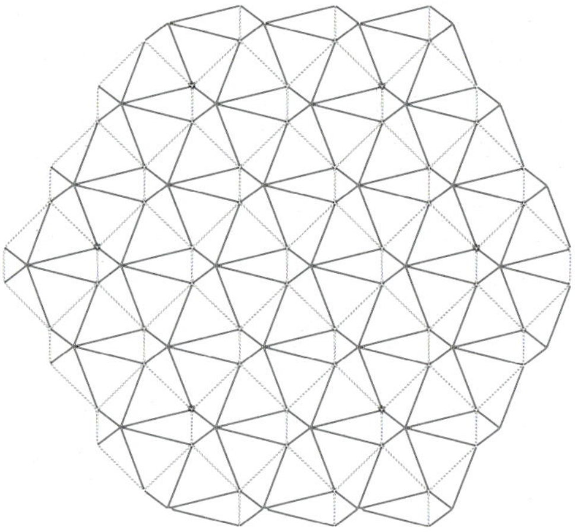

271

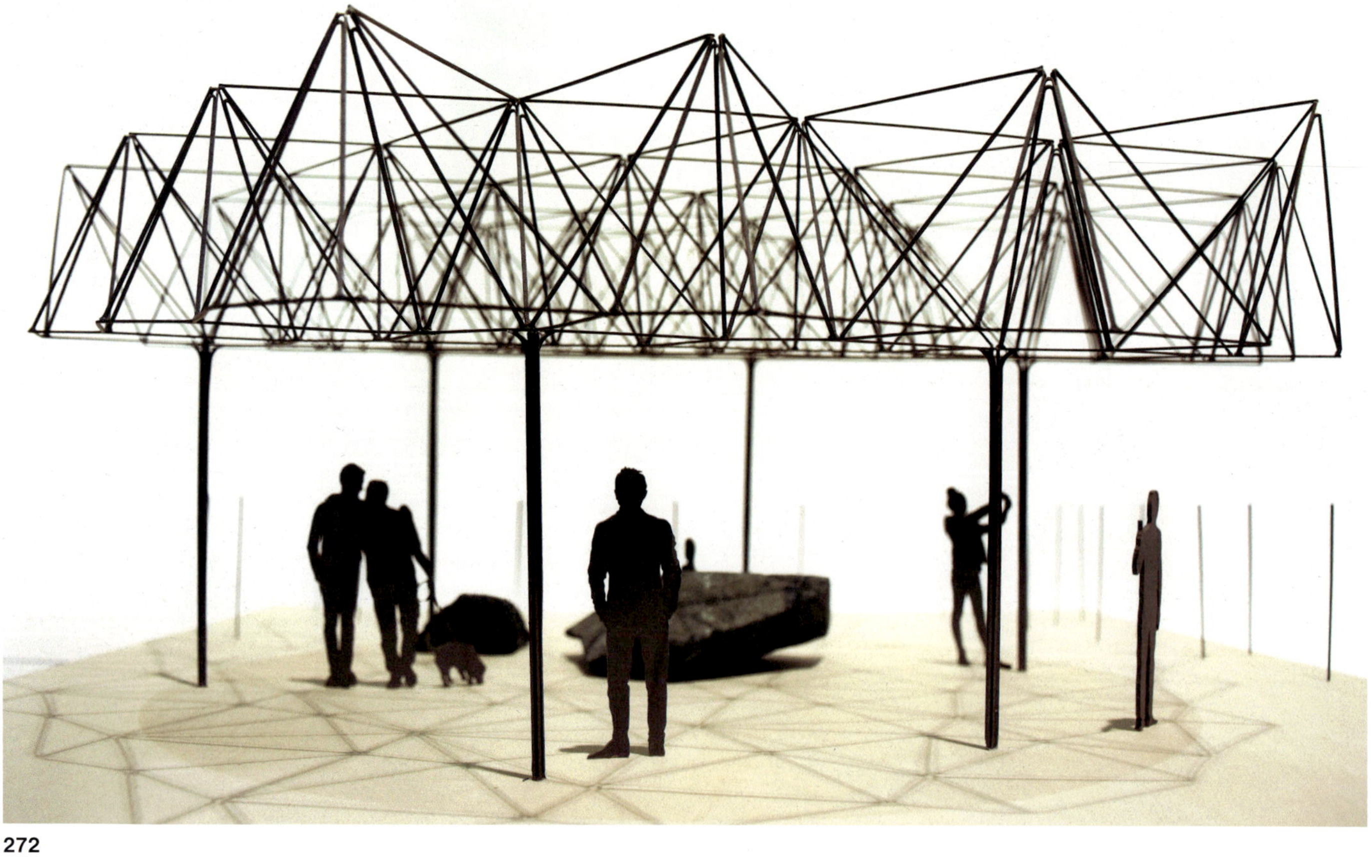

272

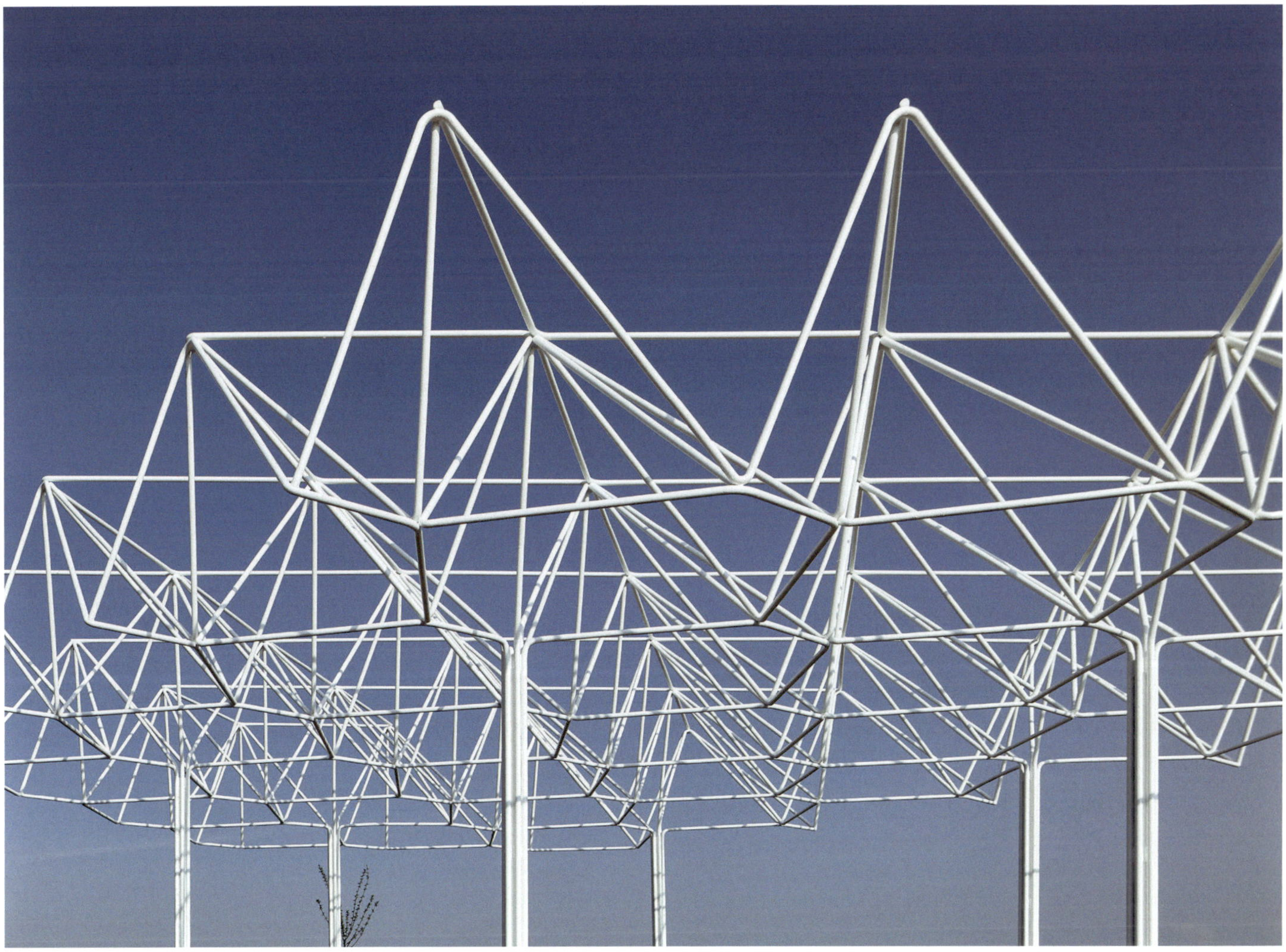

273

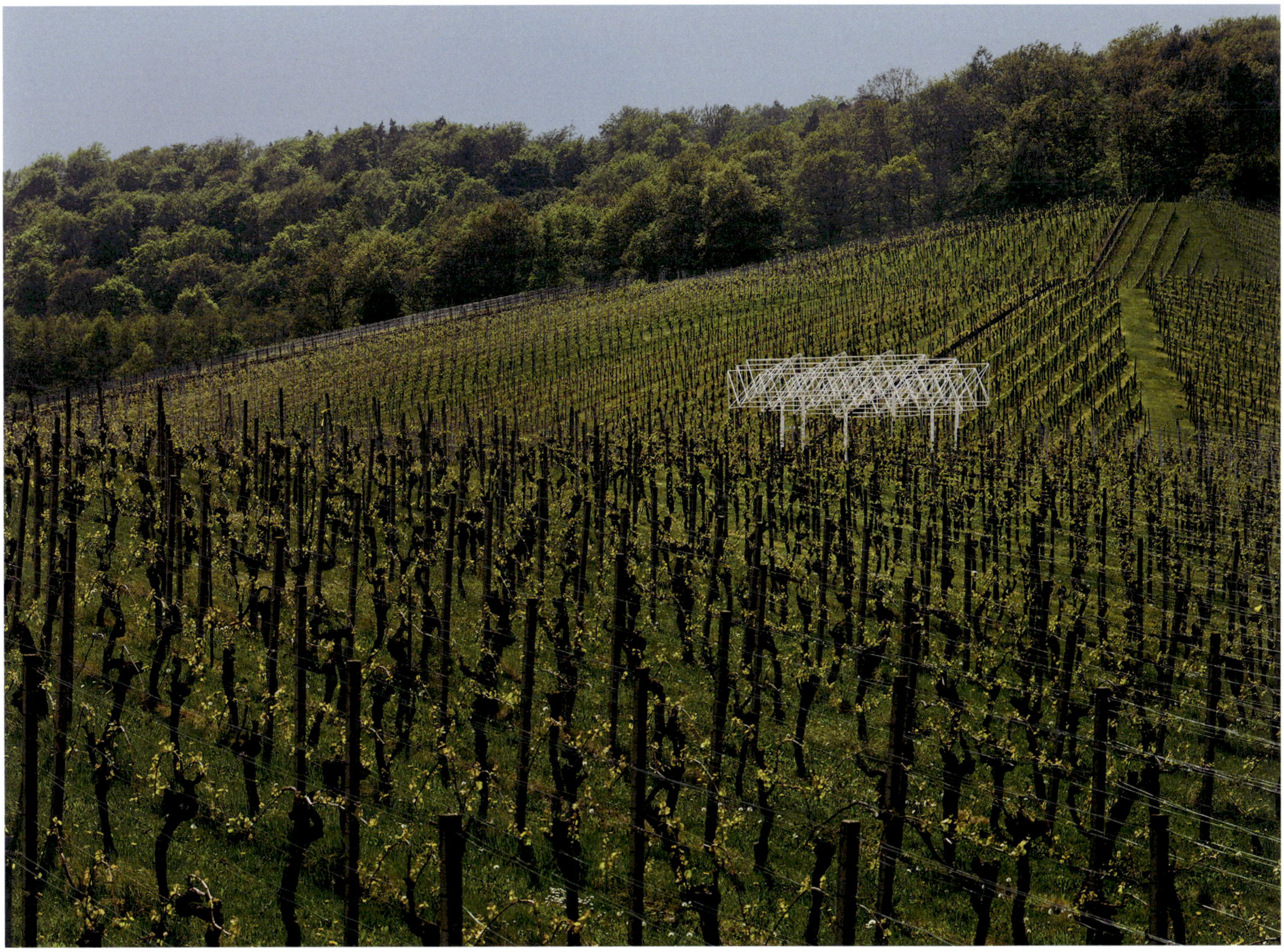

274

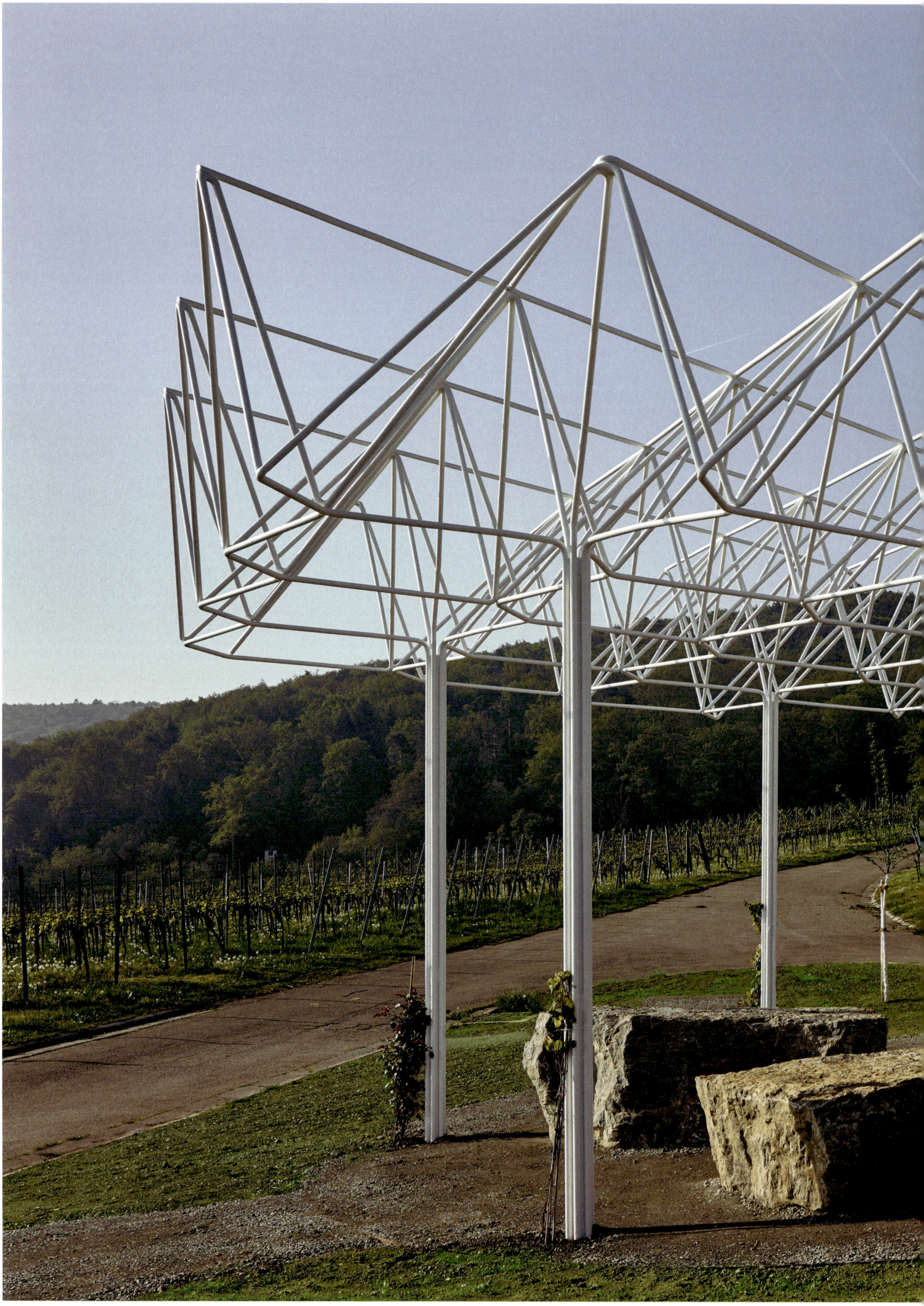

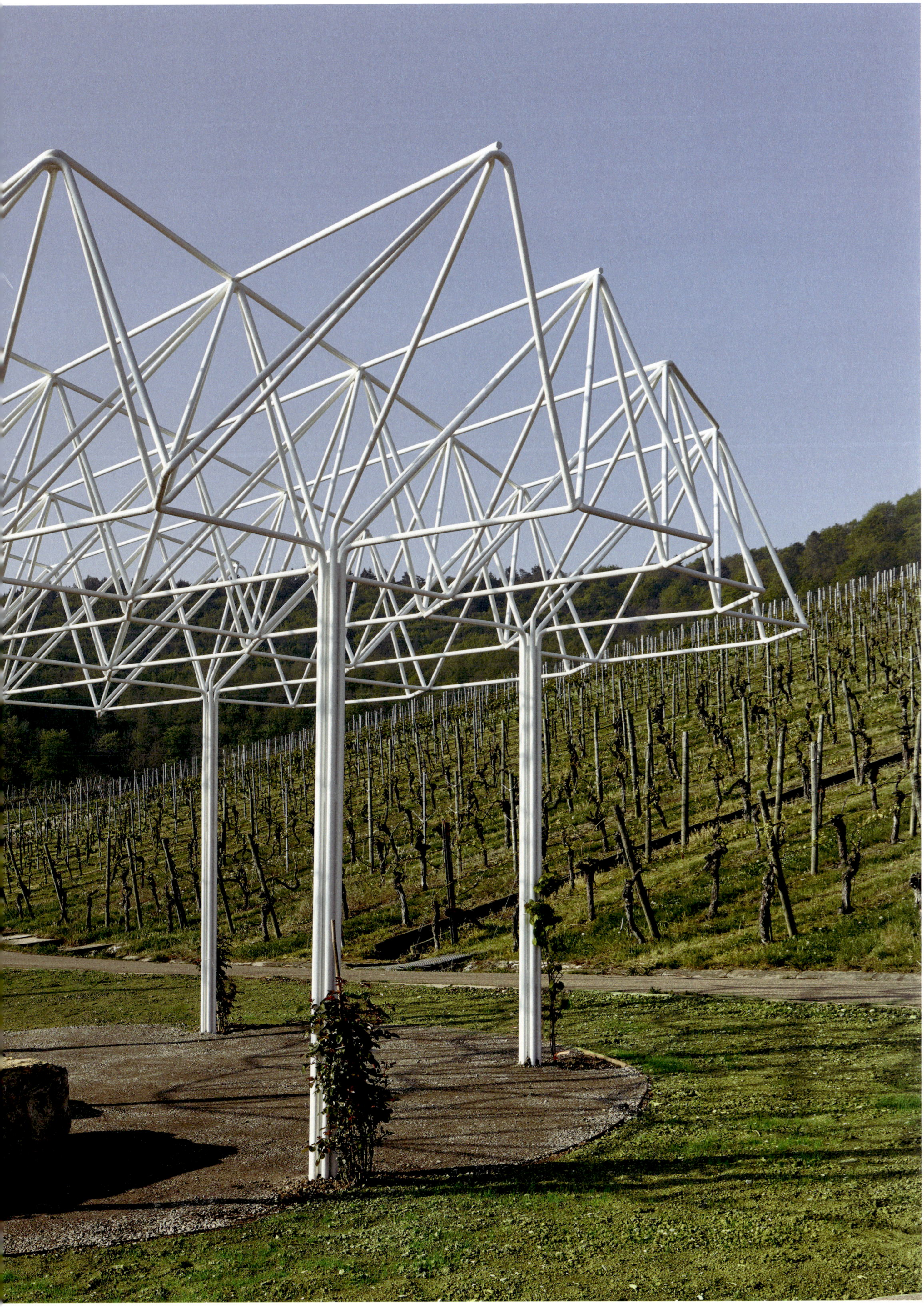

Support program (studios, shop, film, and office) pinwheel around a central double-height hub space for performances and exhibitions. The building glows like a lantern from across the campus. The building is temporary and can later be relocated.

Its frames are structurally stable and can be demounted and moved piece by piece. The value of this transient frame lies both in its reuse value (as a sustainable and useful architectural paradigm) and in its allowance for programmatically dynamic and aesthetically adaptable interior space.

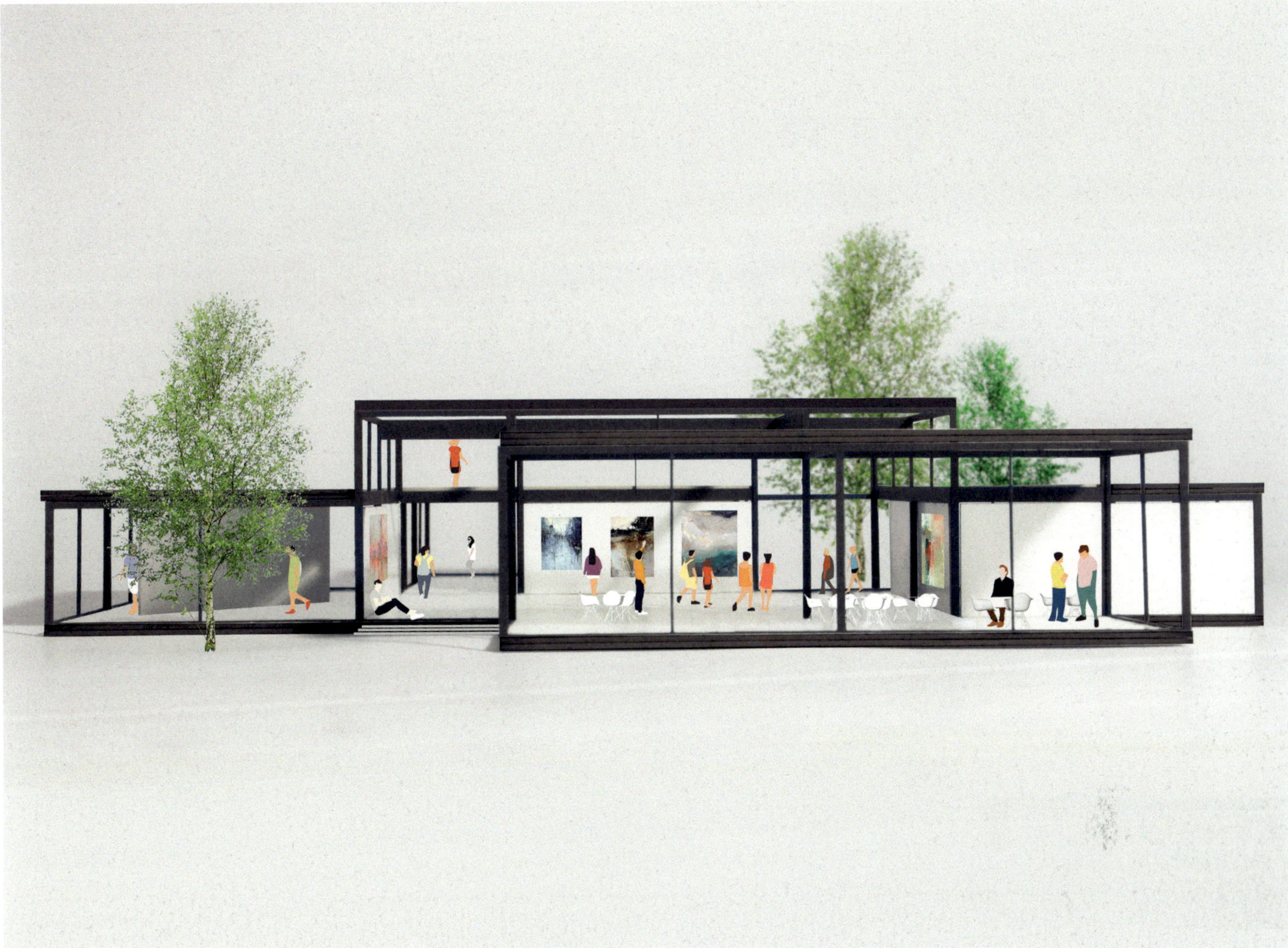

278

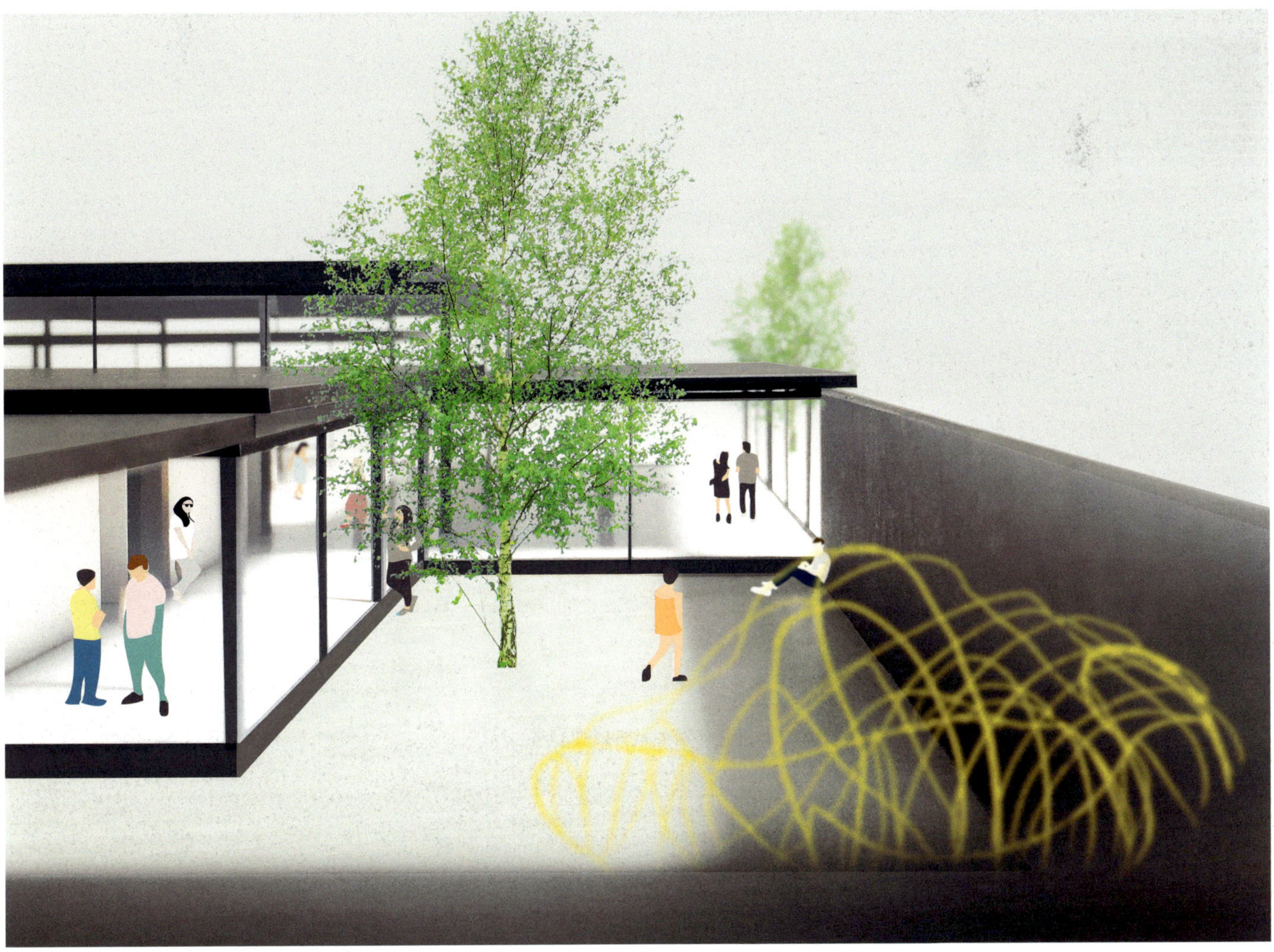

279

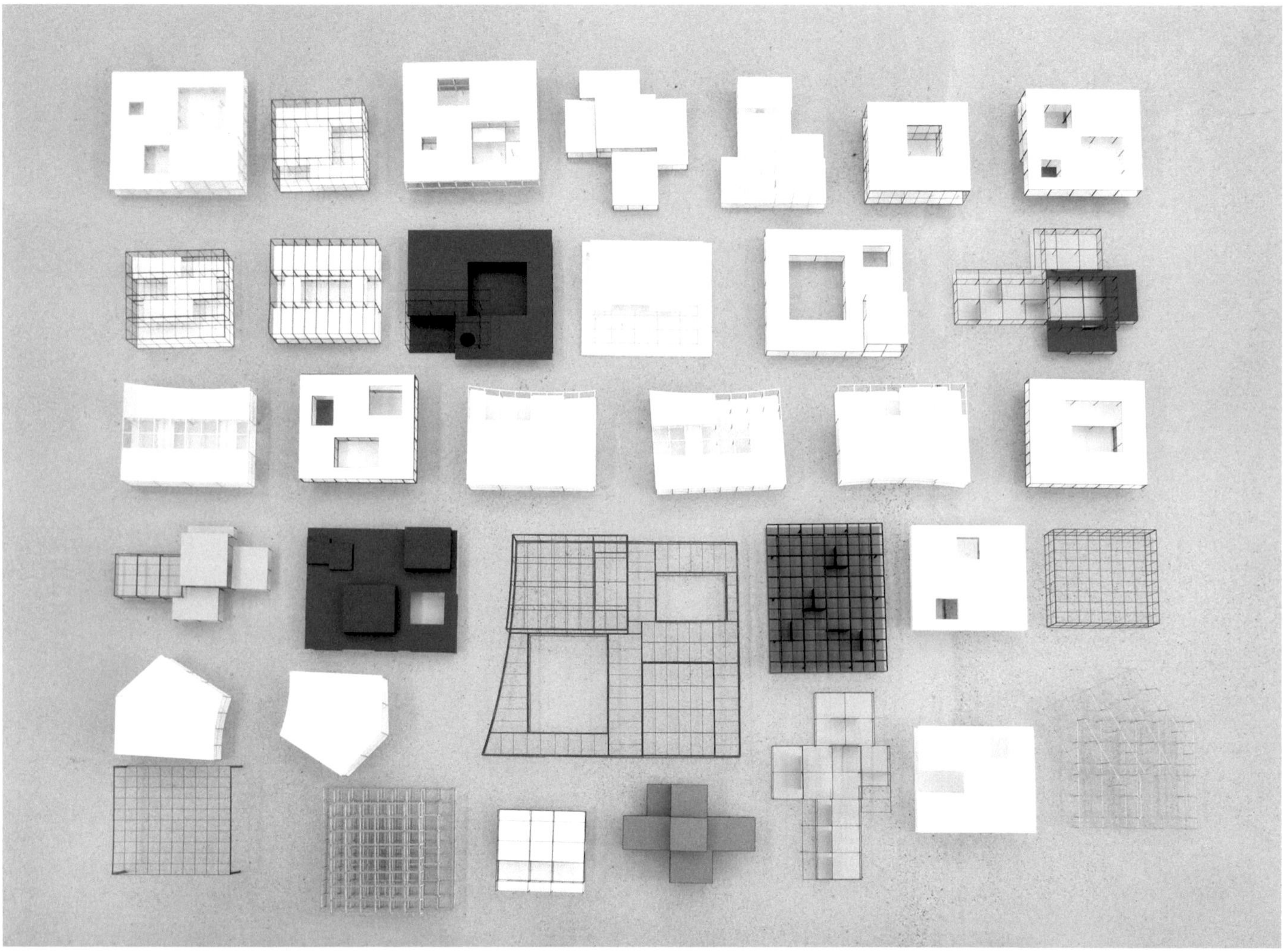

280

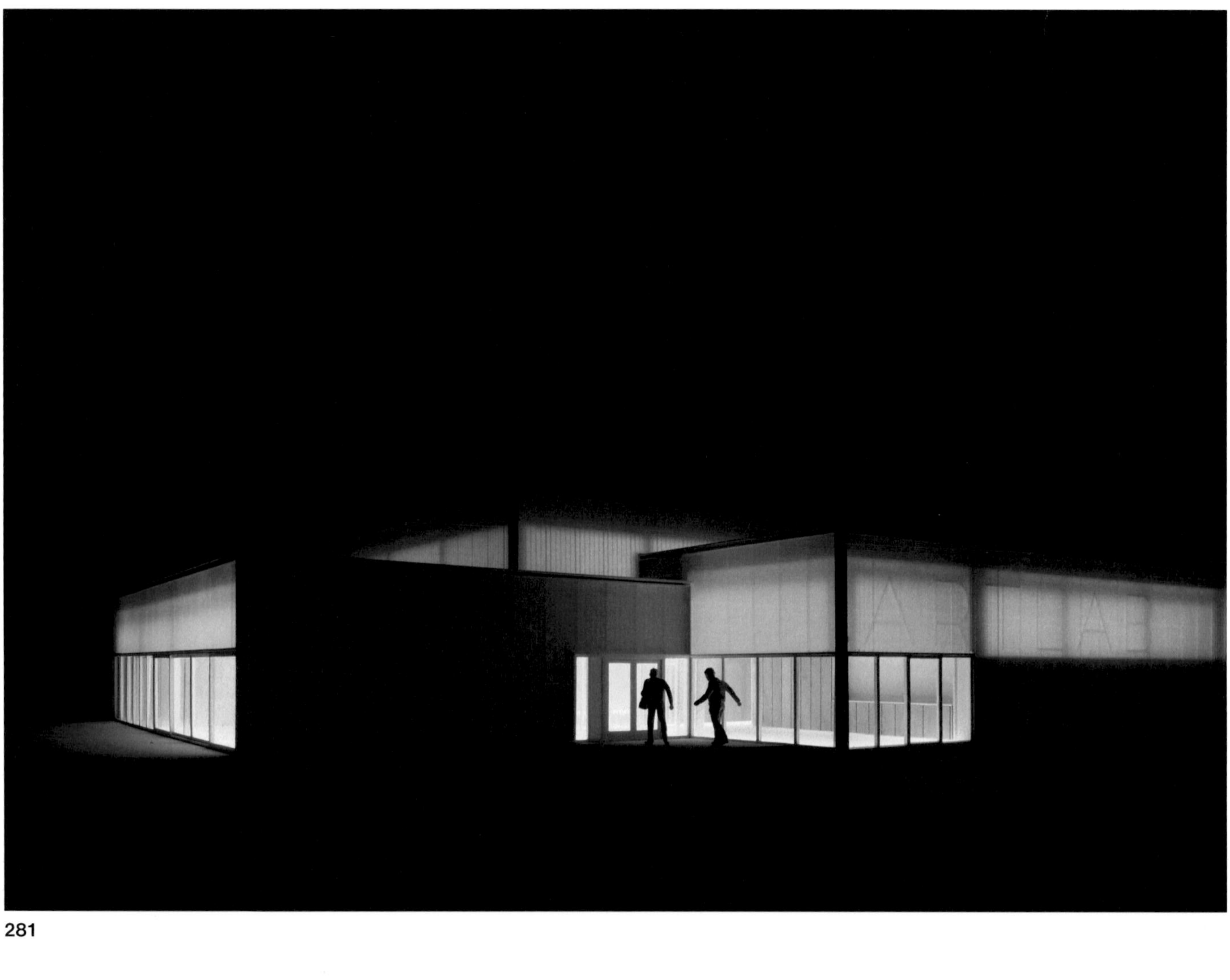

281

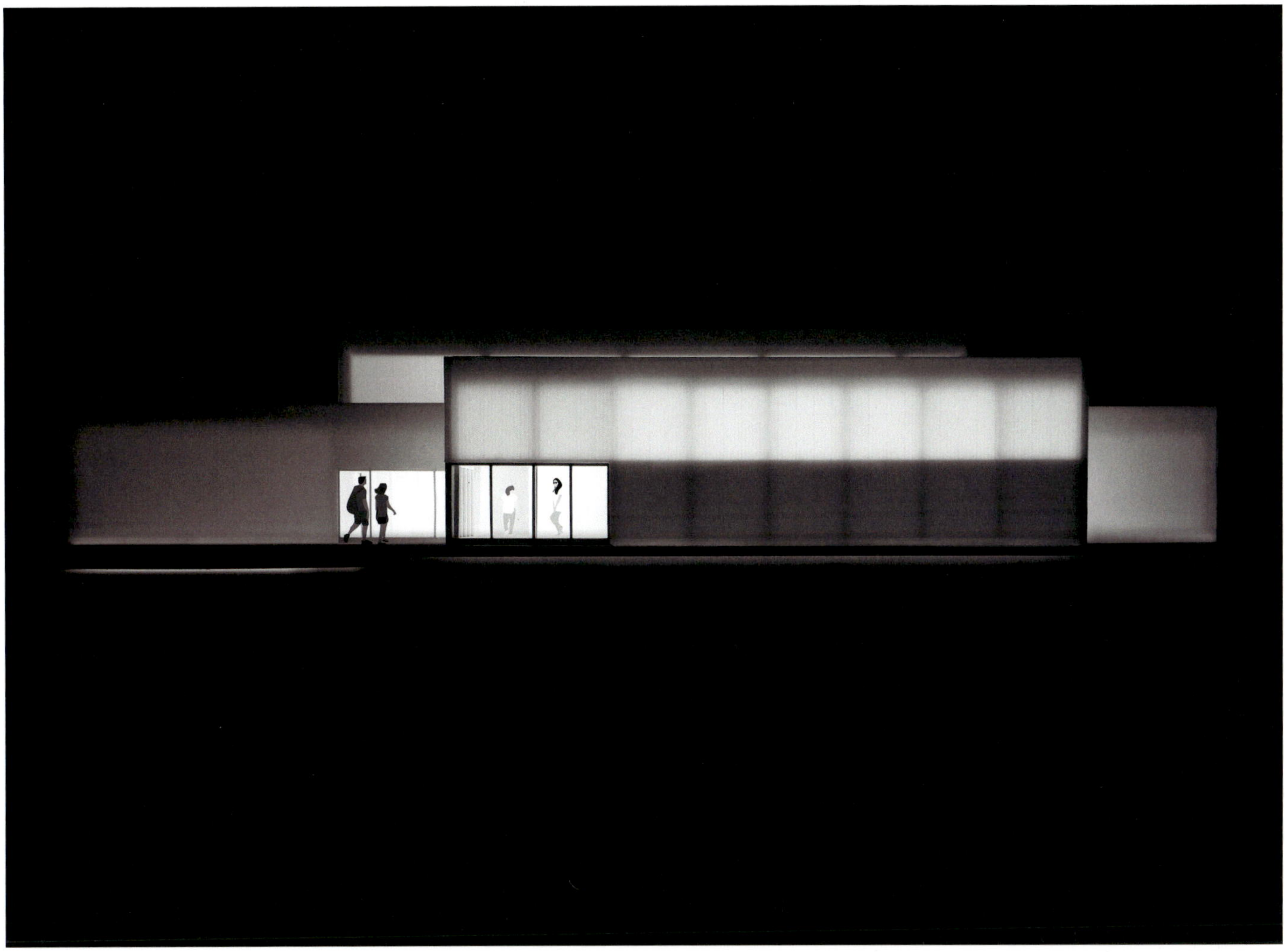

282

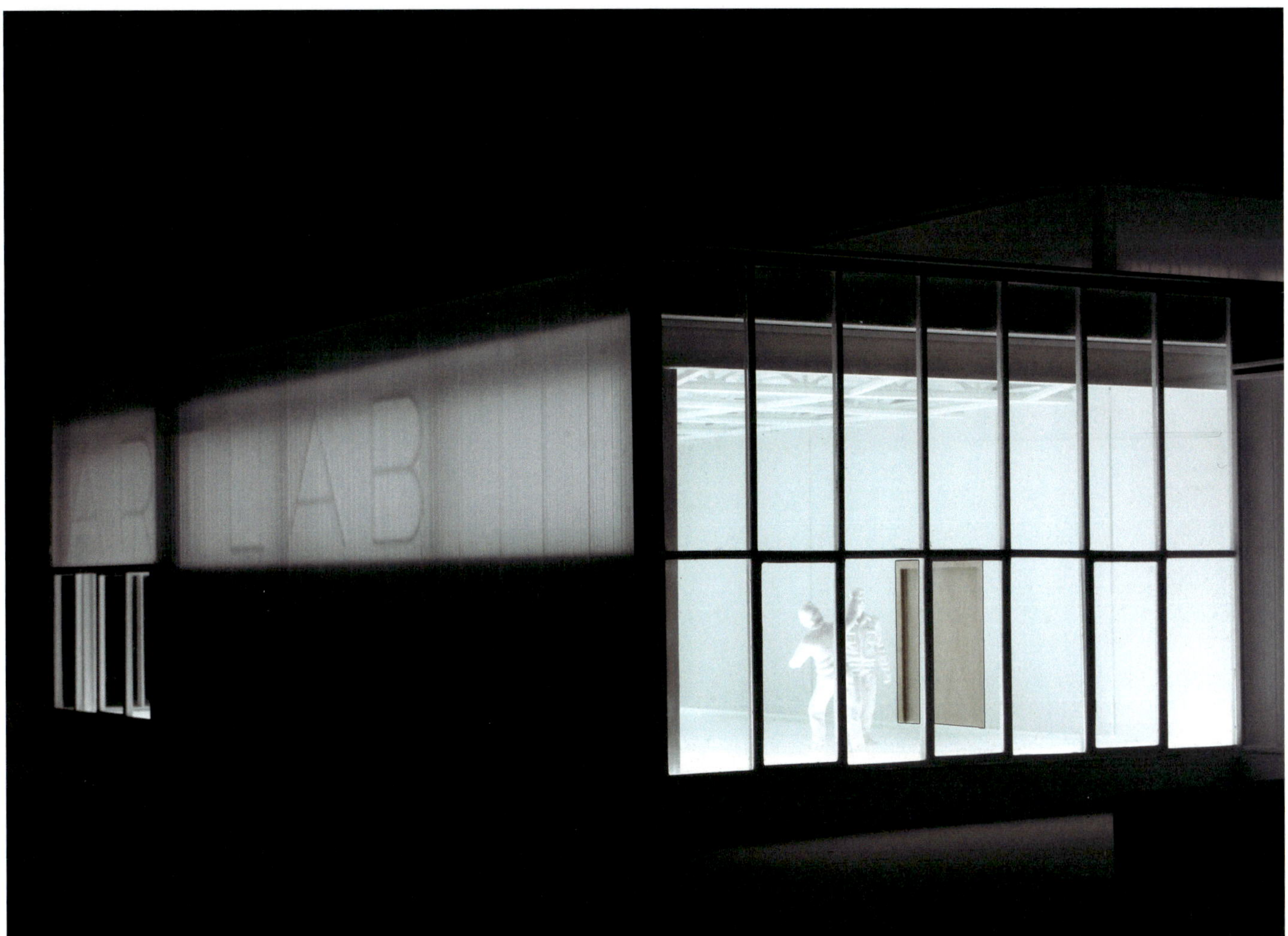

283

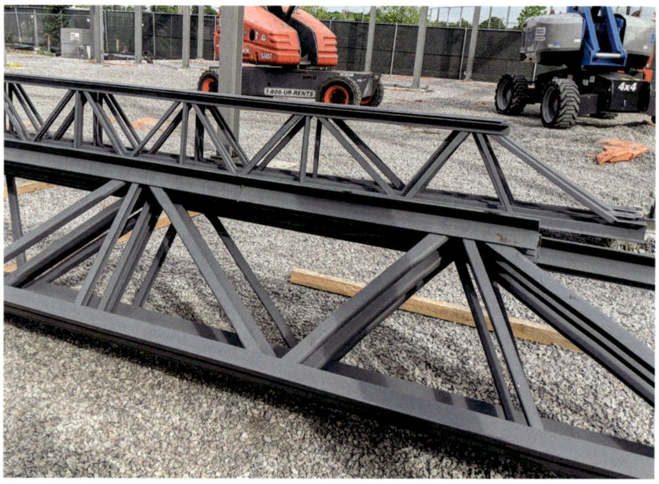

284

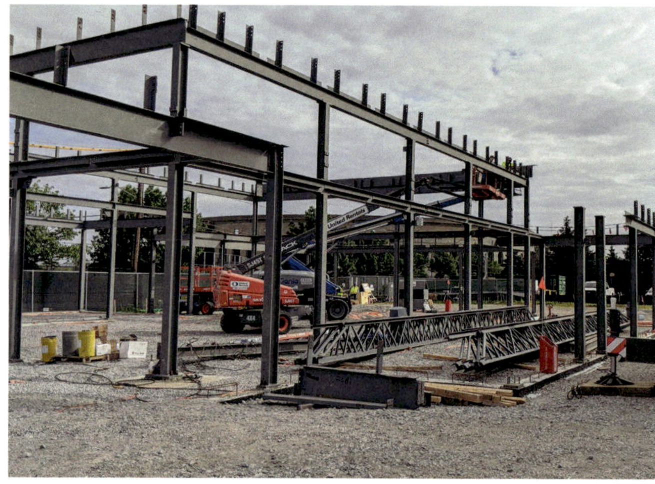

285

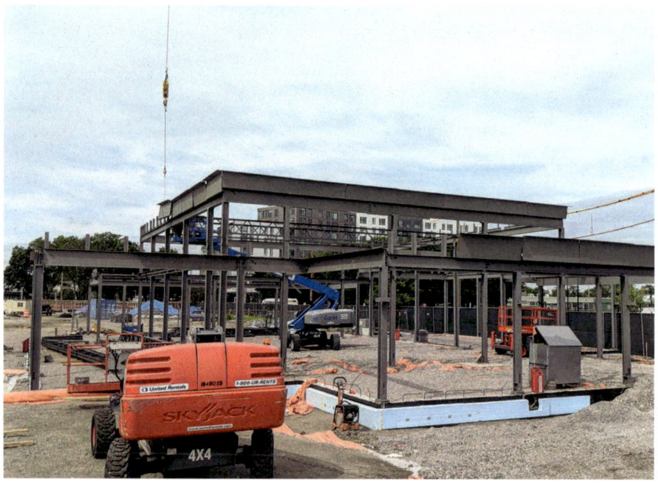

286

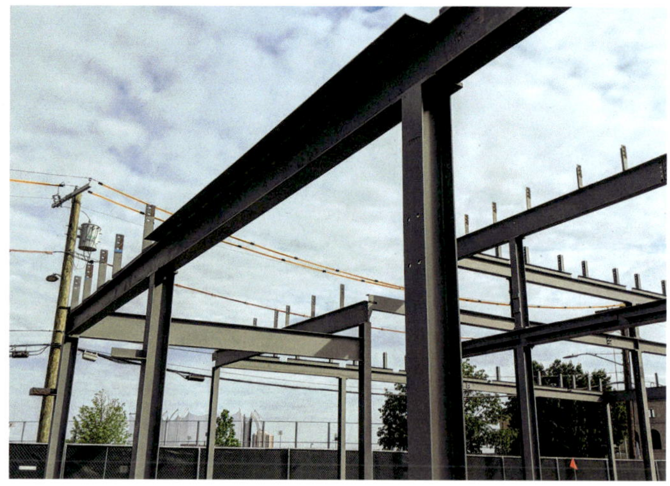

287

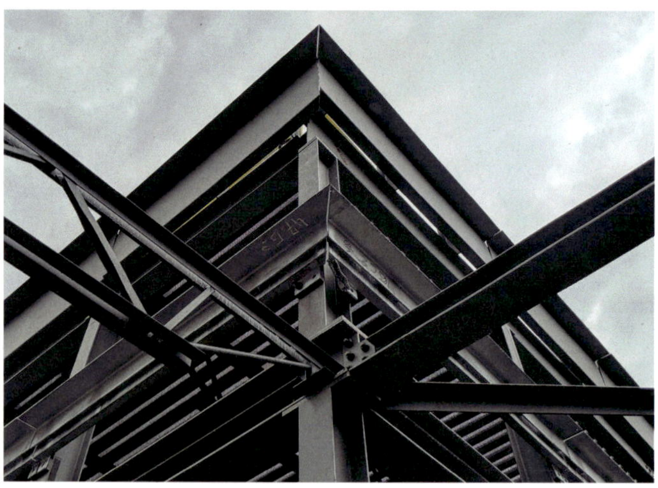

288

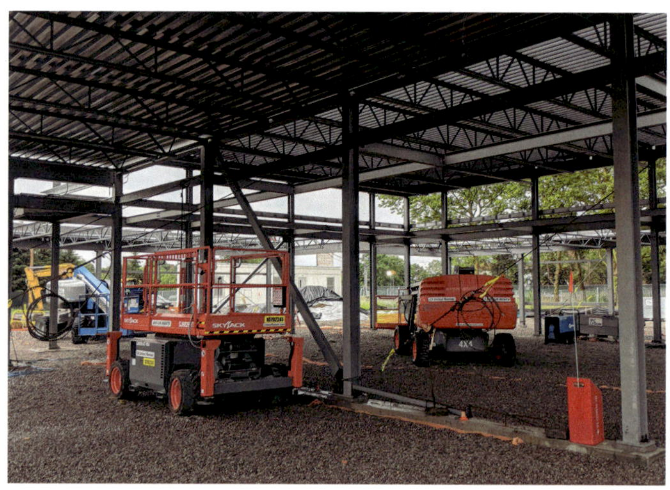

289

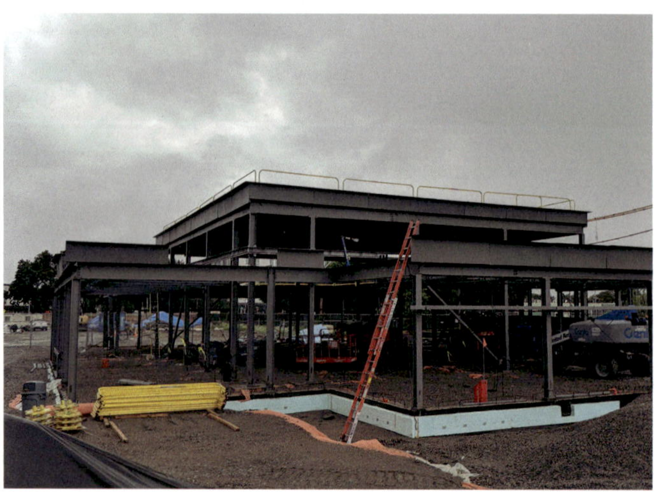

290

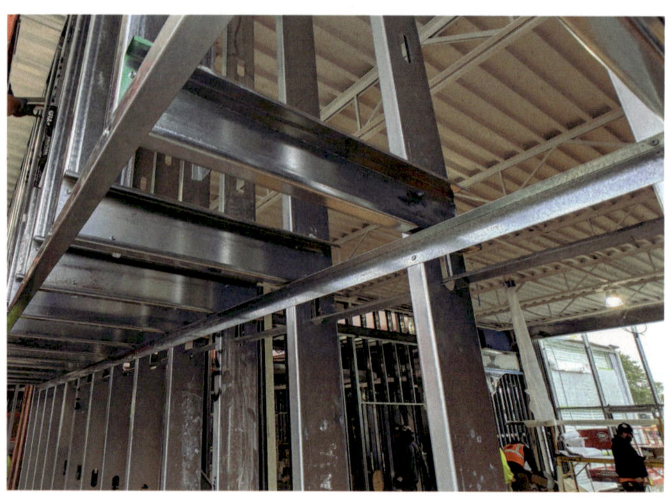

291

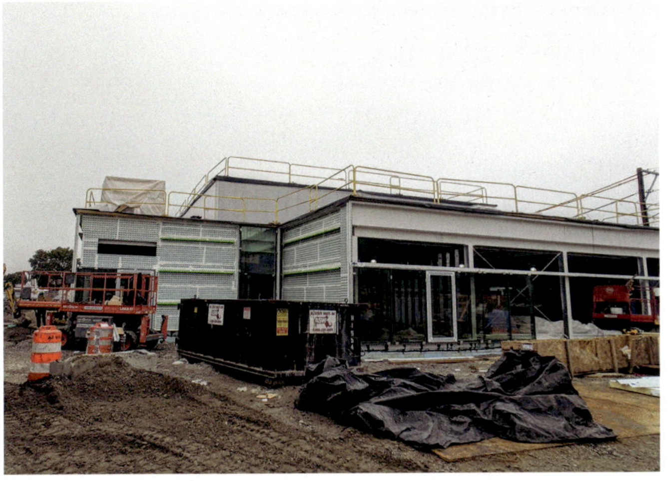

292

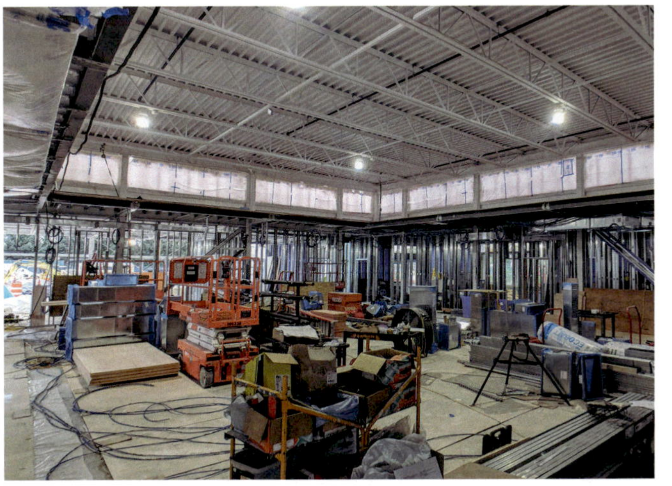

293

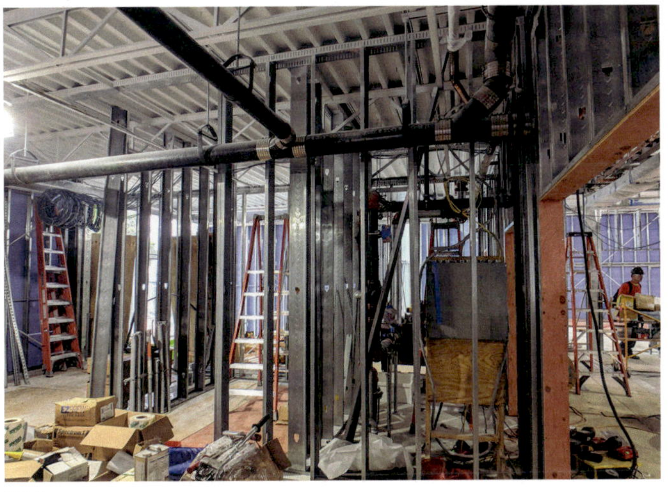

294

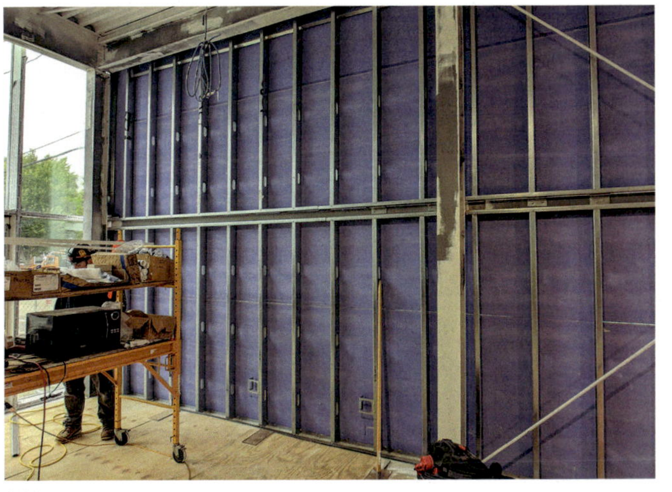

295

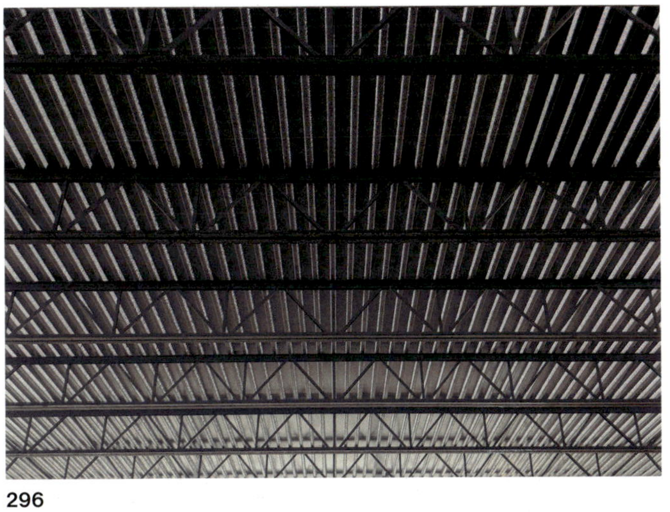

296

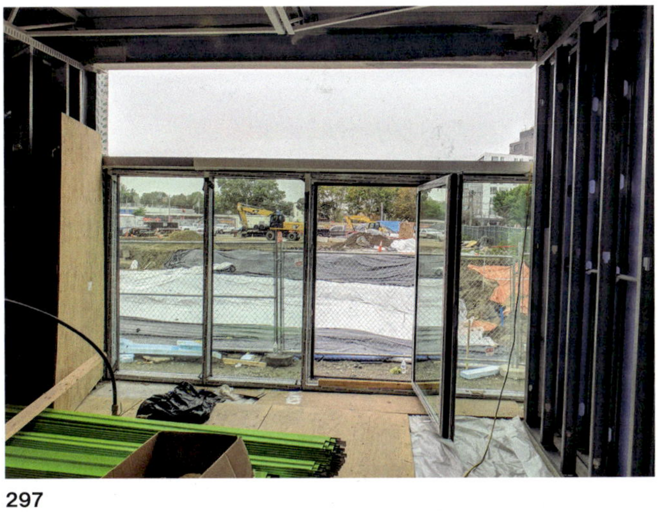

297

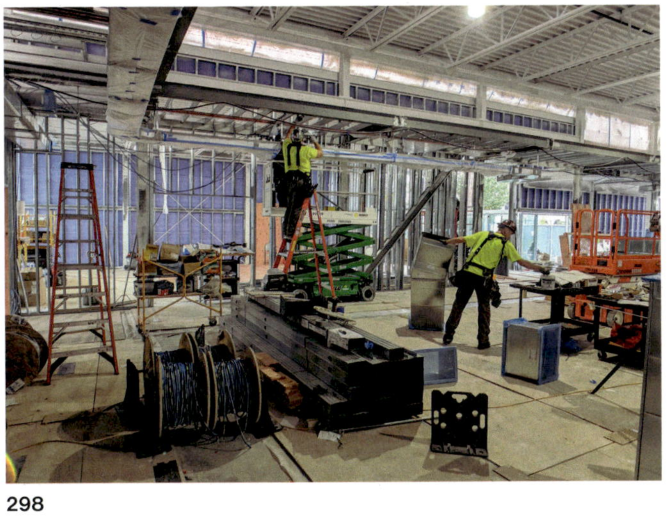

298

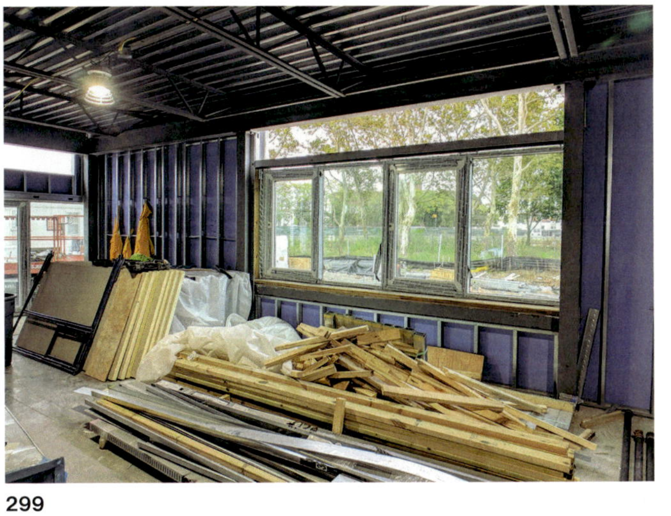

299

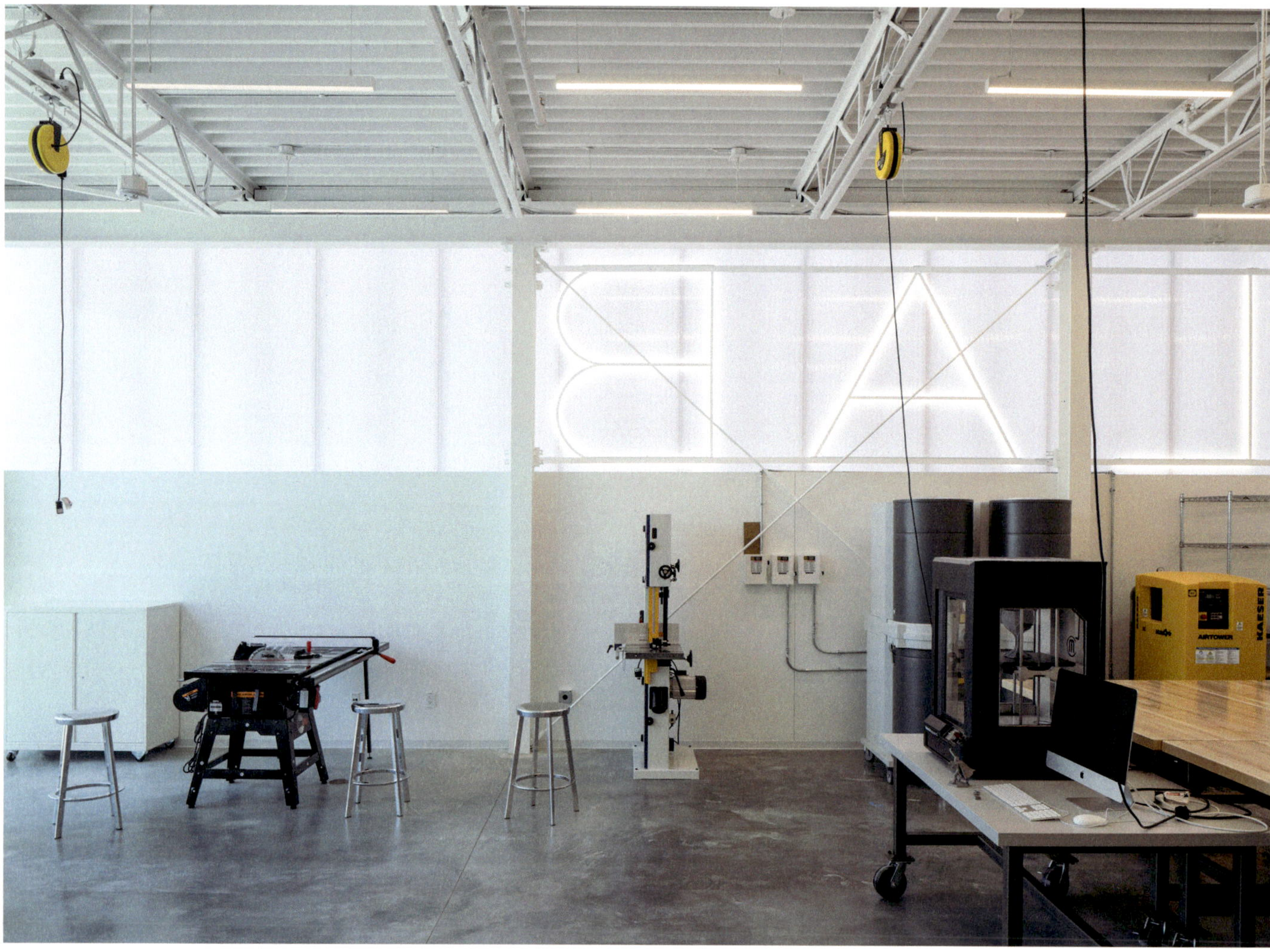

300

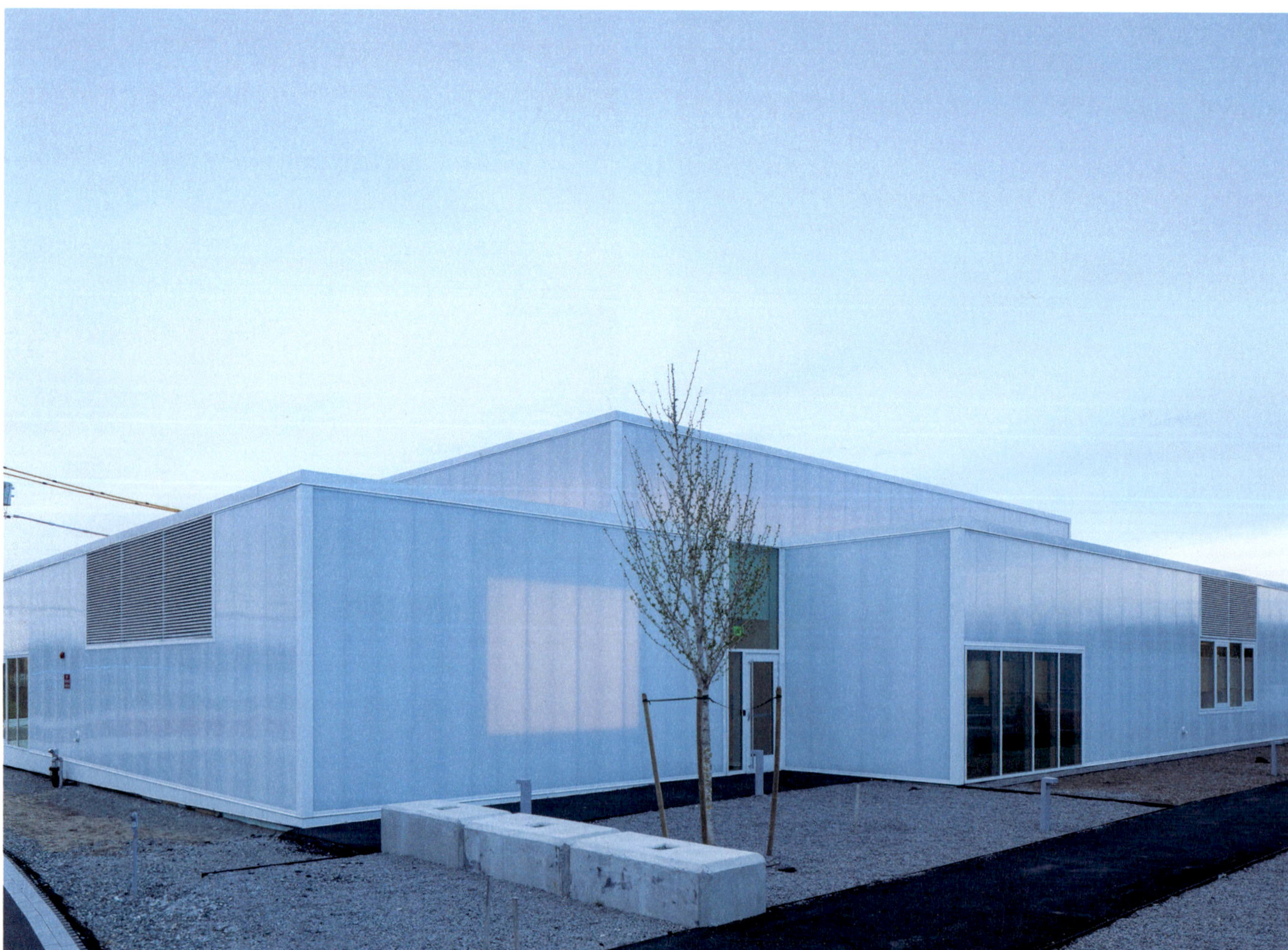

301

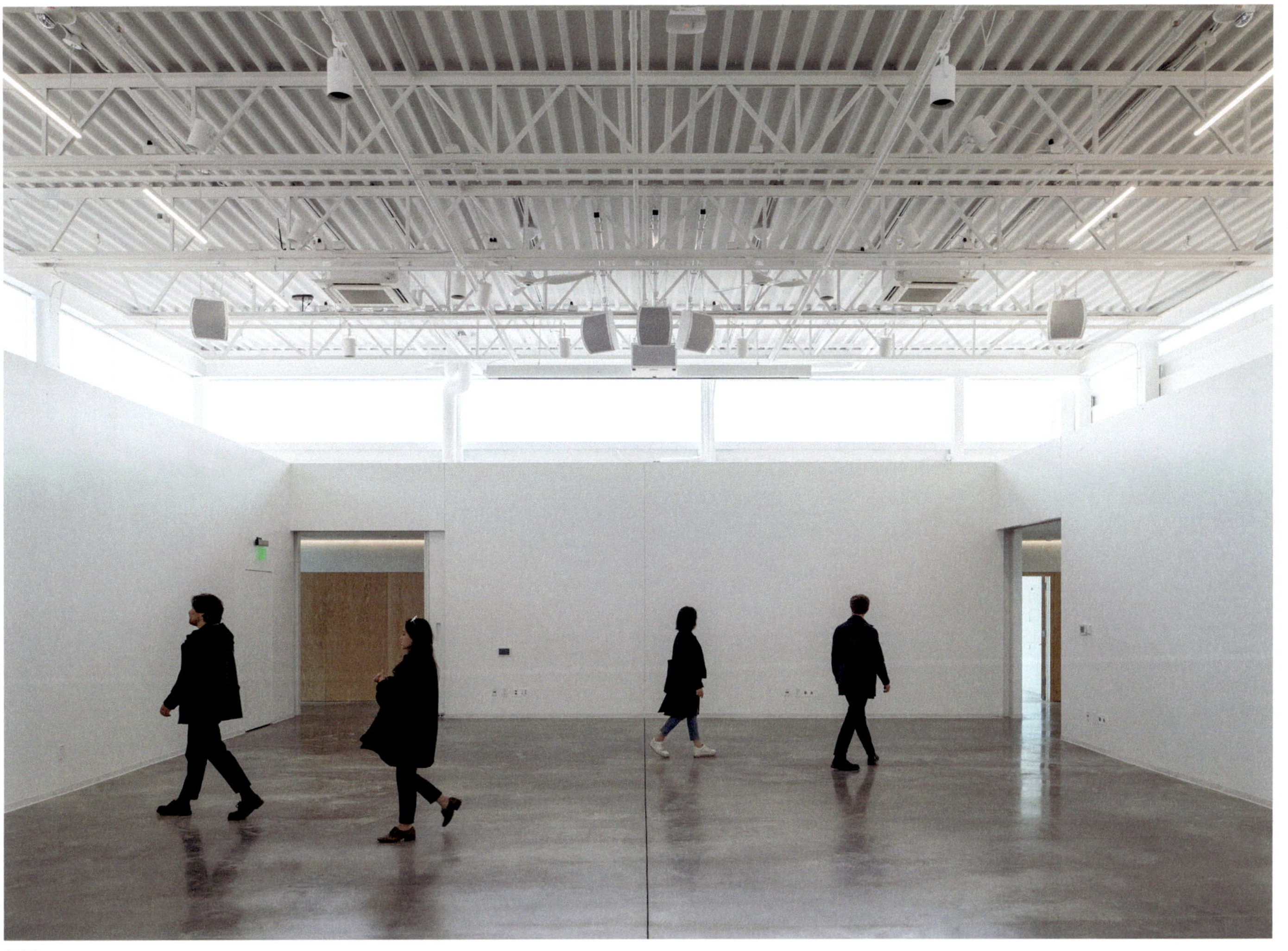

302

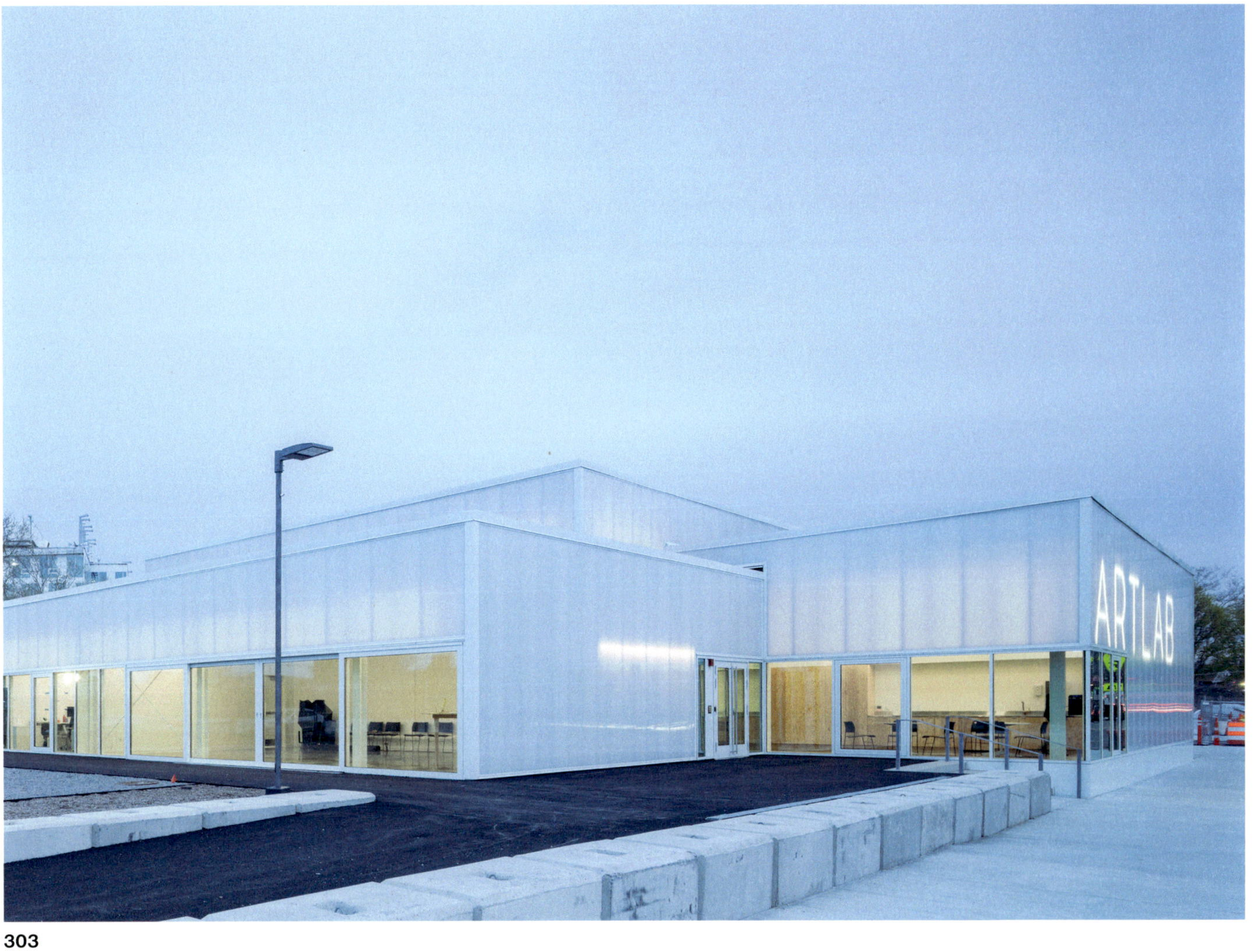

303

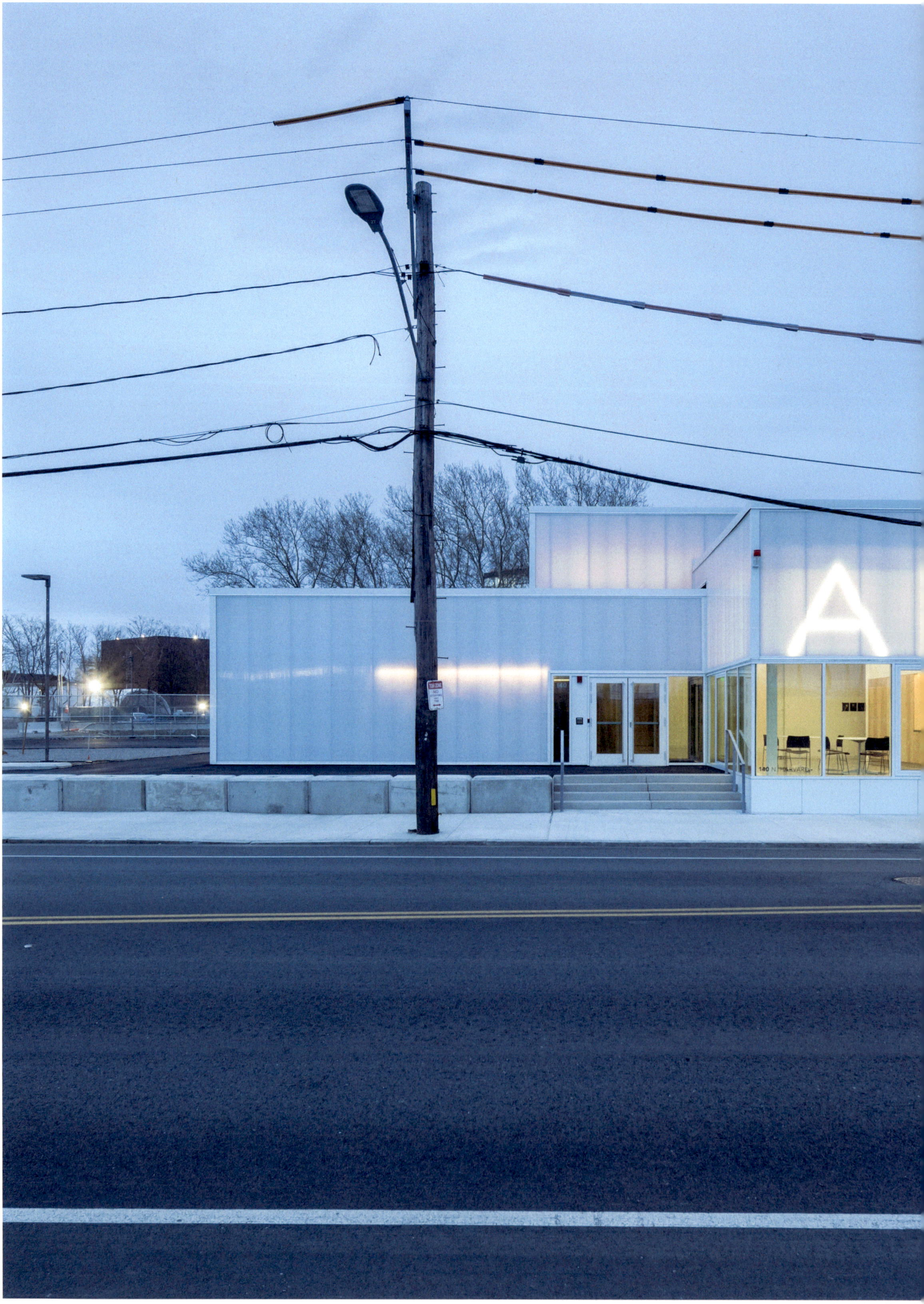
A
140 N

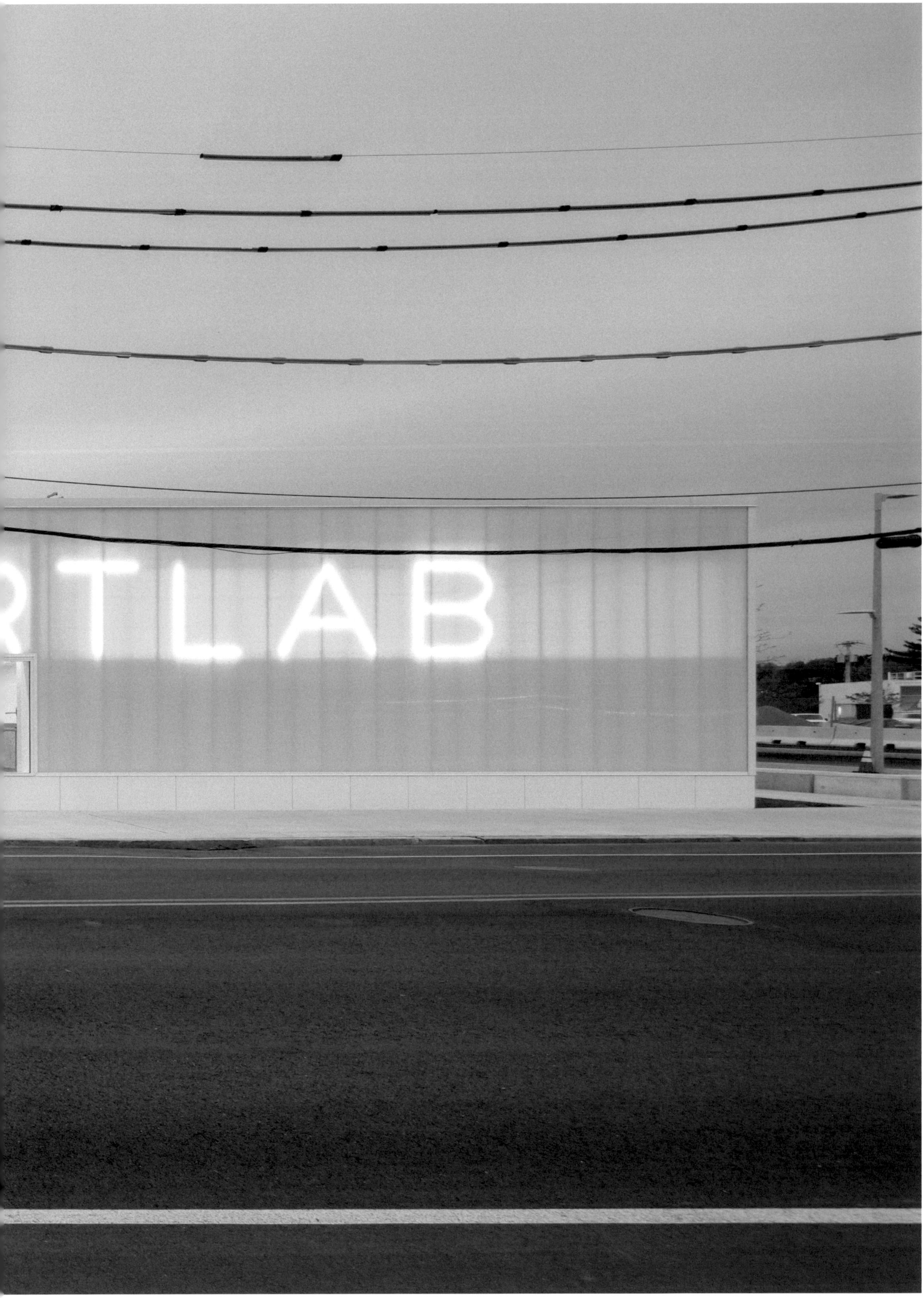
RTLAB

Inspired by steel building in Chicago (including Mies van der Rohe's steel work at IIT), we wanted to revisit the rust belt with a building in steel and core-ten cladding. We therefore harnessed US steel construction capabilities and chose to span the factory hall of TRUMPF Smart Factory with laser-cut Vierendeel trusses.

Located between the highway and a protected wetland, the factory is configured in the form of a figure eight. The sloping roof covers a showroom and lower volumes include a café, offices, and an auditorium. The showroom is spanned by eleven customized laser-cut steel Vierendeel trusses, approximately 45-m long.

Another ambition of ours was to update the Vierendeel truss (invented in 1898 in Belgium) with laser-cutting technology, parametric structural loading, and expressed efficacies. A skywalk – a bridge that runs through the structure – provides a viewing platform from which to observe the machines below.

Structure, here – indeed the frame itself – provides a space that can be occupied, granting onlookers a privileged position to observe the processes below. The view from the skywalk offers a picture of industry 4.0: machines working in concert, connected physically and digitally, requiring little to no human interaction.

In granting this distanced – and spatially superior – position to the onlooker, the implications of the current distance between humans and machines become unavoidably resonant.

309

The core concept driving the development of the exhibition building for the chainsaw manufacturer Stihl was the 'From Tree to Truss.' The chainsaw company owns a forest that could be harvested to construct a large and simple space frame to house exhibition spaces at its factory.

310

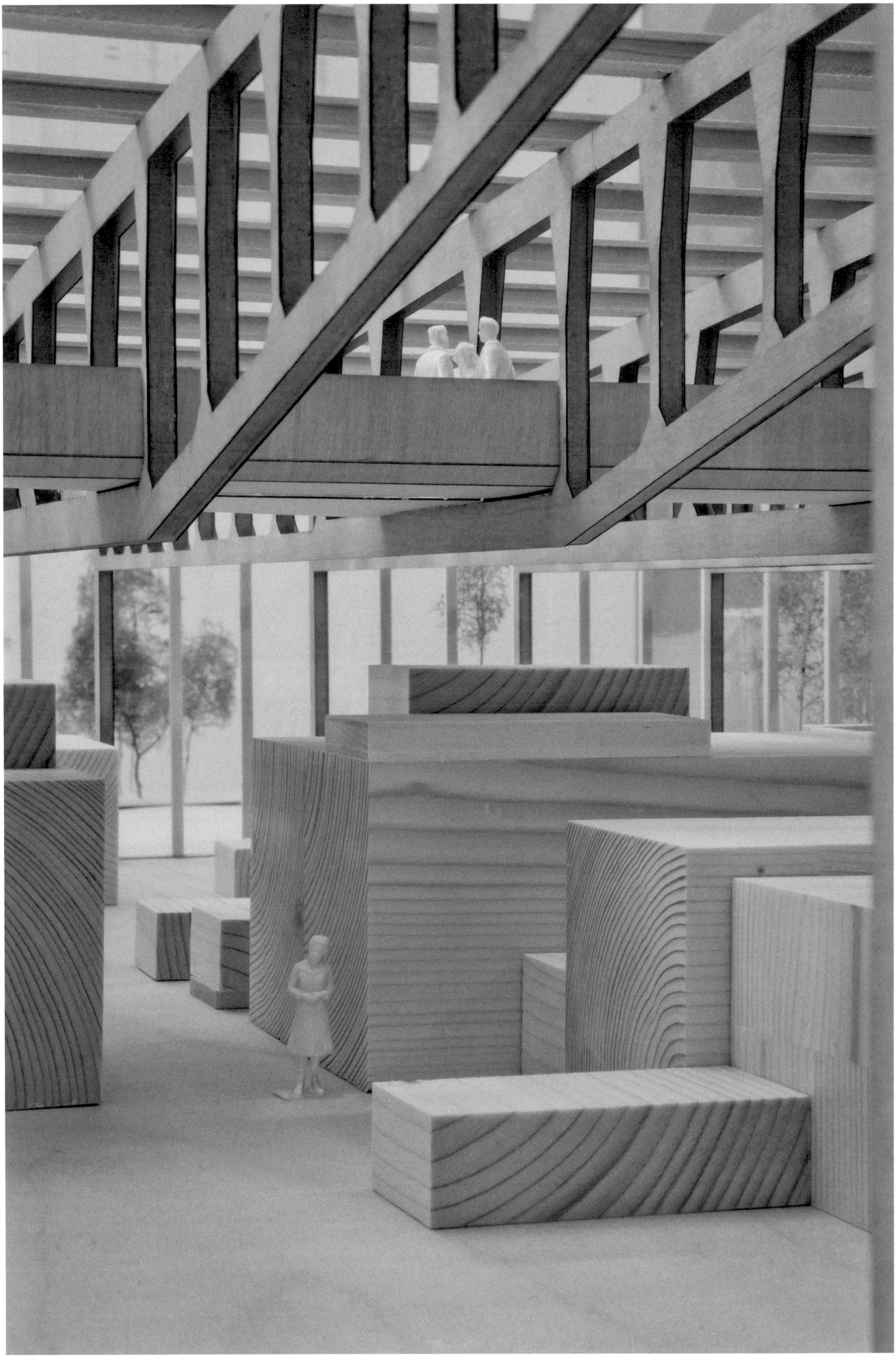

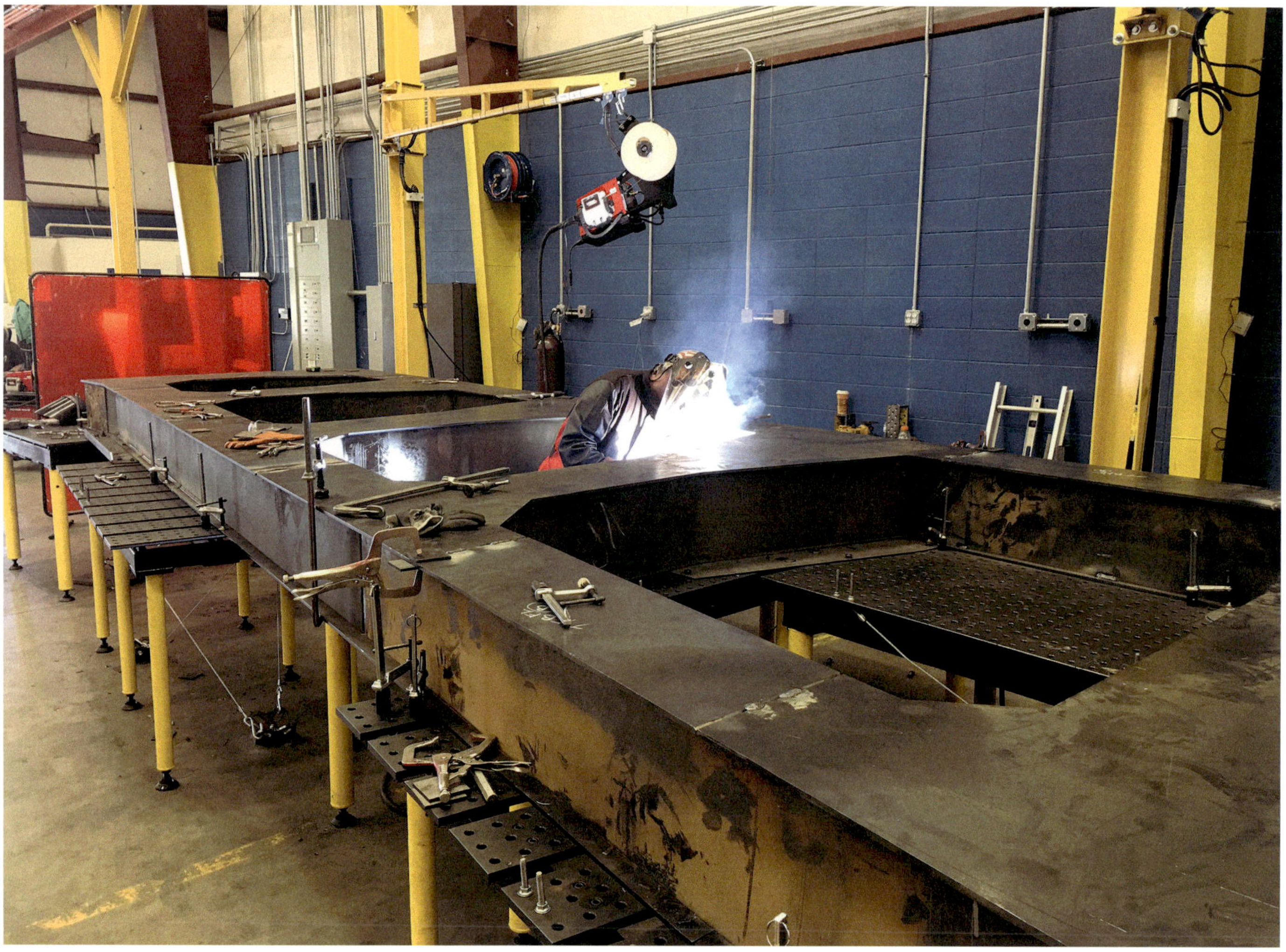

312

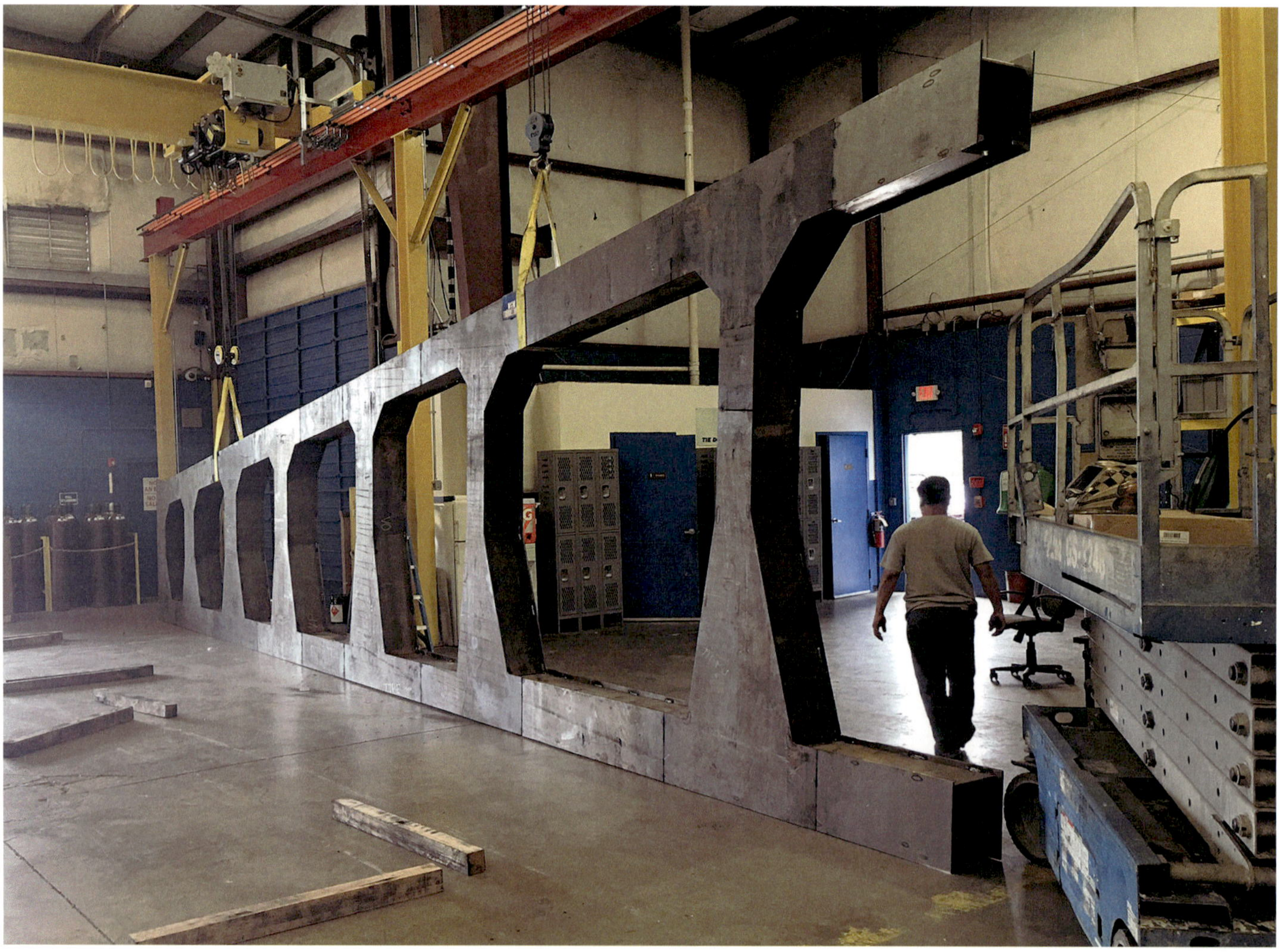

313

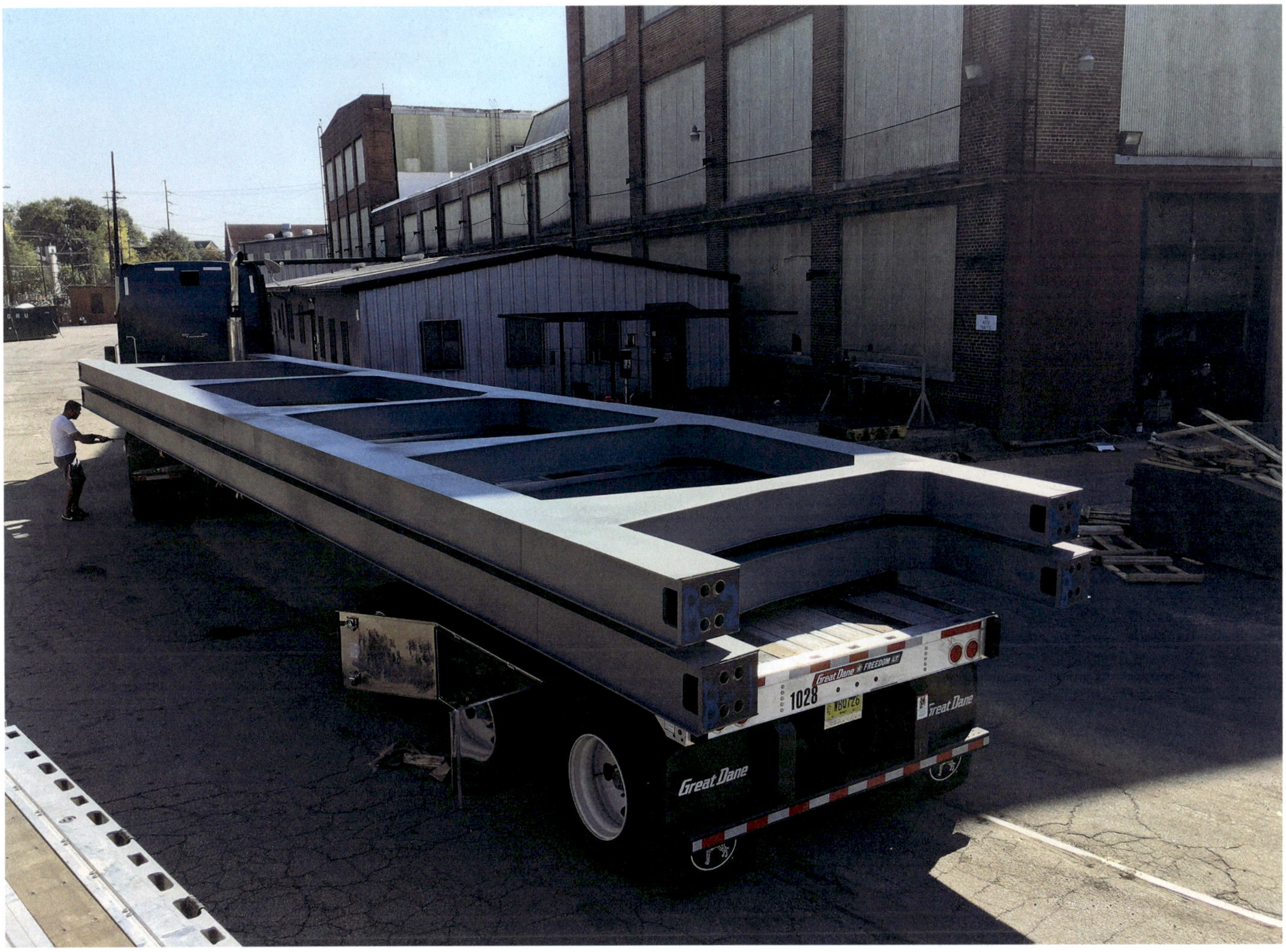

314

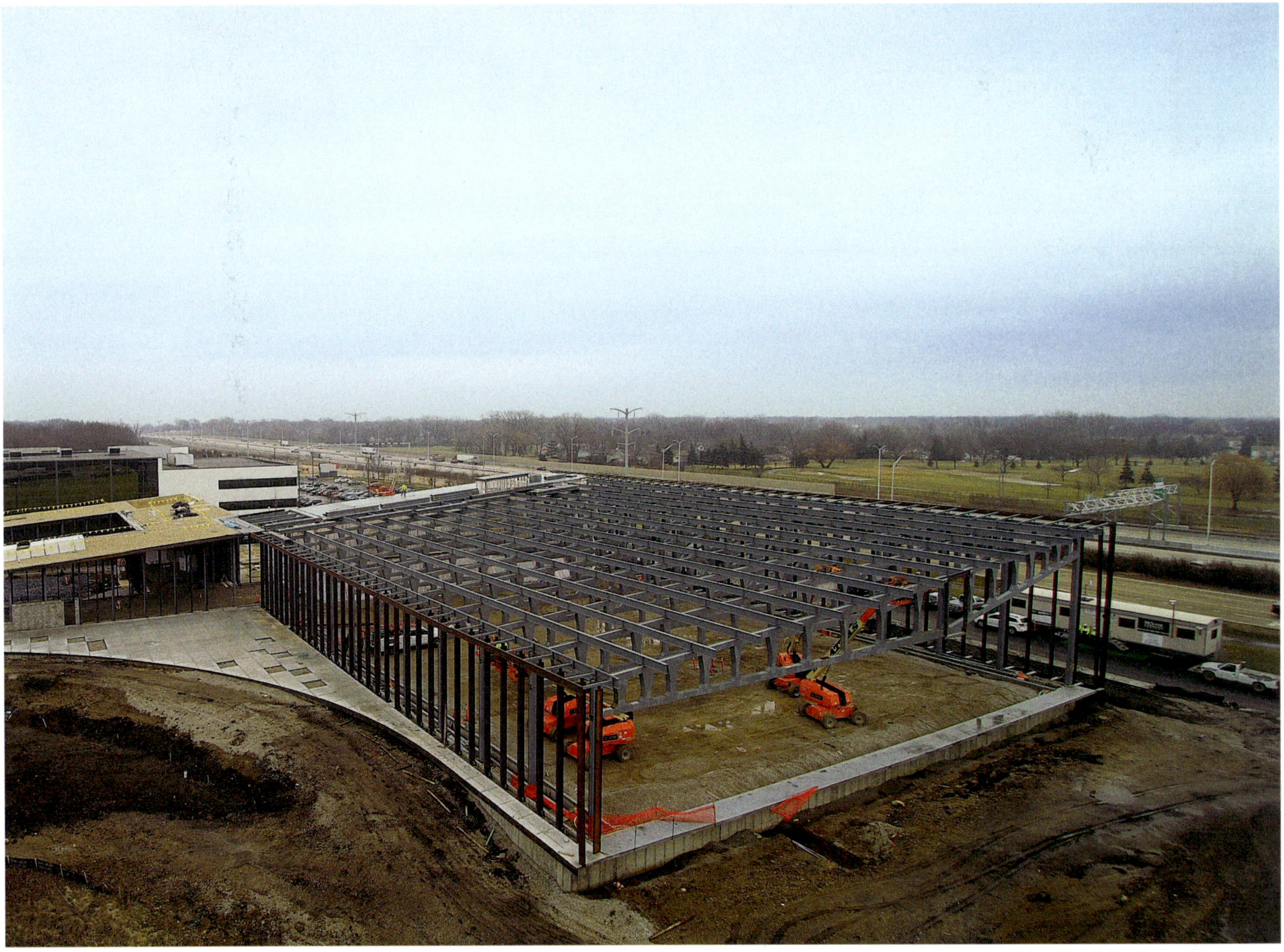
315

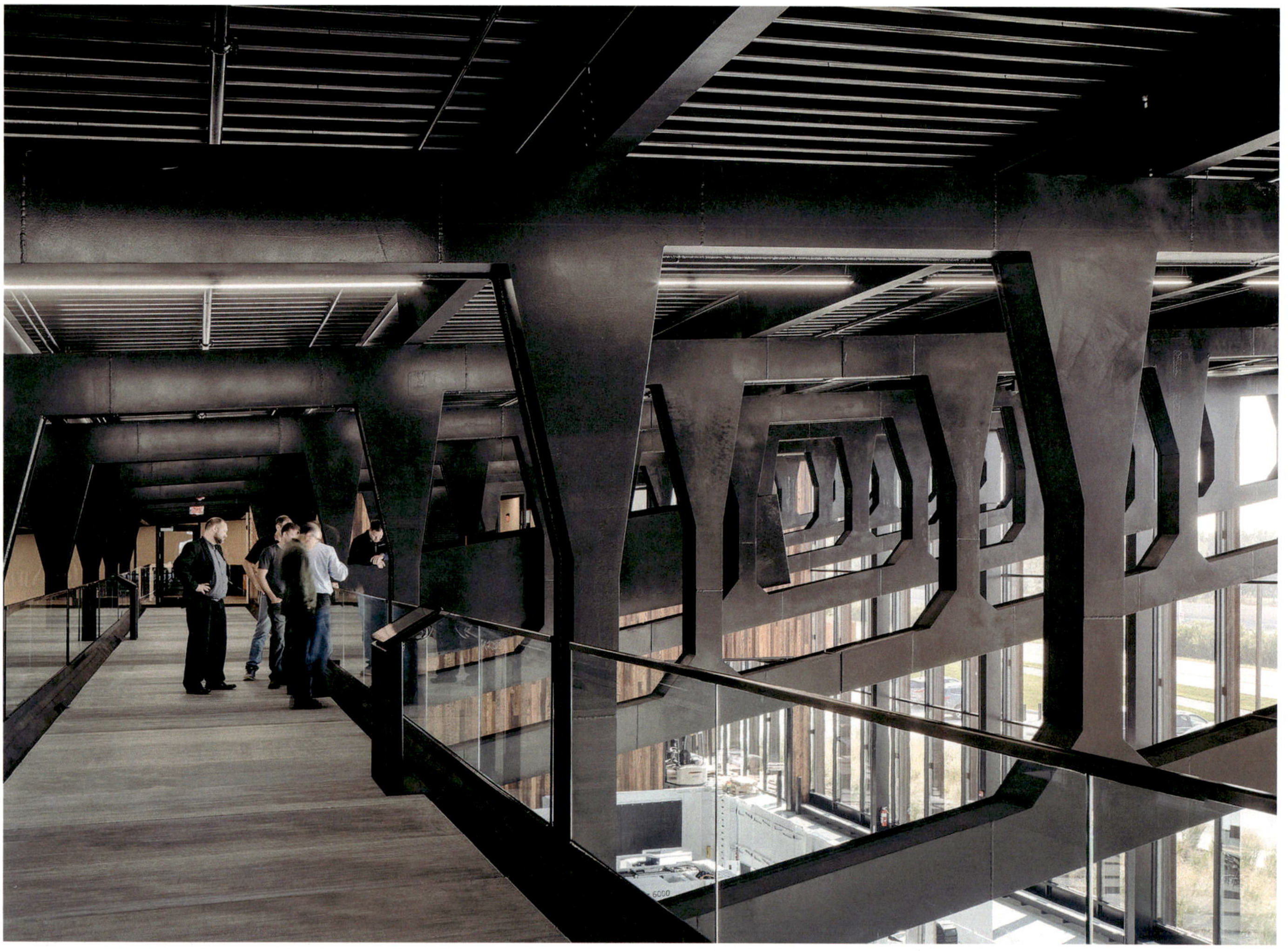

316

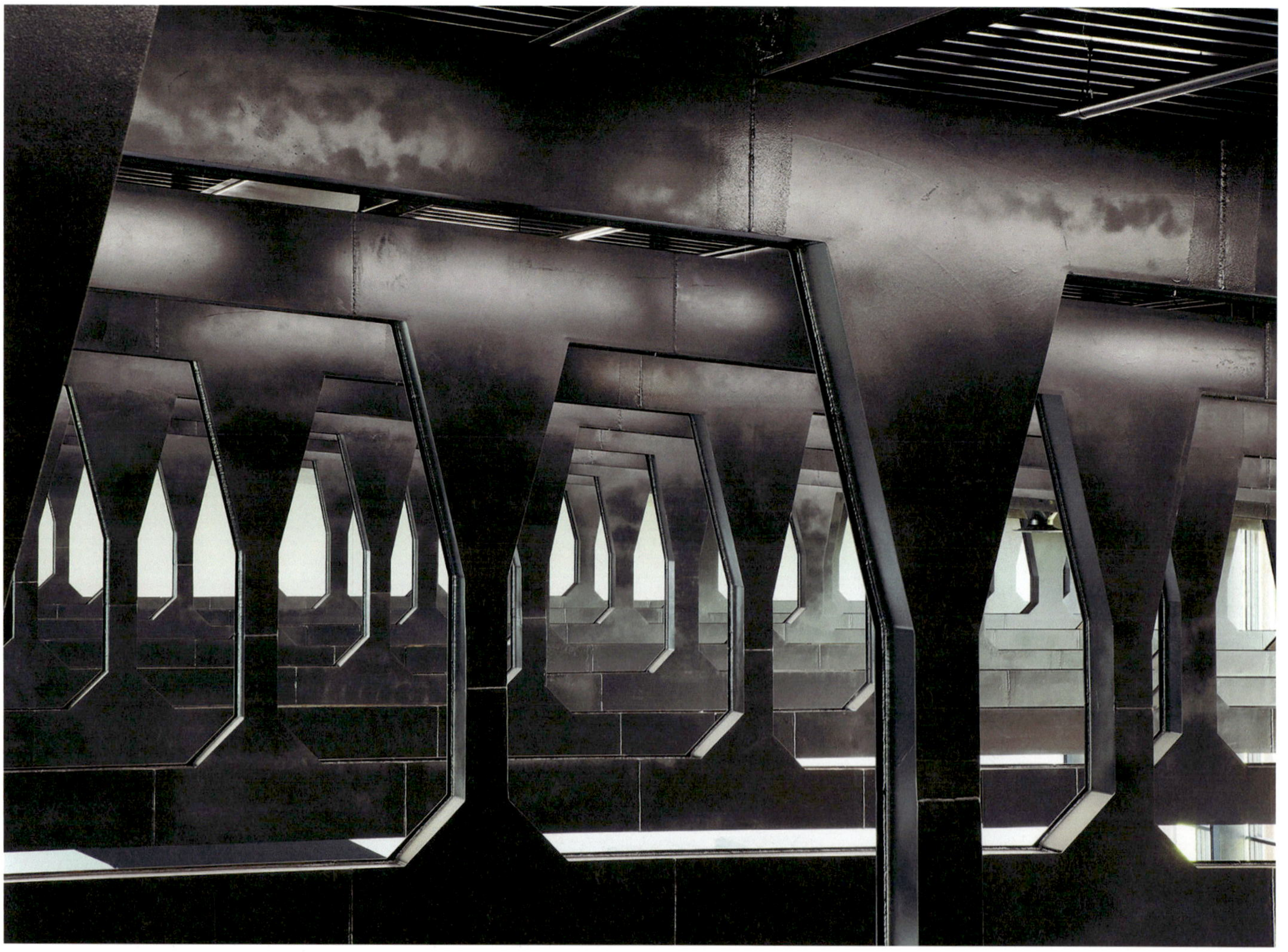

317

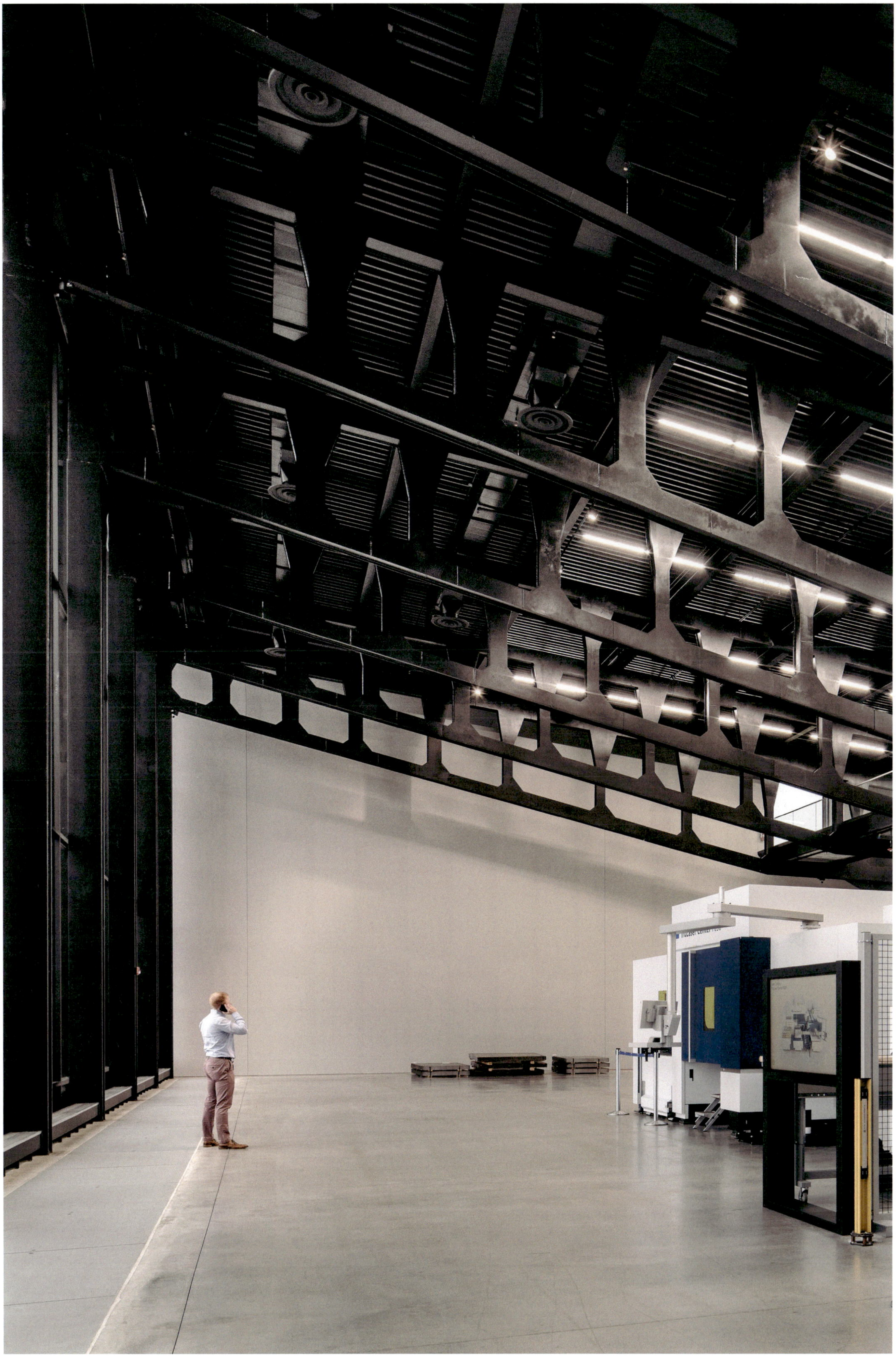

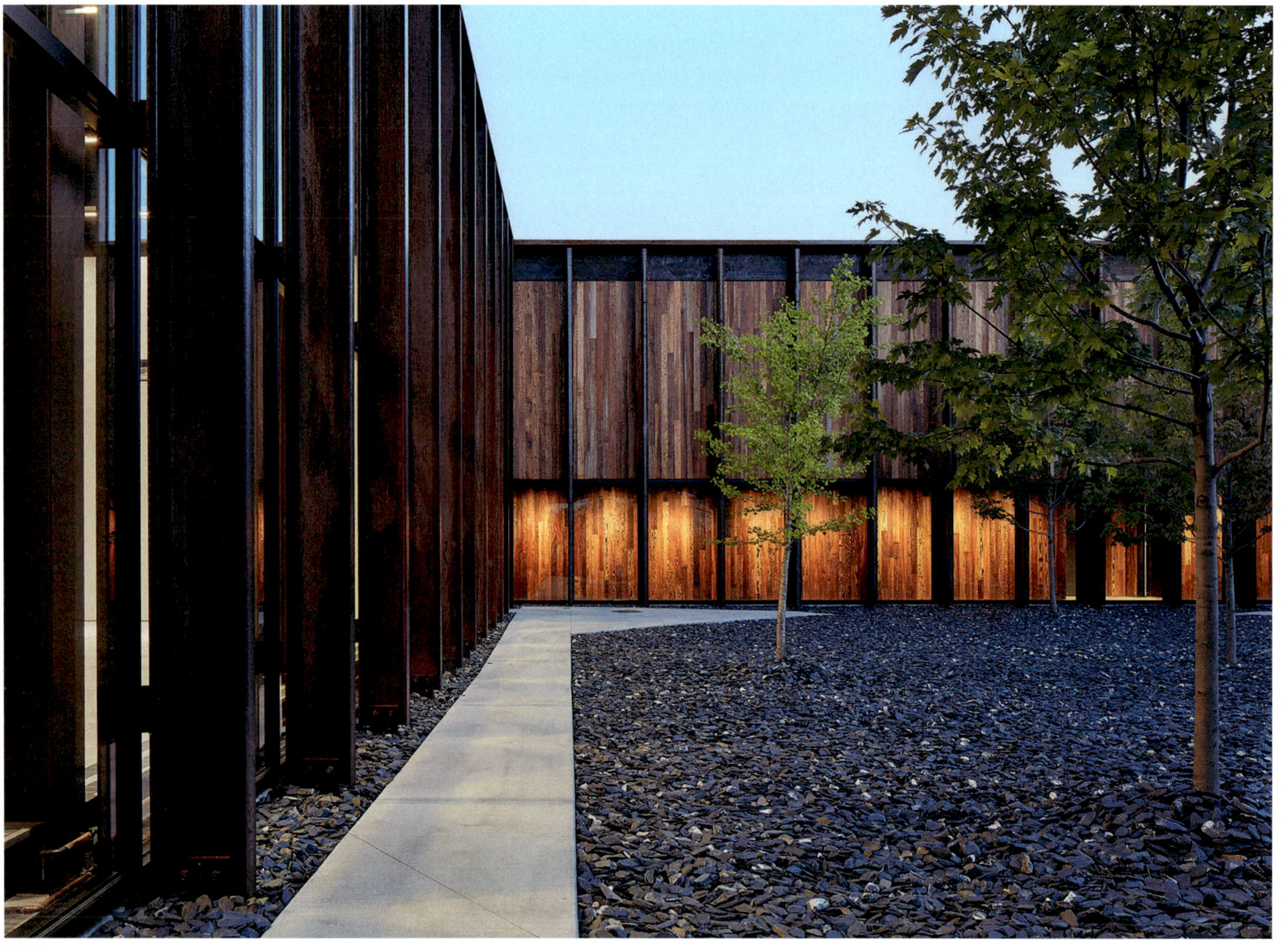
319

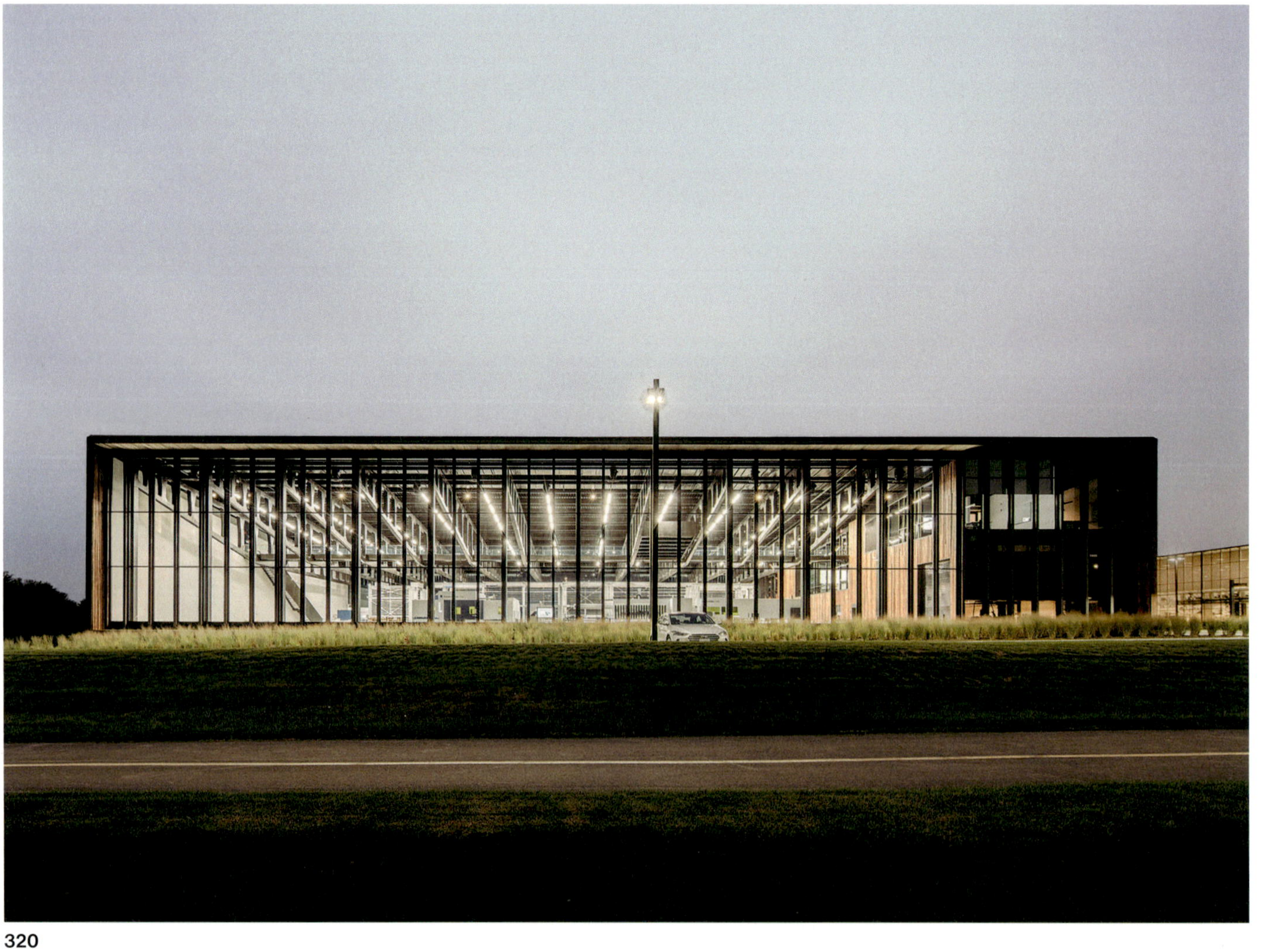
320

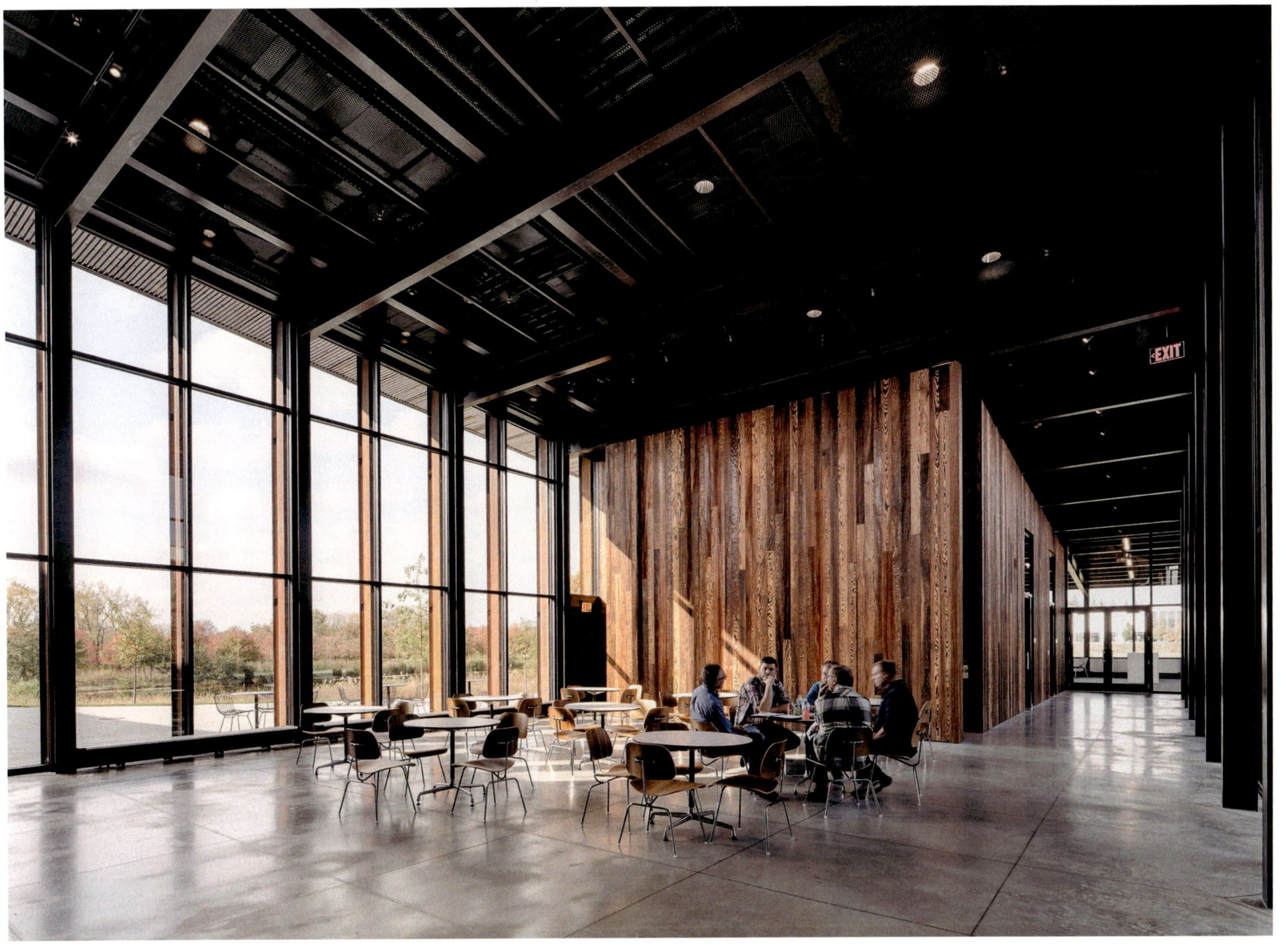

321

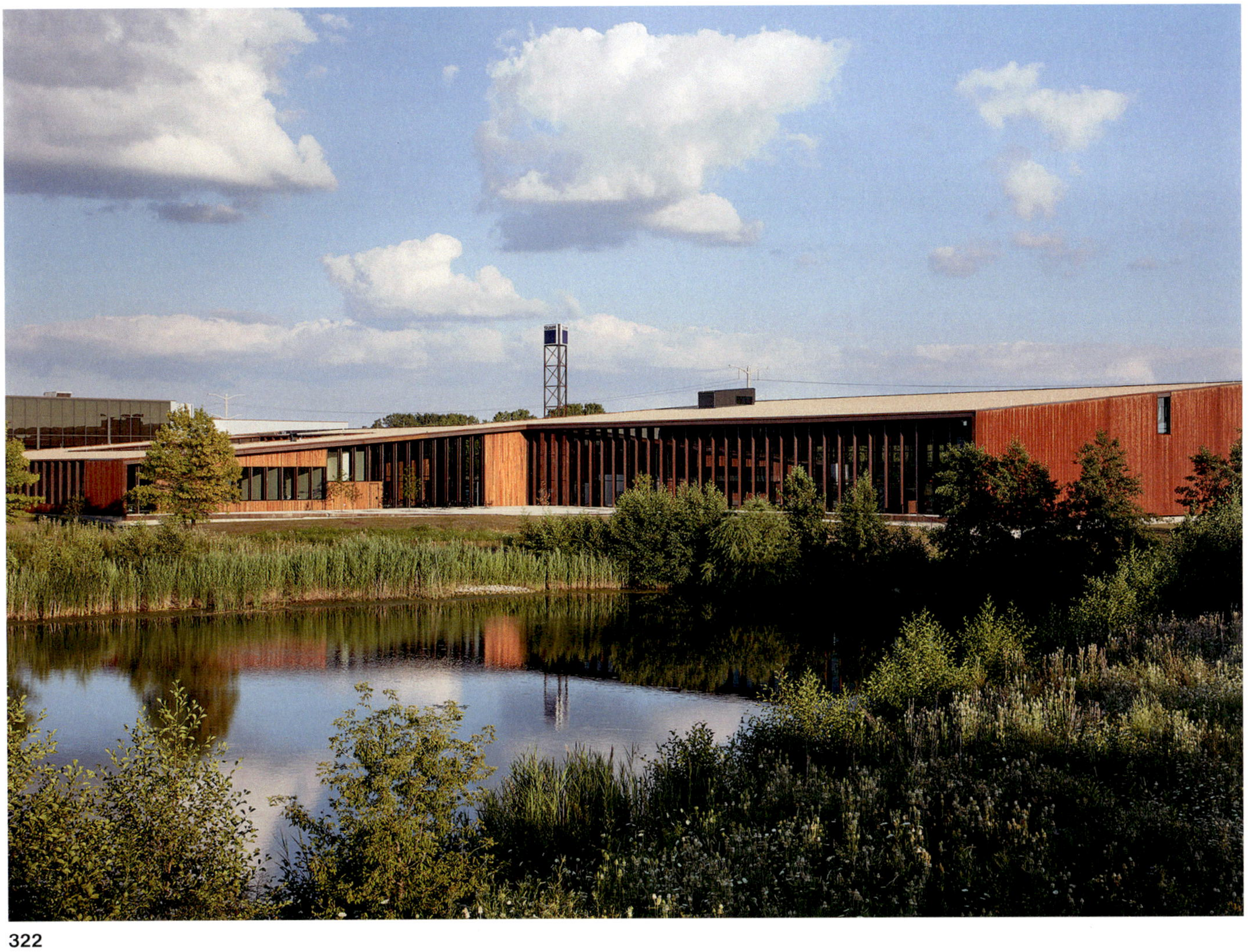

322

The chainsaw, a high-tech and robust tool, was proposed as the primary method for constructing a large and complex building. In this way, the enduring industrial construction type – the space frame – would be the result of an industry providing both the raw material and the tools to transform it.

The implications of industry providing the raw material for its own structure is symbolically significant.

This frame does not unite the exterior and the interior, nor does it particularly accommodate programmatic transformation or grant unexpected or privileged perspectives; this frame, instead, embodies and manifests the nature of the work it houses and its own making.

Over the past few years, Barkow Leibinger has engaged in research projects in the hopes of further interrogating the generative and radical potential of the Frame.

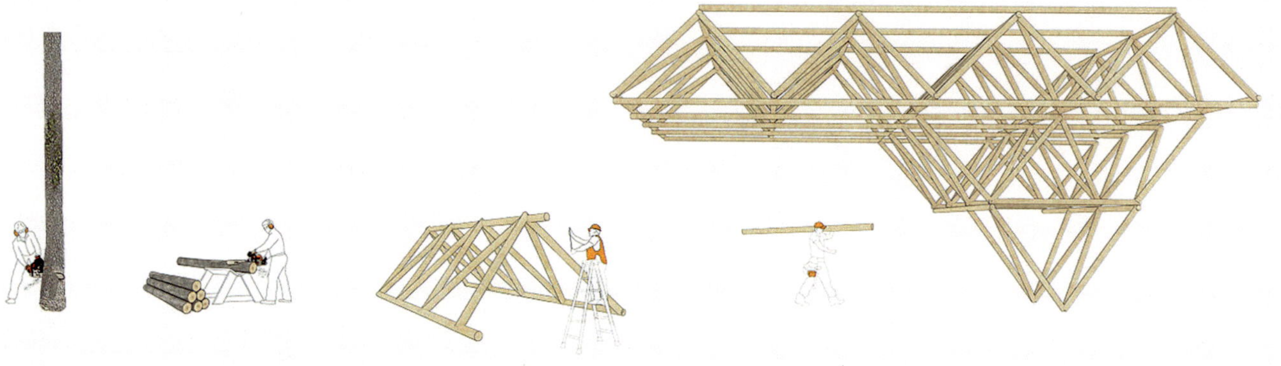

327

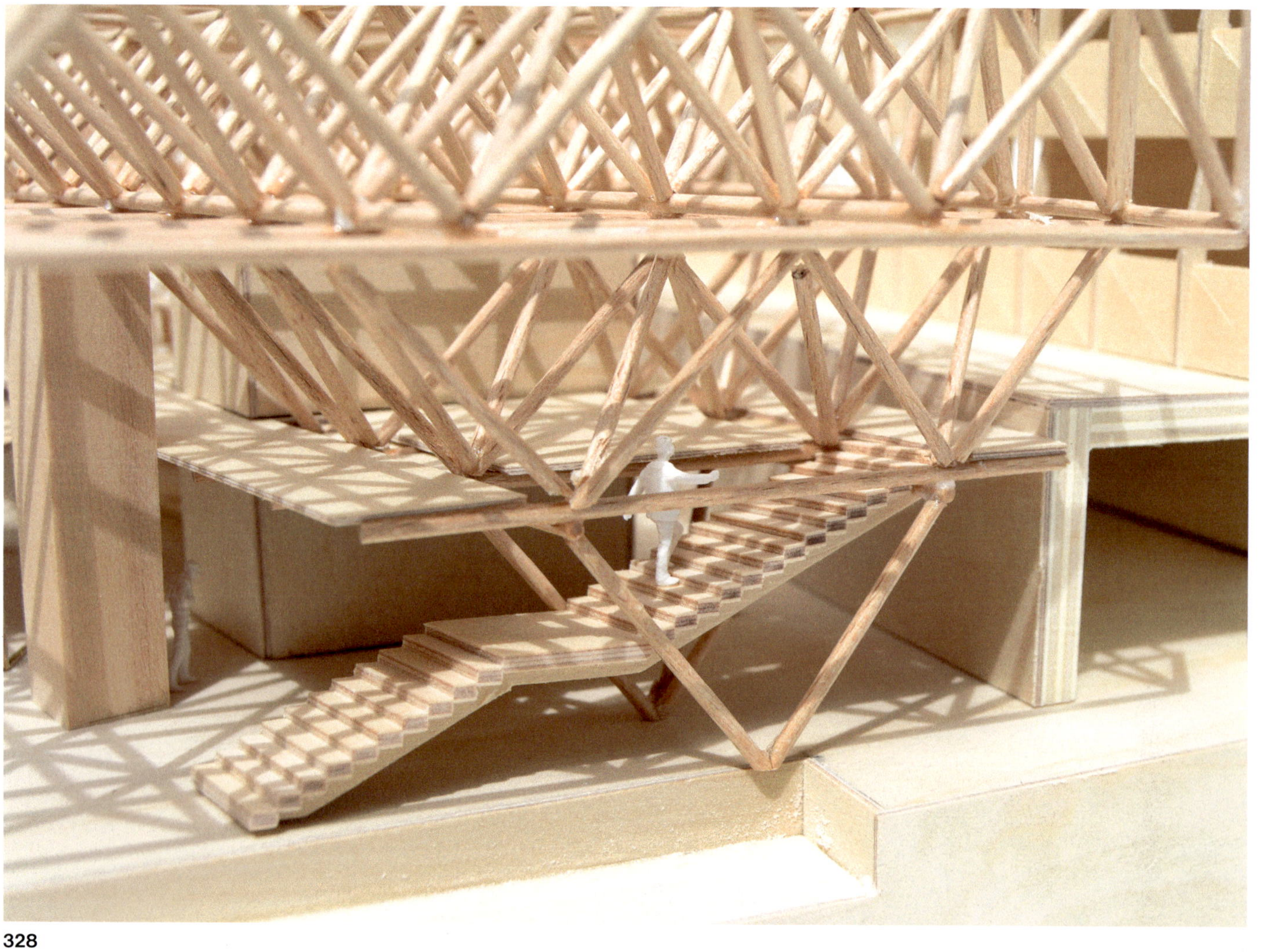

328

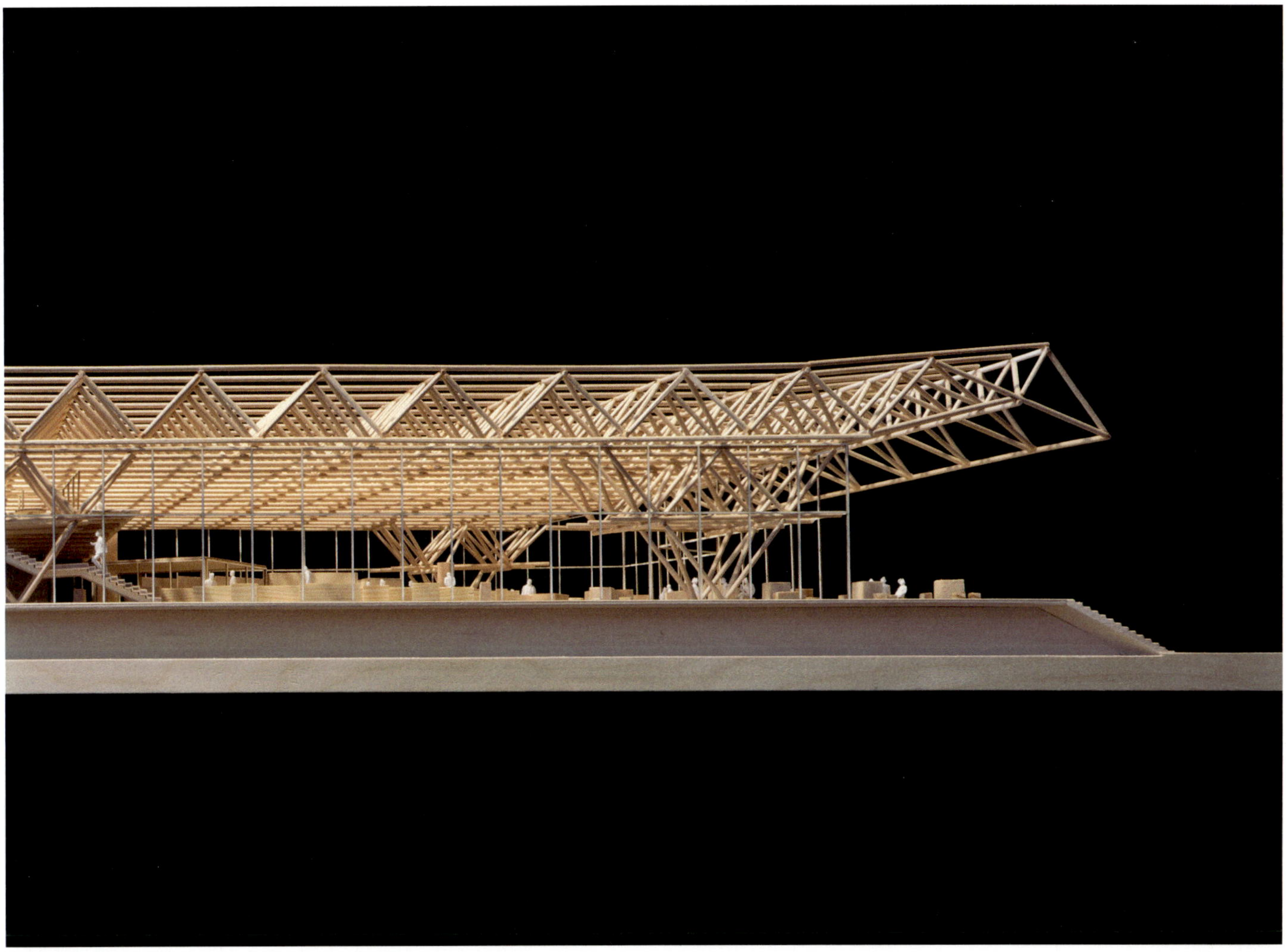

329

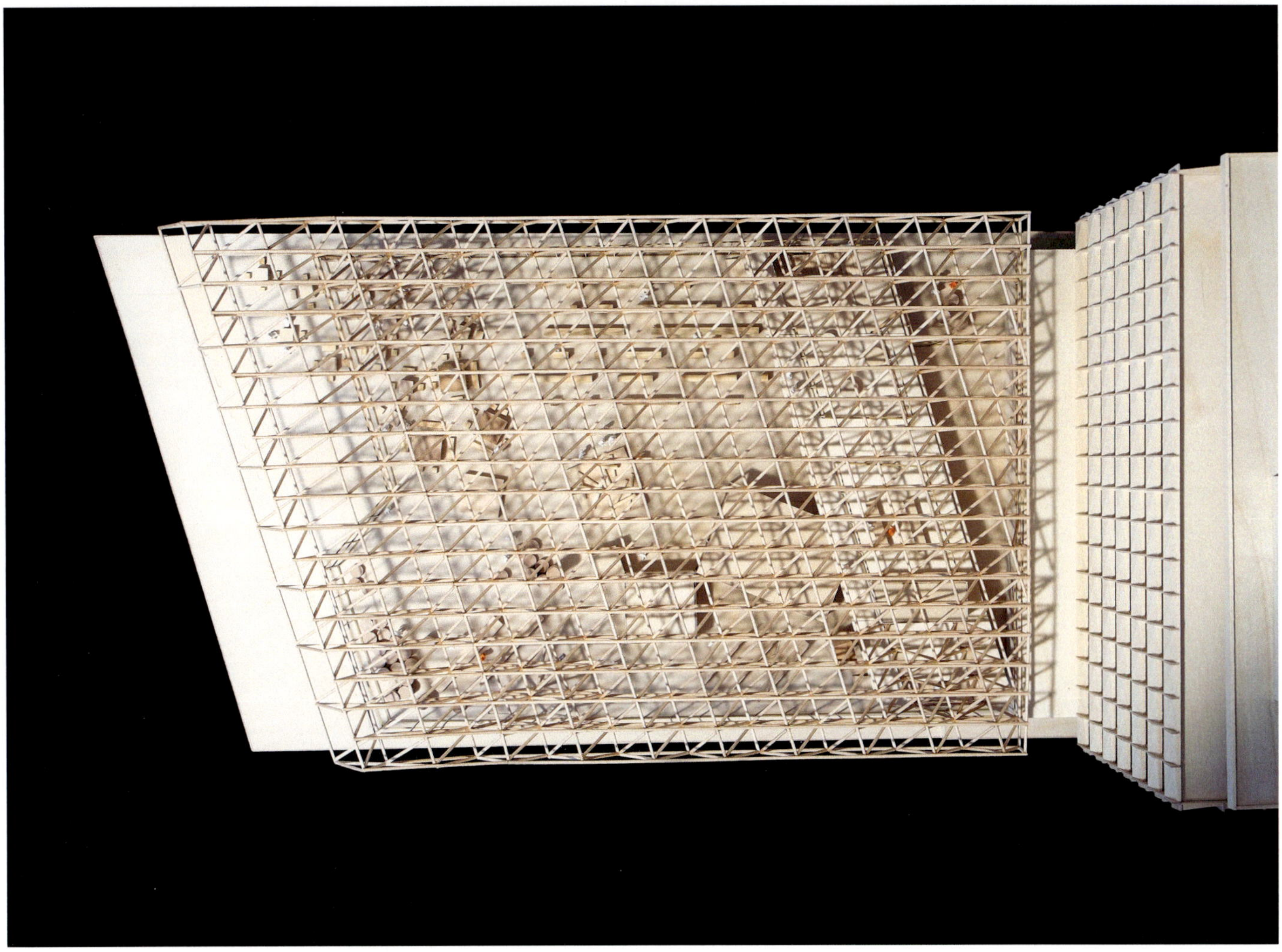

330

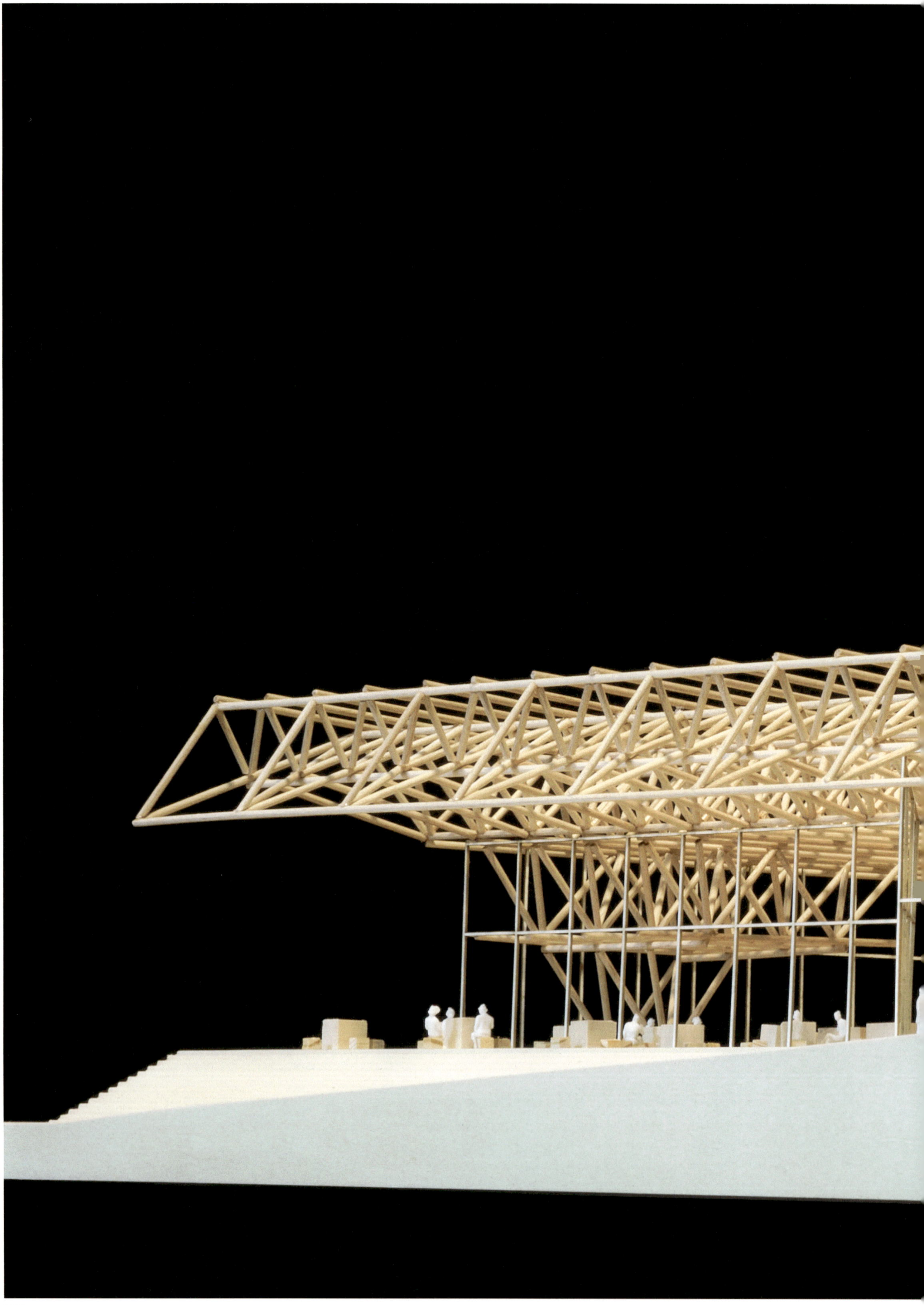

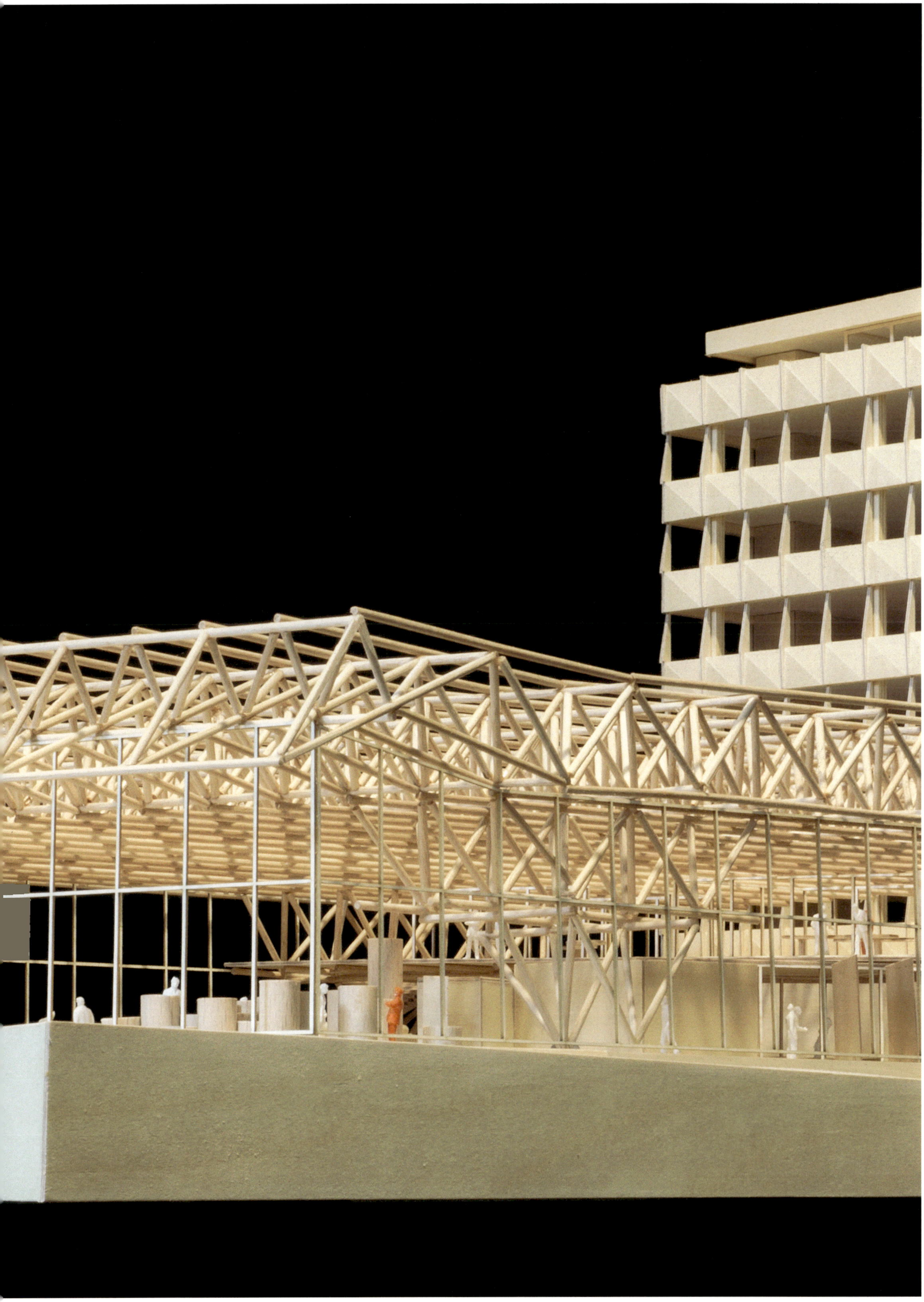

This study was a response to Buckminster Fuller's plywood version of his geodesic dome project. The intention was to use the logic of plywood as an element but to create a more geometrically complex structure than the dome that Fuller promoted.

It was also a shift in frame making from the joining of lineal stick-like parts to the joining of shingle-like surfaces.

The plywood acted as both structure and enclosure – a shift which at once solved the problem of enclosing the space, while changing the space aesthetically and seamlessly.

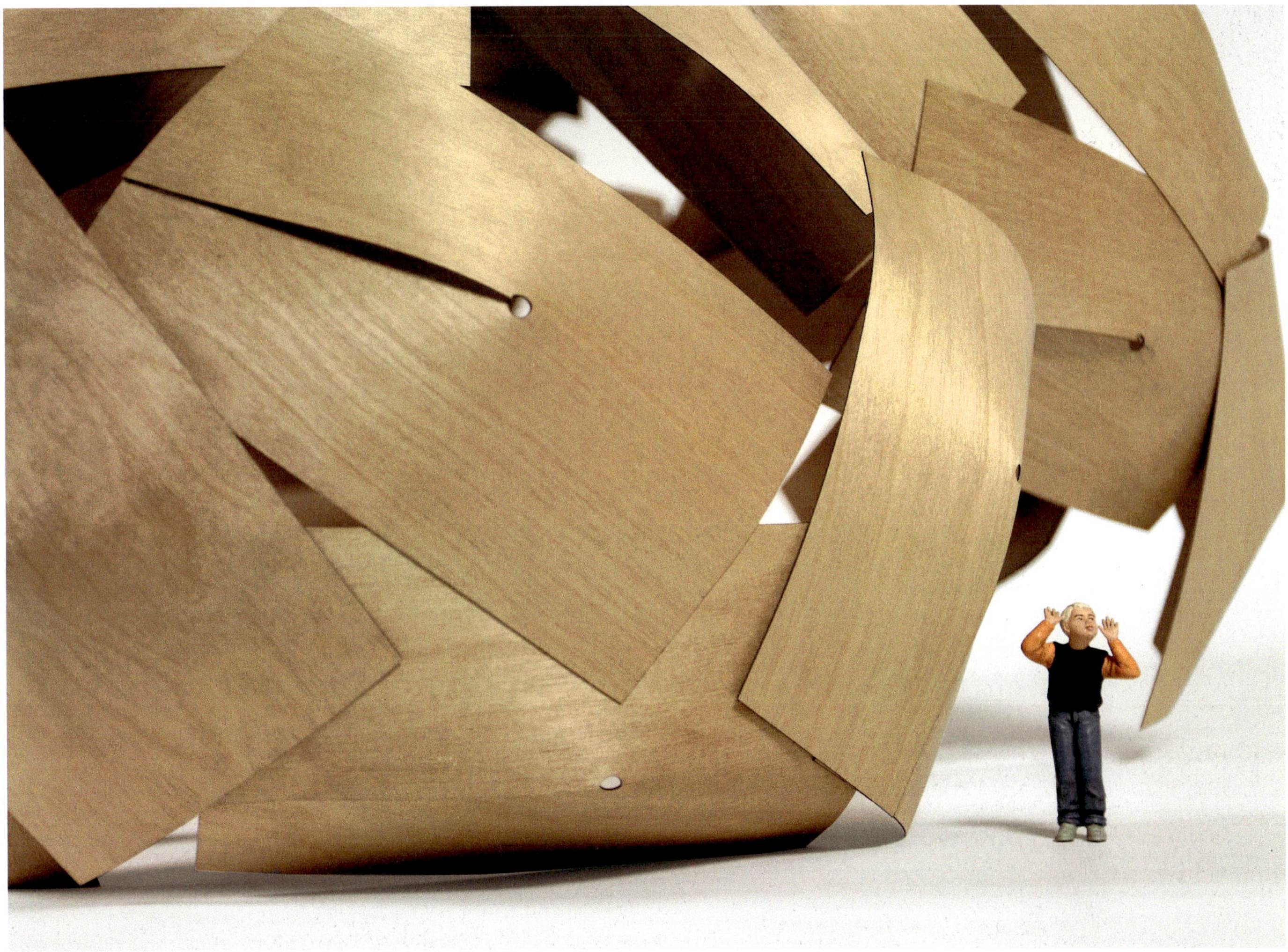

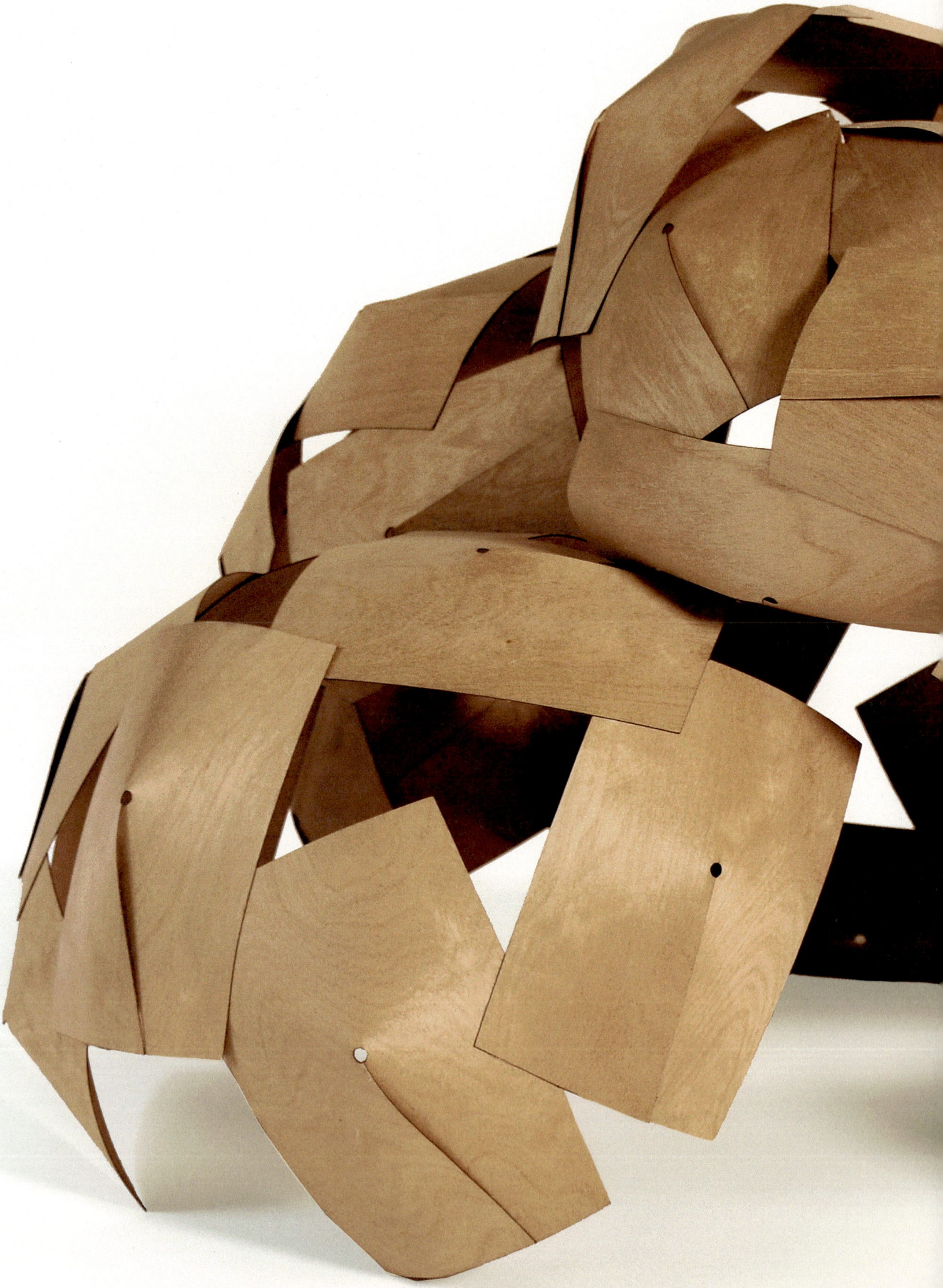

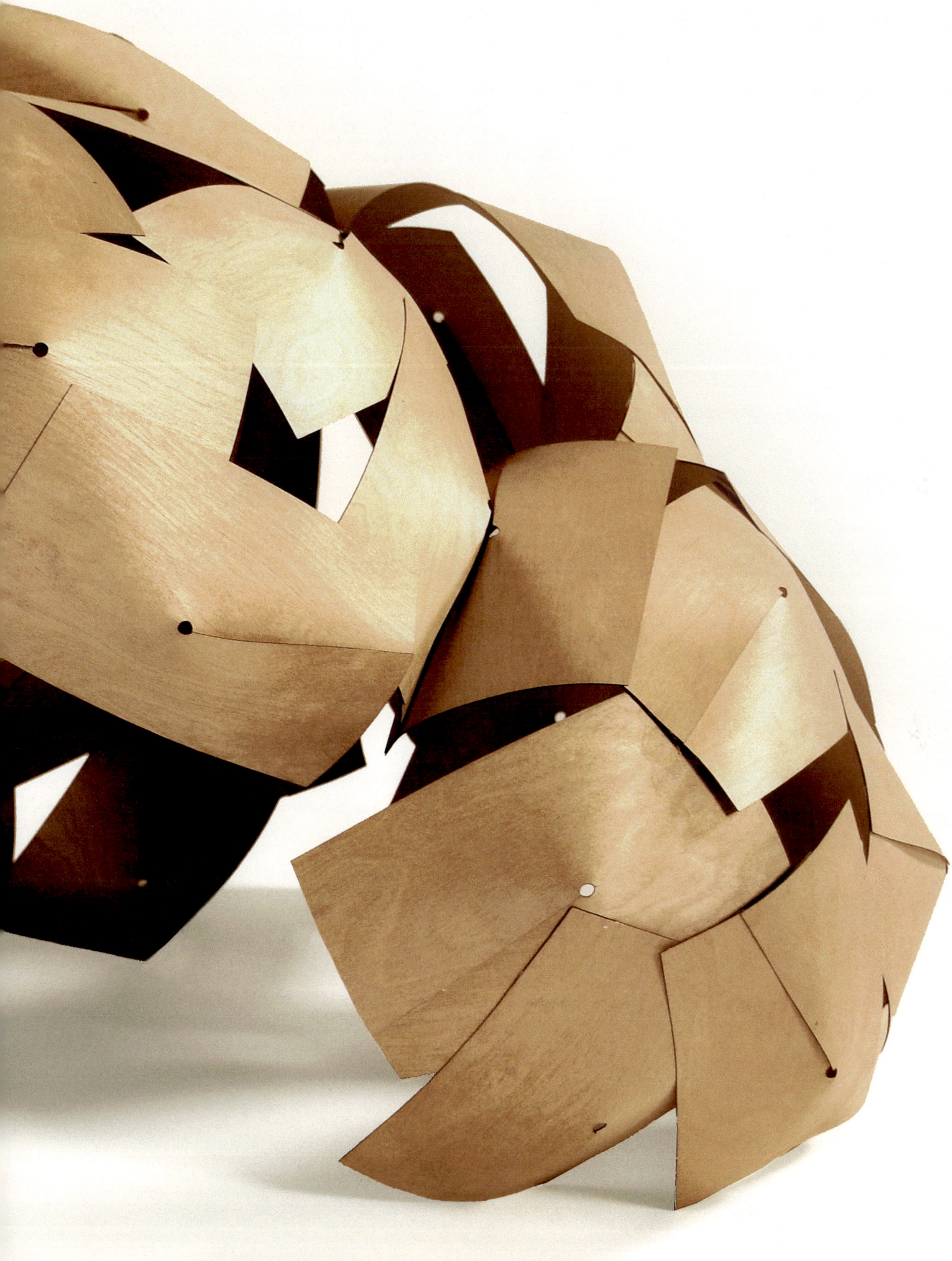

In recent years, we have conceived of a number of updated iterations of the structural space frame concept pioneered by Konrad Wachsmann and furthered by I.M. Pei.

We tested this concept through different materials, including concrete, timber, and metal, using contemporary digital fabrication techniques, with variable structural loading in mind. Our guiding question was: How do modern techniques and digital tooling impact and modify these now out-of-date typologies?

Our study for a bus-stop shelter used a large piece of thick sheet metal. The sheet metal was slotted with one cut in order for the material to fold over itself, forming a canopy-like space that could stand in the simplest possible fashion to provide shelter.

Like the plywood geodesic dome, simple folding made this surface spatial, structural, and capable of generating a shallow enclosure. This version, however, had the added advantages of permanence and weatherproofing.

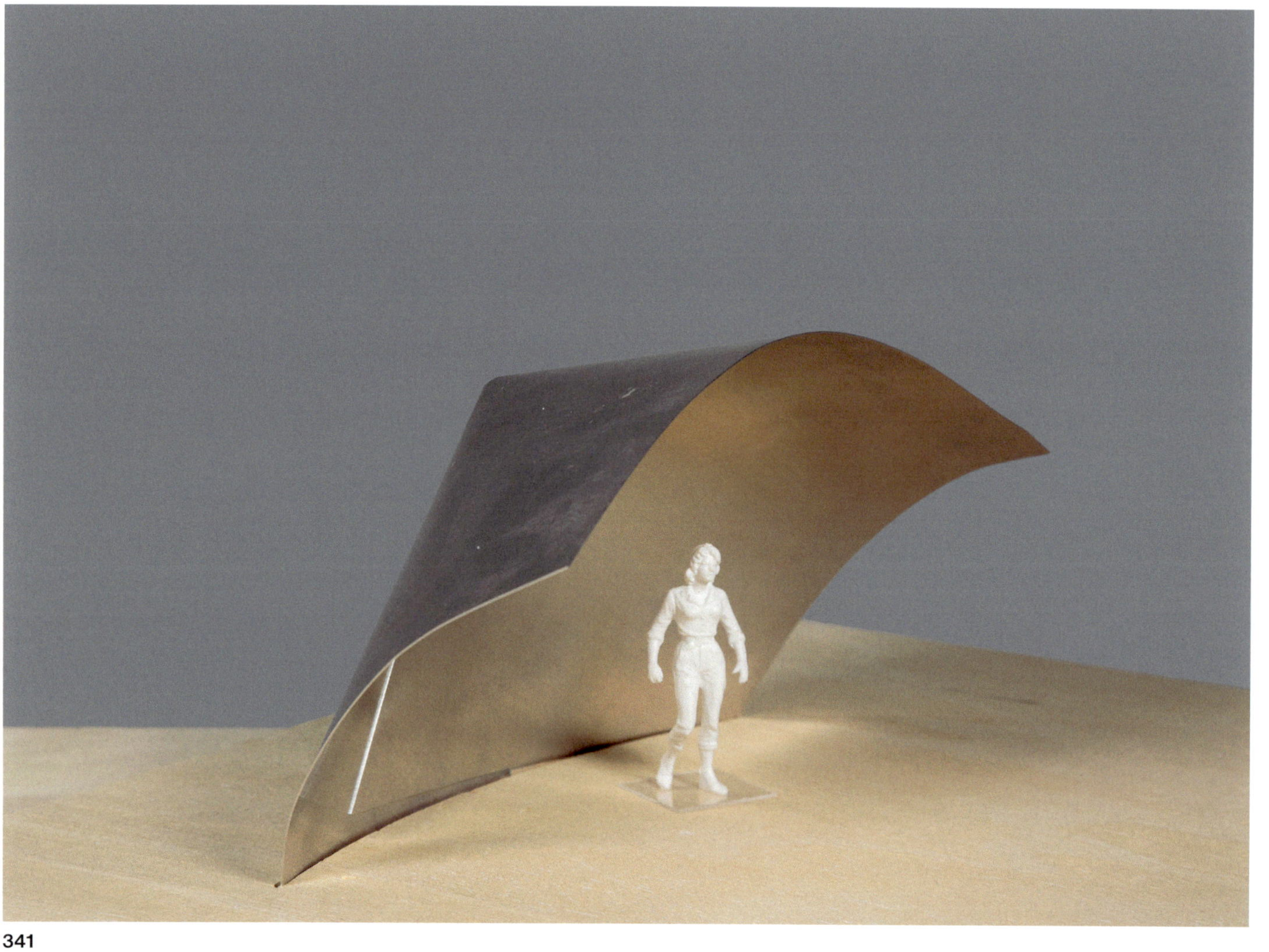
341

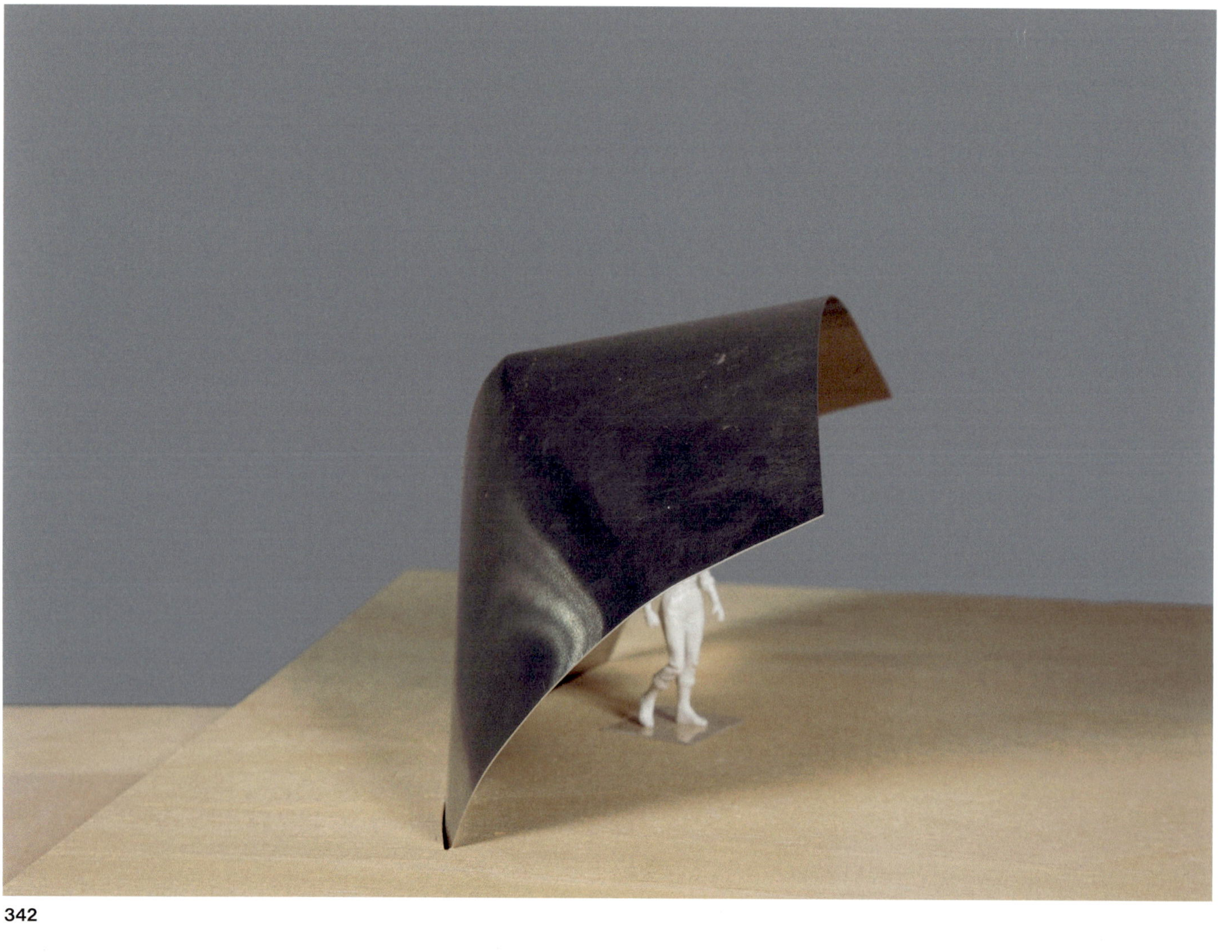
342

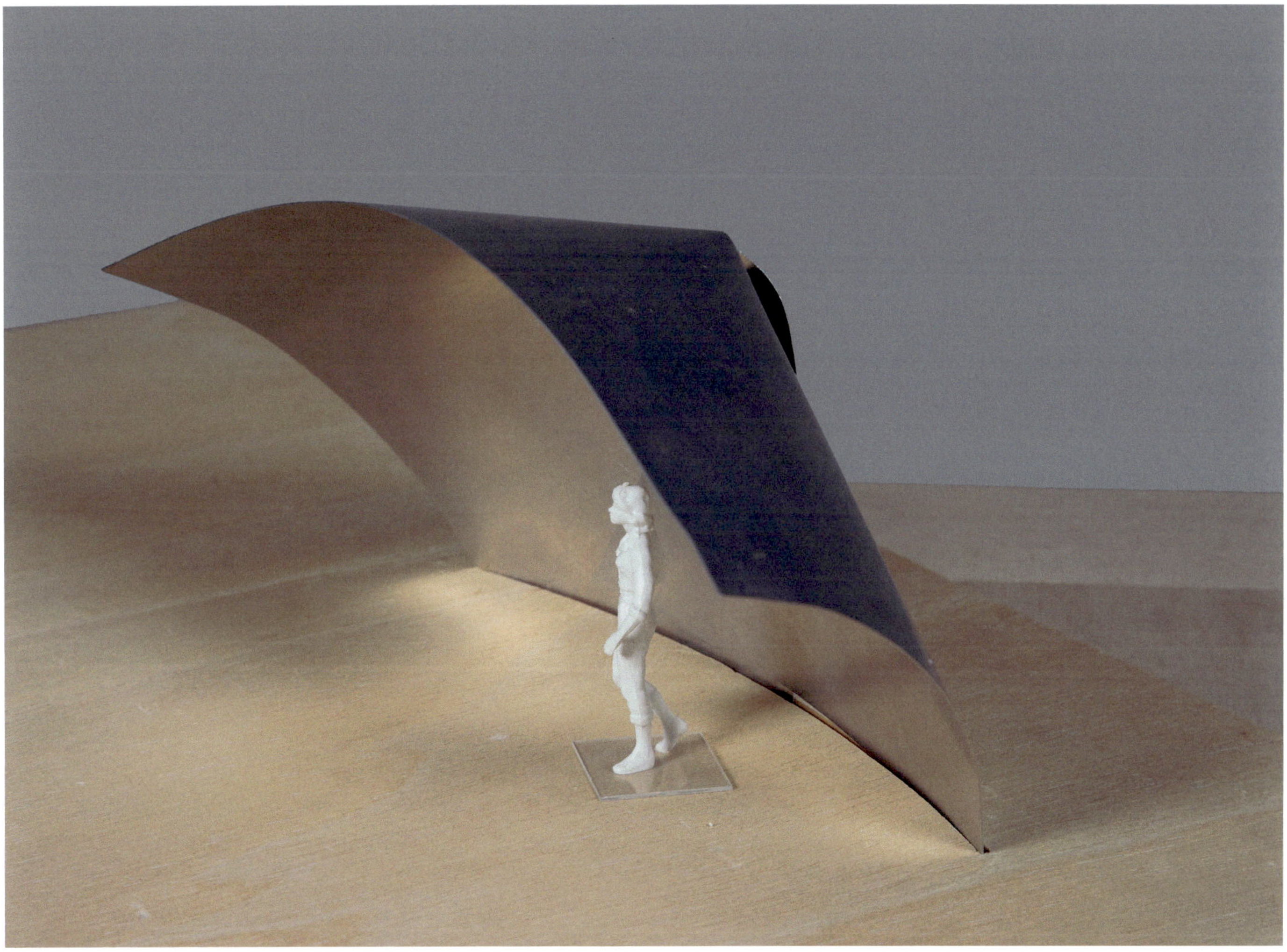

343

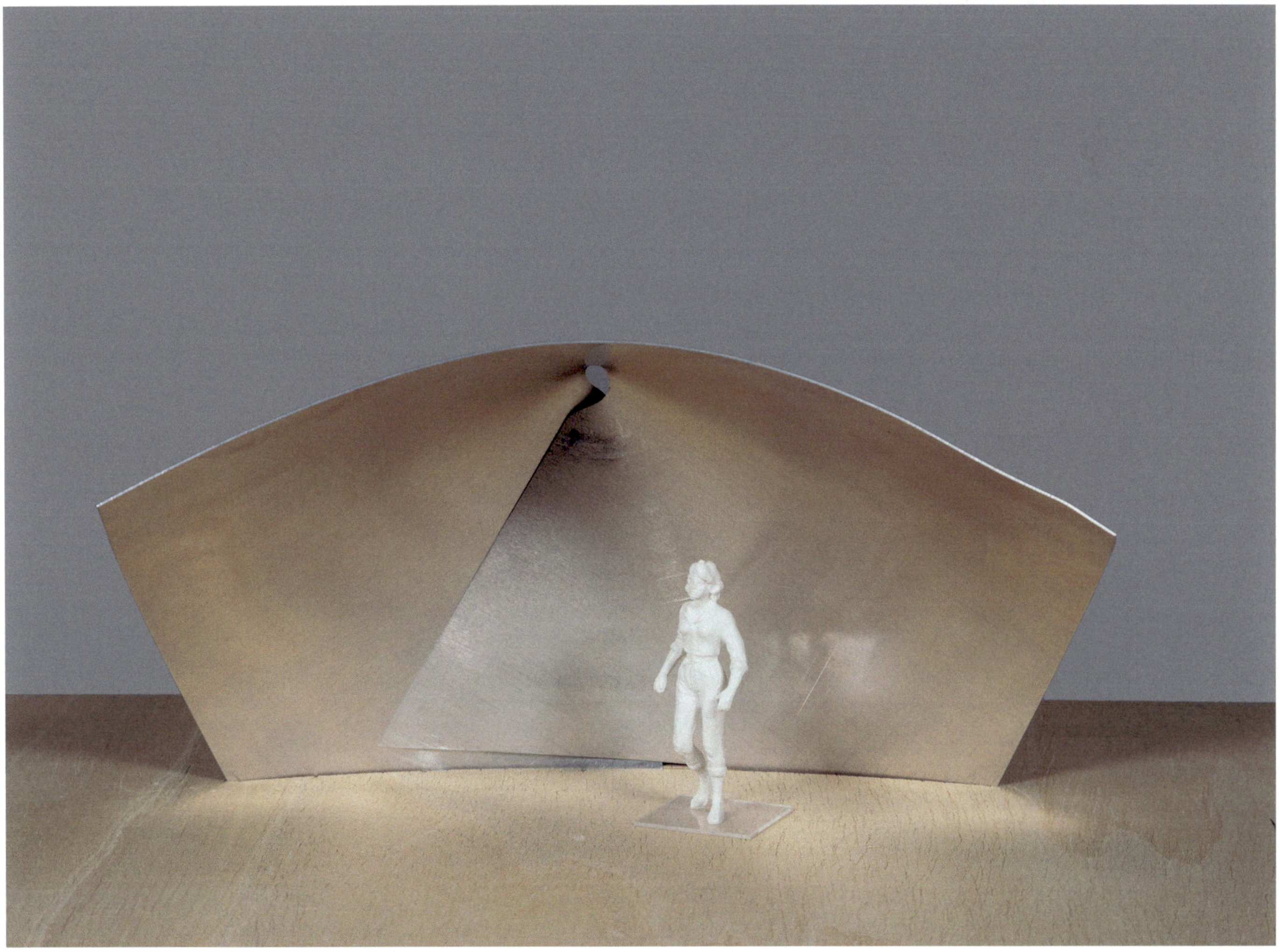

344

Works in Evolving Progress

345

These six work areas constitute the fields of study that have motivated our research over the last few years. It is a list that can expand or recede – it certainly will, as we continue to tinker, play, experiment, and create.

346

Our emphasis is on architectural tectonics: on how materials can be transformed by tools to generate form and space. This is key to our practice. We build active archives as dynamic sources of knowledge from which to draw. This conceptual tinkering forms the very groundwork for how we build our buildings.

Our interest in technology is bound up with a curiosity about architecture and engineering history (how we can revisit it, learn from it, update and modify it for our purposes). The relays between technology and history animate our recent work and form the threads that weave through this edition.

Some themes are old but nevertheless offer us enduring modalities (Casting or Frames), while other themes are seated firmly in the present and engage technological questions (Ultrastructural).

349

One theme is perhaps more phenomenal, interested in the visual and performative effects of architecture (Liminal Facade), while another is concerned with the more enduring aspects of architecture – its geometry and one-thing-after-another nature (Bundles).

350

Finally, one theme embodies a commitment to landscape and the way it forms a dialectic with architecture (Topographical). In all cases, we are dedicated to the evolving and enduring discipline of architecture – its secrets, methods, histories, techniques, and possibilities.

351

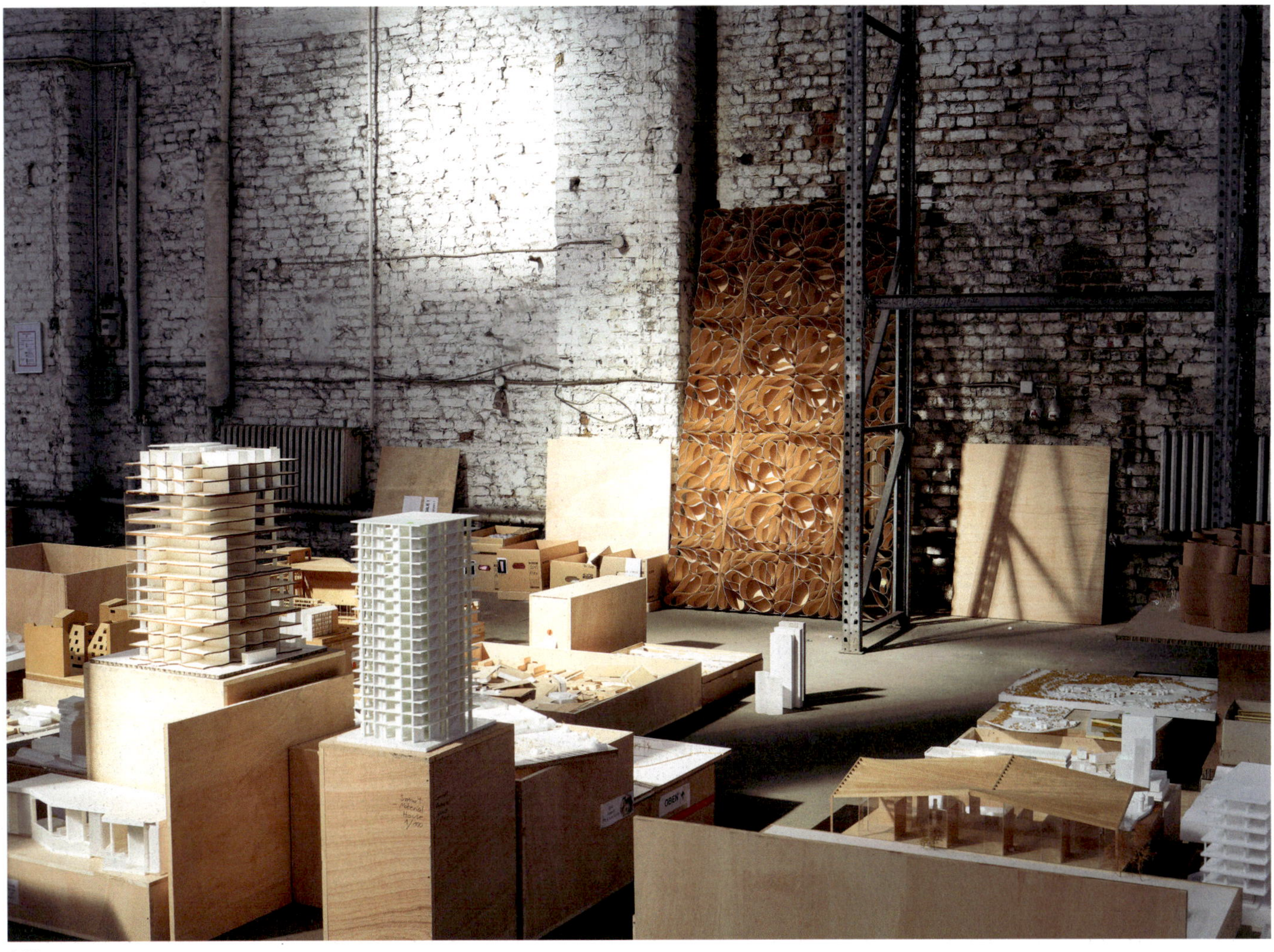

352

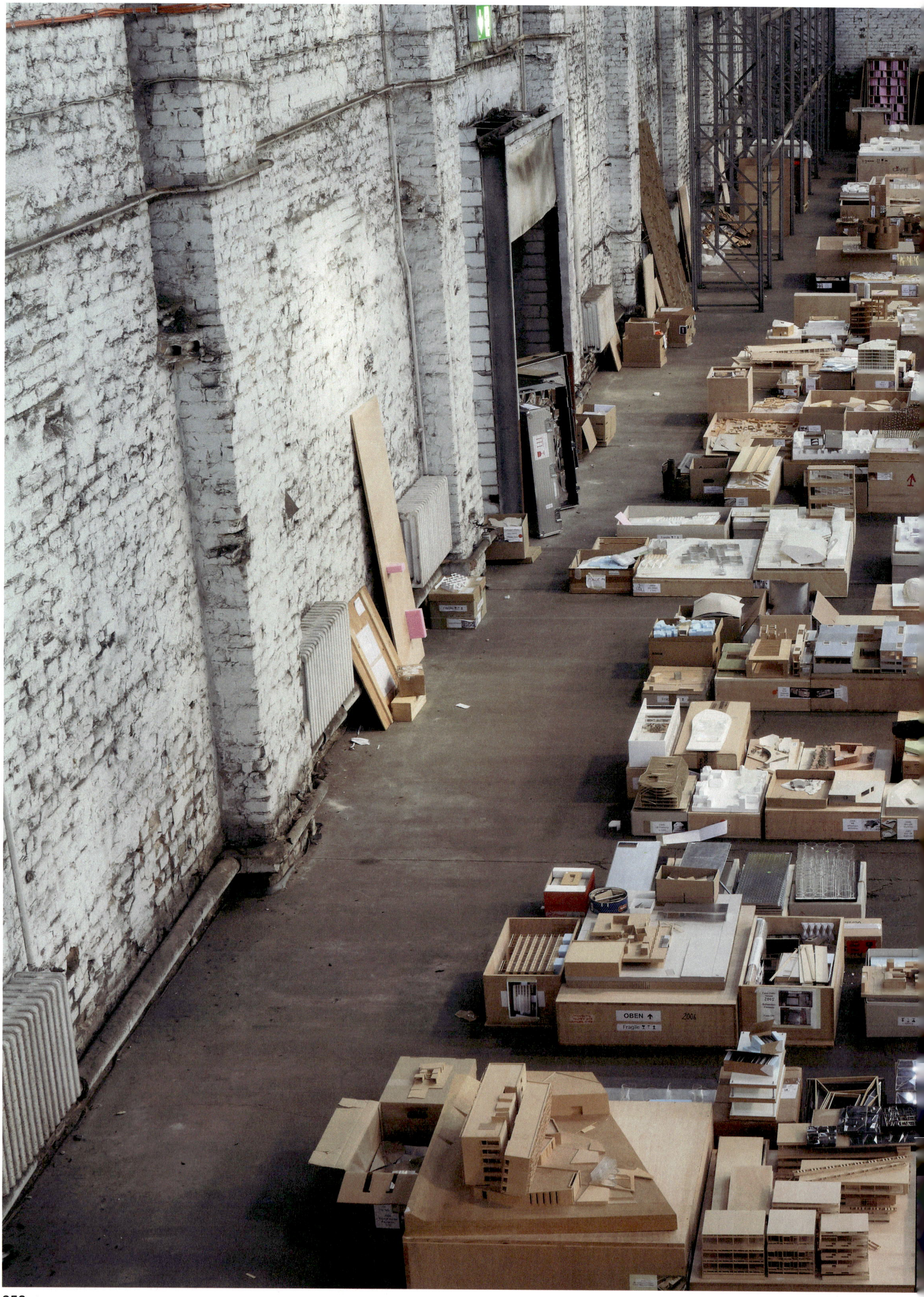
OBEN
Fragile

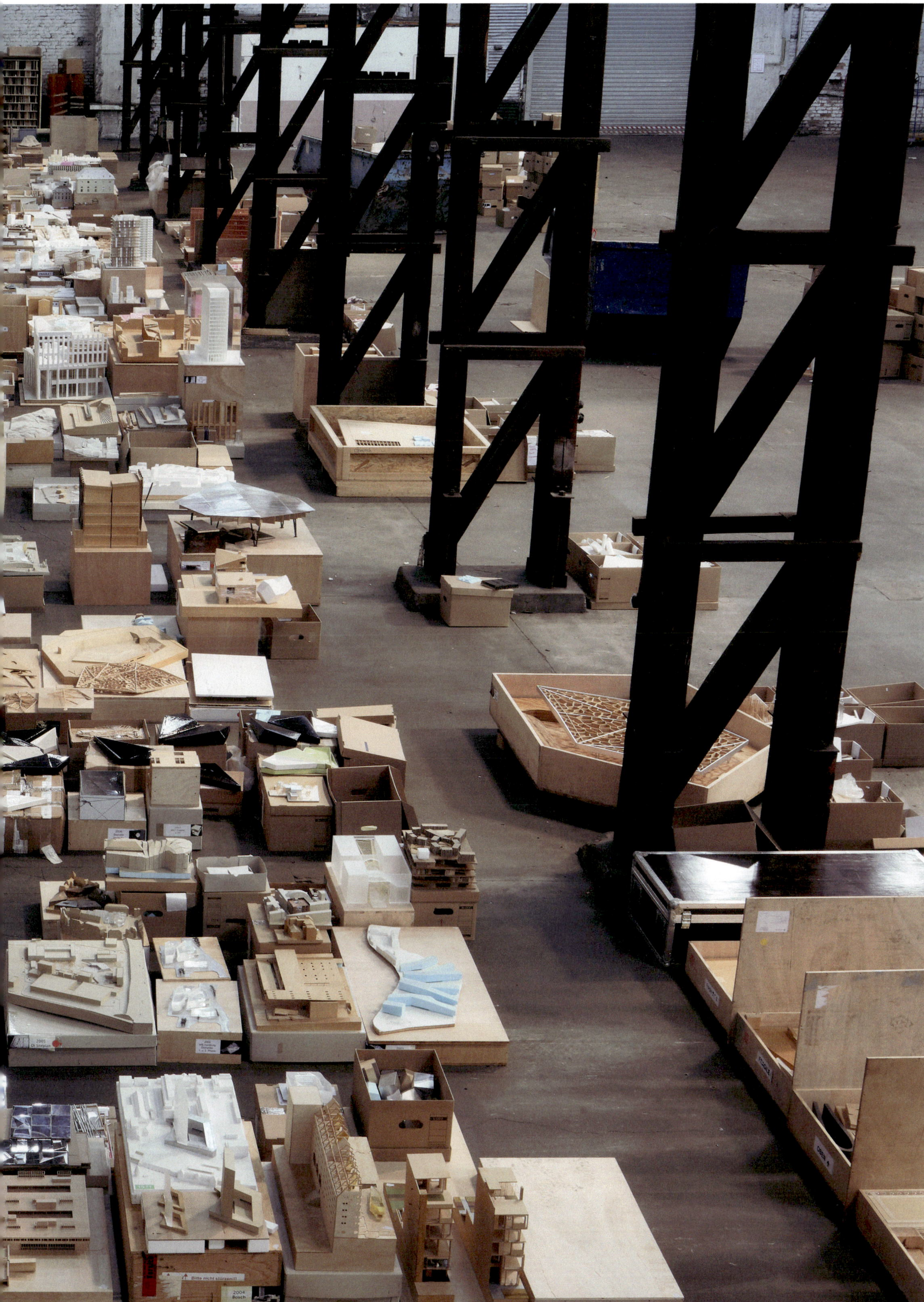

Bitte nicht stürzen!!!
2004
Bosch

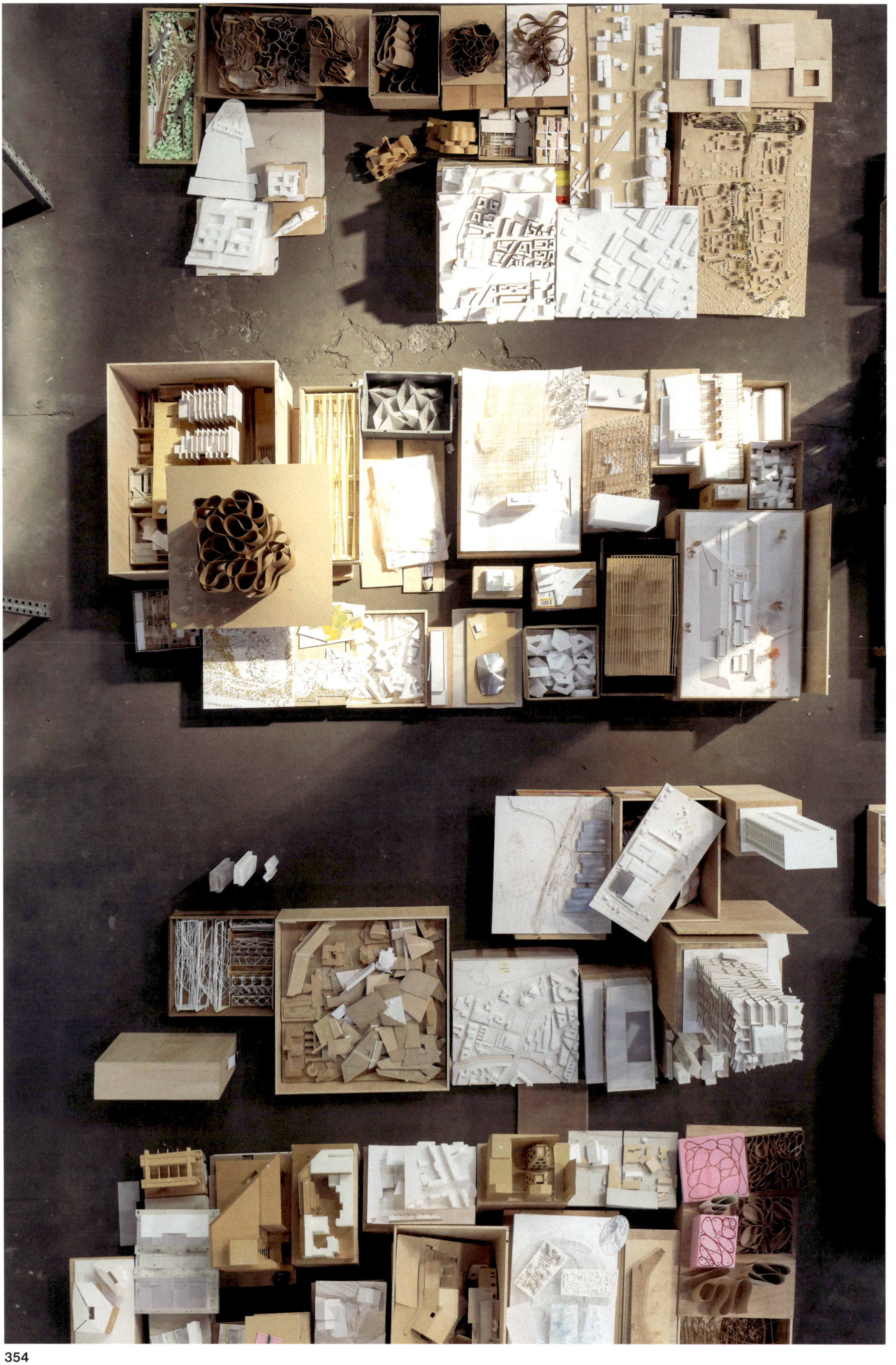

An Act of Active Archiving

It was mid-March 2020. *Revolutions of Choice* would be opening in four months at Haus am Waldsee, and the COVID-19 pandemic held Germany in full lockdown. This moment of great uncertainty and stasis created an opportunity to bring together all the models and other artifacts that had been amassed in the nearly thirty years of Barkow Leibinger's practice in a rather grand, empty warehouse.

In the warehouse, we had the beautiful experience of seeing an archive unfold and start to speak, unraveling its secrets. Cardboard boxes, wooden crates, models, and other objects were placed together, creating an absurd city, a gridded urban landscape, each block consisting of many different projects that – through sheer coincidence – created new pairings, giving way to new possibilities of interpretation. Here, a scenario unfolded in which time stretched infinitely. As the office's archivists tried to structure the projects chronologically – with exceptions due to size – walking the hundred-meter-long fantastic city became a walk through Barkow Leibinger's design experiments over the last decades. There were glitches. Sometimes a box would be temporarily stacked on a model, sometimes a model would sit on a box-cum-pedestal. Sometimes the chronology was broken. Various heights and scales of the models created new, unintended silhouettes of buildings and sculptures – each oblivious to the constraints and contexts the models were initially intended for. Being in that warehouse with all that material was a truly elevating experience. Questions of scale suddenly became secondary.

The model's scales became somewhat arbitrary. It was hard to differentiate whether one was looking at a broken ceramic experiment, the scaled-down detail of a facade element, or an entire building. This also made the connections between the different projects – beyond form and material – more apparent. Things that were quite small dominated the landscape to a certain degree, as they were being preserved as material studies and 1:1 prototypes, while large buildings were only present in tiny scaled-down versions of themselves. Not only scale but also repetition affected perception: various versions and iterations of the same model demystified the 'one' to turn it into one of many. What usually would have been overlooked became very present, while what was normally very present was displayed less prominently.

Beyond the uplifting sensation of roaming through this fantastic fictional city in the warehouse, the setup allowed for access to Barkow Leibinger's work that gave way to new insights. Sorting things was possible, of course, but *how* didn't feel existential: the themes and categories used could be non-exclusive, overlapping, repetitive even. Because what we could now see clearly was that all of Barkow Leibinger's works were were informed by each other. Reconnecting their body of work in this 'act of active archiving' allowed *Revolutions of Choice* to be transformed from a comprehensive retrospective into a forward-looking activation of the office's archive.

Ludwig Engel
Futurologist and urbanist, co-curator
of the exhibition *Revolutions of Choice*,
Haus am Waldsee

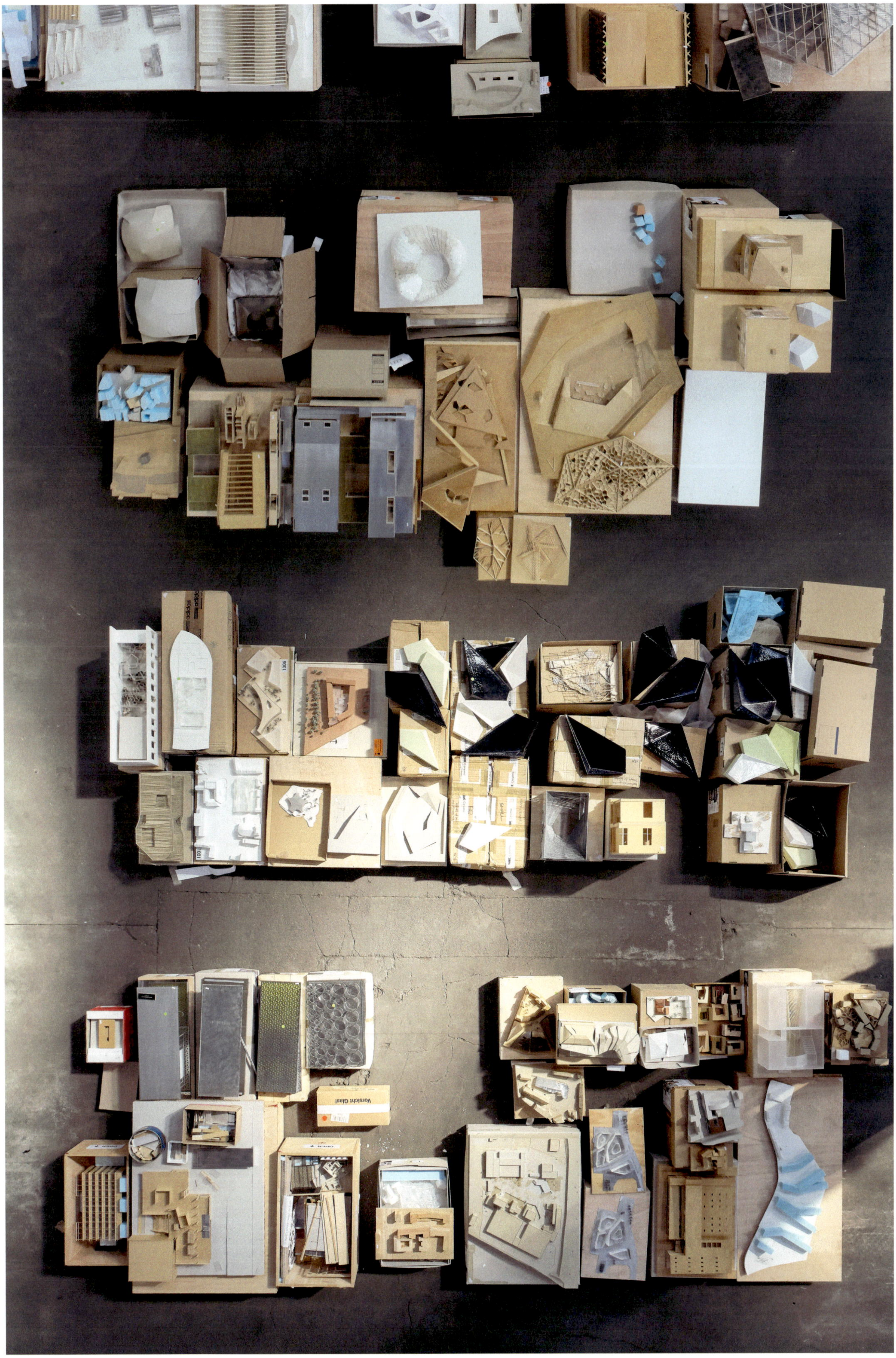

Bundles

MAKE NEW HISTORY – CHICAGO ARCHITECTURE BIENNIAL 2017

23–24: model studies
25–27: exhibition models
28–31: model studies

CHANDELIER AT TRUMPF SMART FACTORY, CHICAGO

34: mock-up
35: plan study
36: model study
37: 1:1 mock-up
38–39: chandelier

SERPENTINE SUMMER HOUSE, LONDON & BERLIN

52–55: model studies
56: plan
57: 1:50 study model
58–59: sketches
60: 1:50 study model
61: steel framing (StageOne Workshop, York)
62: site installation (London)
63: completed pavilion (London)
64: aerial view (London)
65: pavilion detail (Berlin)
66: aerial view (Berlin)
67: completed pavilion (Berlin)

THICKET, BERLIN & PRINCETON

71: installation detail (Berlin)
72–75: concept drawings
76–78: installation views (Berlin)

Casting

CAMPUS CANTEEN II, DITZINGEN

86–91: concrete sand casting
92: concrete model detail
93, 94, 96: 1:100 roofscape model
95: 1:100 ceiling study model
97–100: hanging models

CERAMIC WORKSHOPS, BERLIN

104: clay slab study
105: pottery wheel
106: clay coil study
107: initial studies
108: ceramic extrusions
109: 1:1 facade mock-up
110–114: double wall section studies
115–116: ceramic tubes
117: stacked candelabra study

TWO WALLS, BERLIN

124–125: cast infra-lightweight concrete and stone aggregate studies
126: cast infra-lightweight concrete and recycled rubble study
127–129: CNC-milled Styrodur model studies
129–130: infra-lightweight concrete and plaster model studies
131–133: off-site prefabrication, formwork, and casting
134: site installation at Haus am Waldsee
135–136: completed installation

Topographical

FIDELIO, VIENNA

145: wireframe drawing
146: model components
147–148: 1:50 study model
149: factory mock-up and scaffolding (Poland)
150: completed set and view from stage, Theater an der Wien (Vienna)
151–152: *Fidelio* set and performance

ESTREL TOWER, BERLIN

155: 1:500 site model (competition phase)
156: paper concept model (ballroom)
157: paper concept models
158: Tangram process diagram
159: atrium model detail (competition phase)

HIGH-RISE LIVE/WORK KURFÜRSTENSTRASSE, BERLIN

162: stone cutting
163: axonometric urban block diagram
164: stone cutting
165: corner perspective

Liminal Facade

FACTORY PARKING STRUCTURE, DITZINGEN

173: prefabricated steel structure
174: axonometric facade study/laser tube cutting
175: axonometric facade corner study
176: 1:20 corner study model
177–178: completed project elevation views
179: completed project

TECHNOLOGY CENTER, WARSAW

182–183: facade details
184: interior mezzanine view
185: 1:1 facade mock-up

HIGHRISE JANNOWITZBRÜCKE, BERLIN

190–191: 1:200 model
192–195: 1:200 model details

BERTHA BUILDING, BERLIN

199: facade components
200–201: facade model studies
202: completed project
203: facade detail

Ultrastructural

PROTOTYPE PAVILION

211: Tony Smith, Bultman Chapel, axonometric and plan
212: collage
213: sketch collage
214–215: model views

FELLOWS PAVILION, AMERICAN ACADEMY, BERLIN

220–222: steel prefabrication and site installation
223: completed project

URBAINABLE, BERLIN

230: line drawing
231–236: suspension drawings
237: installation drawing
238: vertical garden sketch
239: gradient concrete formwork
240: gradient concrete sample
241: gradient concrete section model
242–245: installation details

EDUCATION CENTER, DITZINGEN

250–253: plan variations
254: wireframe model
255–256: 1:50 model studies
257–260: interior views

Frames

REMSTAL PAVILION, FELLBACH

267: drawing of Konrad Wachsmann's Grapevine Structure
268–271: structural variations
272: 1:50 model
273: detail roof structure
274–275: completed pavilion

HARVARD ARTLAB, ALLSTON

278–279: model collages
280: model studies
281–283: 1:50 model/ lighting studies
284–299: construction process
300: completed project, workshop
301: completed project, rear entrance
302: completed project, event space
303: completed project, main entrance
304: completed project

TRUMPF SMART FACTORY, CHICAGO

311: 1:50 model detail/ catwalk
312–314: truss prefabrication
315: site installation
316: completed project, catwalk
317: completed project, truss
318: completed project, machine showroom
319: completed project, courtyard
320: completed project, highway view
321: completed project, cafeteria
322: completed project

STIHL HEADQUARTERS, WAIBLINGEN

327: from tree to truss... production diagram
328–331: 1:100 model details

PLYWOOD GEODESIC CLUSTER STUDY

335: model detail
336: model

FOLDED METAL BUS STOP

341–344: 1:50 model

Works in Evolving Process

WAREHOUSE MODEL ARCHIVE, BERLIN

352: models
353–355: drone views

Many thanks to the entire
Barkow Leibinger team
who helped us realize
these projects, especially:

Sebastian Awick
Martina Bauer
Johannes Beck
Jordan Berta
Kateryna Bilyk
Jan Blifernez
Oliver Bucklin
Fabian Busse
Gustav Düsing
Ulrich Fuchs
Ayax Abreu Garcia
Cynthia Grieshofer
Heiko Krech
Frederic Lilja
Reidar Mester
Andreas Moling
Klaus Reintjes
Ina Reinecke
Vincenzo Salierno
Konrad Scholz
Pablo Sequero
Richard Sharam
Antje Steckhan
Volker Steinbrenner
Bogdan Strugar
Robert Tzscheutschler
Blake Villwock
Annette Wagner
Lukas Weder
Tobias Wenz
Jens Weßel
Jane Wong
Linda Zhang

Many thanks also to
all our collaborators and
supporters, especially:

Akademie der Künste
American Academy Berlin
ARCH+
Iwan Baan
Katja Blomberg
Pascal Bünning
Carson Chan
Seanna Drew
Ludwig Engel
David Franck
Konstantin Grcic
Steve Hall
Haus am Waldsee
Thorsten Helbig
Julia Peyton Jones
Nils Koenning
Stefan Korte
Zak Kyes
Konrad Wachsmann Archiv
Regula Lüscher
Niklas Maak
Nina von Maltzahn
Simon Menges
Monath und Menzel
Mohsen Mostafavi
Stefan Müller
NBK Keramik
Hans Ulrich Obrist
Monica Ponce de Leon
Max Rebensburg
Corinne Rose
Gary Smith
Tony Smith Estate
Werner Sobek, Institut
für Leichtbau Entwerfen
und Konstruieren (ILEK),
Universität Stuttgart
Brett Steele
Theater an der Wien
Ray Wachsman
Christoph Waltz

Managing Editor:
Tim Berge

Editing:
Leah Whitman-Salkin

Design:
Zak Group

Printing:
Printmanagement Plitt,
Germany

First published by
Verlag der Buchhandlung
Walther und Franz König
Ehrenstraße 4, D-50672 Köln

Bibliographic information
published by the Deutsche
Nationalbibliothek. The
Deutsche Nationalbibliothek
lists this publication in the
Deutsche Nationalbibliografie;
detailed bibliographic data
are available in the Internet
at http://dnb.d-nb.de.

Vertrieb/Distribution:

Europe
Buchhandlung Walther König
Ehrenstraße 4
D-50672 Köln
Tel: +49 (0) 221 20 59 6 53
verlag@buchhandlung-
walther-koenig.de

UK & Ireland
Cornerhouse Publications Ltd.
2 Tony Wilson Place
UK-Manchester M15 4FN
Tel: +44 (0) 161 212 3466
publications@cornerhouse.org

Outside Europe
D.A.P./Distributed Art
Publishers, Inc.
75 Broad Street, Suite 630
USA-New York, NY 10004
Tel: +1 212 627 1999
orders@dapinc.com

ISBN 978-3-7533-0161-7

Photo Credits

All images © Barkow Leibinger
except:

25–27: Simon Menges
38–39: Simon Menges
62: Ina Reinecke/Barkow Leibinger
63–64: Iwan Baan
65–67: Simon Menges
71, 76, 78: Ina Reinecke/Barkow Leibinger
77: Harry Schnitger
100, 129: Roman März
135–136: Simon Menges
150–152: Iwan Baan
155: Werner Huthmacher
159: Monath Menzel
164: Zeidler Wimmel
165: Ponnie Images/Barkow Leibinger
173, 177–179: Stefan Müller
182–184: David Franck
190–195: Pascal Bünning
202–203: Corinne Rose
211: Tony Smith, Bultman Chapel, axonometric and plan, graphite pencil on paper, 8-1/2×11 inches. Collection: Anthony F. Bultman IV © Tony Smith Estate/Artists Rights Society (ARS), New York
223: Stefan Müller
240: Institut für Leichtbau Entwerfen und Konstruieren (ILEK), Universität Stuttgart
241–245: Stefan Korte
267: Akademie der Künste, Berlin, Konrad Wachsmann Archiv, nr. 164 F.1
273–275: Stefan Müller
284–299: Sasaki
300–304: Iwan Baan
311: Ina Reinecke/Barkow Leibinger
312–315: McShane Construction
316, 318, 320, 321: Simon Menges
317, 319, 322: Steve Hall, Hall+Merrick Photographers
353–355: Nils Koenning